T0259039

BIOAVAILABILITY
Physical, Chemical, and Biological Interactions

Edited by

Jerry L. Hamelink, Ph.D.
Dow Corning Corporation

Peter F. Landrum, Ph.D.
Great Lakes Environment Research Lab

Harold L. Bergman, Ph.D.
Department of Zoology and Physiology
University of Wyoming

William H. Benson, Ph.D.
School of Pharmacy
The University of Mississippi

Proceedings of the Thirteenth Pellston Workshop
Pellston, Michigan, August 17-22, 1992

SETAC Special Publications Series

Series Editors

Dr. T.W. La Point
The Institute of Wildlife and Environmental Toxicology, Clemson University

Dr. B.T. Walton
Environmental Sciences Division, Oak Ridge National Laboratory

Dr. C.H. Ward
Department of Environmental Sciences and Engineering, Rice University

Publication sponsored by the Society of Environmental Toxicology
and Chemistry (SETAC) and the SETAC Foundation for Environmental Education

CRC Press
Taylor & Francis Group
Boca Raton London New York

CRC Press is an imprint of the
Taylor & Francis Group, an **informa** business

CRC Press
Taylor & Francis Group
6000 Broken Sound Parkway NW, Suite 300
Boca Raton, FL 33487-2742

© 1994 by Taylor & Francis Group, LLC
CRC Press is an imprint of Taylor & Francis Group, an Informa business

No claim to original U.S. Government works

Visit the Taylor & Francis Web site at
http://www.taylorandfrancis.com

and the CRC Press Web site at
http://www.crcpress.com

The SETAC Special Publications Series

The SETAC Special Publications Series was established by the Society of Environmental Toxicology and Chemistry to provide in-depth reviews and critical appraisals on scientific subjects relevant to understanding the impacts of chemicals and technology on the environment. The series consists of single- and multiple-authored or edited books on topics reviewed and recommended by the SETAC Board of Directors for their importance, timeliness, and contribution to multidisciplinary approaches to solving environmental problems. The diversity and breadth of subjects covered in the series reflects the wide range of disciplines encompassed by environmental toxicology, environmental chemistry, and hazard and risk assessment. Despite this diversity, the goals of these volumes are similar; they are to present the reader with authoritative coverage of the literature, as well as paradigms, methodologies, controversies, research needs, and new developments specific to the featured topics. All books in the series are peer reviewed for SETAC by acknowledged experts.

The SETAC Special Publications are useful to environmental scientists in research, research management, chemical manufacturing, regulation, and education, as well as to the students considering careers in these areas. The series provides information for keeping abreast of recent developments in familiar areas and for rapid introduction to principles and approaches in new subject areas.

Bioavailability: Physical, Chemical, and Biological Interactions presents the collected papers stemming from a SETAC-sponsored Pellston workshop on developing a mechanistic understanding of the bioavailability of toxic chemicals. The workshop was held in Pellston, Michigan, August 17–22, 1992. The workshop focused on the array of physical-chemical factors influencing the effective exposure concentrations of chemicals. Hence, the papers herein critically evaluate the major abiotic and biotic controlling factors that affect bioavailability in aquatic environments.

Preface

This book is the result of the 13th "Pellston Workshop," the last nine of which were held under the auspices of the Society of Environmental Toxicology and Chemistry (SETAC). Unlike the previous eight meetings, this workshop was actually held at the University of Michigan Biological Station on Douglas Lake in Pellston, Michigan, where the workshop series began in 1977. Like all the previous workshops, participation was limited to invited experts from government, academia, and industry who were selected because of their experience with the workshop topic. The workshop provided a structured environment for the exchange of ideas and debate such that consensus positions would be derived and documented for some of the issues surrounding the fate and effects of chemical contaminants in aquatic environments.

This workshop continued to build on the concept that hazard assessment requires us to understand the relationship between the measured environmental concentrations of chemical contaminants and the true, biologically available exposure concentrations that have adverse effects on aquatic life in that environment. This is exceptionally challenging because the absolute amount of toxic chemical measured in any particular environmental compartment is seldom equivalent to the effective exposure concentration. Rather, the exposure concentration depends on an array of physical-chemical interactions that constitute the environment occupied by any particular organism. Hence, this book is intended to provide a synopsis on the major factors which are believed to control the bioavailability of toxic chemicals over time and space to different kinds and sizes of organisms in aquatic environments.

Acknowledgments

The 1992 Pellston Workshop, *Bioavailabilty: A Mechanistic Understanding*, and this publication were made possible by financial support from:

- American Cyanamid Company
- Battelle
- Dow Chemical Company
- Dow Corning Corporation
- E.I. DuPont DeNemours and Company
- EXXON
- Hoechst-Celanese Corporation
- ICI Americas, Inc.
- National Water Research Institute of Canada
- Procter & Gamble Company
- Sandoz Agro, Inc.
- U.S. Environmental Protection Agency
- U.S. National Oceanic and Atmospheric Administration

The excellent support provided by the SETAC/SETAC Foundation Office staff and volunteers contributed to the success of the workshop.

Special thanks go to Linda Longsworth and Gail Kummers, SETAC, and the spouse volunteers Jill Hamelink, Jan Dickson, Sue Giesy, and Annie Bergman for all their hard work.

Foreword

This workshop was a continuation of a series of successful workshops called the Pellston Workshop Series. Since 1977, twelve workshops have been held at Pellston and several other locations to address a variety of emerging issues in environmental toxicology and chemistry. These workshops were as follows:

- *Estimating the Hazard of Chemical Substances to Aquatic Life,* held in Pellston, MI, June 13–17, 1977, and published by the American Society for Testing and Materials as STP 657 in 1978.
- *Analyzing the Hazard Evaluation Process,* held in Waterville Valley, NH, August 14–18, 1978, and published by the American Fisheries Society in 1979.
- *Biotransformation and Fate of Chemicals in the Aquatic Environment,* held in Pellston, MI, August 14–18, 1979, and published by The American Society of Microbiology in 1980.
- *Modeling the Fate of Chemicals in the Aquatic Environment,* held in Pellston, MI, August 16–21, 1981, and published by Ann Arbor Sciences in 1982.
- *Environmental Hazard Assessment of Effluents,* held in Cody, WY, August 23–27, 1982, and published by SETAC and Pergamon Press in 1985.
- *Fate and Effects of Sediment-Bound Chemicals in Aquatic Systems,* held in Florissant, CO, August 11–18, 1984, and published by SETAC and Pergamon Press in 1987.
- *Research Priorities in Environmental Risk Assessment,* held in Breckenridge, CO, August 16–21, 1987, and published by SETAC in 1987.
- *Biomarkers: Biochemical, Physiological, and Histological Markers of Anthropogenic Stress,* held in Keystone, CO, July 23–28, 1989, and published by SETAC and Lewis Publishers in 1992.
- *Population Ecology and Wildlife Toxicology of Agricultural Pesticide Use: A Modeling Initiative for Avian Species,* held in Kiawah Island, SC, July 22–27, 1990. Proceedings published by SETAC and Lewis Publishers in 1994 (*Wildlife Toxicology and Population Modeling: Integrated Studies of Agroecosystems*).
- *A Technical Framework for [Product] Life-Cycle Assessments,* held in Smuggler's Notch, VT, August 18–23, 1990. Proceedings published by SETAC in January 1991, with a second printing in September, 1991.
- *Aquatic Microcosms for Ecological Assessment of Pesticides,* held in Wintergreen, VA, October 7–11, 1991. Interim Report published by SETAC February, 1992. Proceedings published by SETAC and Lewis Publishers in 1993 (*Aquatic Mesocosm Studies in Ecological Risk Assessment*).
- *A Conceptual Framework for LCA Impact Assessment,* held in Sandestin, FL, February 1–6, 1992. Proceedings published by SETAC in 1993.

Information about the availability of Workshop Proceedings can be obtained by contacting:

SETAC Foundation for Environmental Education
1010 North 12th Avenue
Pensacola, FL 32501–3307
U.S.A.
TEL 904–469–9777
FAX 904–469–9778

About the Editors

Jerry L. Hamelink, Ph.D., is an environmental research scientist. As such, he is interested in the environmental fate and effects of all kinds of chemicals.

Dr. Hamelink received his B.S., with honors in 1963, and Ph.D. in Fisheries and Wildlife in 1969 from Michigan State University. His graduate studies on DDT pioneered the concept of bioconcentration—partitioning directly from the water—and kindled his fascination with bioavailability.

He has worked in both academia and industry, and was employed by Dow Corning Corporation, the silicone company, when the workshop reported herein was planned, held, and reported. He is a life member of the American Fisheries Society (AFS) and an active member of SETAC; he served as the Awards Committee chair from 1984 to 1990 and on the Board of Directors from 1989 to 1992.

Peter F. Landrum, Ph.D., is a research chemist for the National Oceanic and Atmospheric Administration at the Great Lakes Environmental Research Laboratory, Ann Arbor, MI. He received a Bachelor of Science degree in Chemistry from California State College, San Bernardino in 1974. He received a Ph.D. degree in Pharmacology and Toxicology from the University of California, Davis in 1979. He pursued research on the transport and fate of polycyclic aromatic hydrocarbons in stream systems at the Savannah River Ecology Laboratory as a University of Georgia employee from 1979 to 1981.

In 1981, Dr. Landrum moved to Ann Arbor, MI and began his position at the Great Lakes Environmental Research Laboratory. His research has focused on the bioavailability of organic contaminants to aquatic organisms. This work has evolved from a focus on water column issues to sediment-associated contaminants. His current research focuses on bioavailability of sediment-associated contaminants, contaminant food chain transfer, development of whole sediment bioassay techniques, use of body residues as a measure of dose, the behavior of compounds in mixtures in both bioavailability and toxicity, and the role of the zebra mussel in contaminant cycling in the Great Lakes.

Dr. Landrum was also active in teaching toxicology at Eastern Michigan University from 1984 to 1991. Dr. Landrum is active in the Society of Environmental Toxicology and Chemistry and served on the Board of Directors from 1990 to 1993. Dr. Landrum is a fellow of the Cooperative Institute for Limnology and Ecosystem Research and has pursued cooperative research with scientists from Clemson University, Michigan State University, Ohio State University, and University of Michigan. Dr. Landrum has over 70 publications and serves on the editorial boards for *Chemosphere* and *Critical Reviews in Environmental Science and Technology* and is Associate Editor for the *Journal of Great Lakes Research.*

Harold L. Bergman, Ph.D., is Professor of Zoology and Physiology and Director of the Red Buttes Environmental Biology Laboratory at the University of Wyoming.

Dr. Bergman received B.A. and M.S. degrees in biology at Eastern Michigan University and a Ph.D. in fisheries biology at Michigan State University. Prior to joining the University of Wyoming faculty in 1975, he was a fishery biologist at the Great Lakes Fishery Laboratory, U. S. Fish and Wildlife Service, Ann Arbor, Michigan, and Research Associate in the Environmental Sciences Division at Oak Ridge National Laboratory, Oak Ridge, Tennessee.

He has authored or co-authored over 60 research articles on diverse topics related to his principal research interests in fish physiology and aquatic toxicology. Professional society memberships include the American Association for the Advancement of Science, American Fisheries Society, North American Benthological Society, Sigma Xi, and the Society of Environmental Toxicology and Chemistry (SETAC), which he served as President in 1984–1985.

William H. Benson, Ph.D., is Professor of Pharmacology and Toxicology and serves as Coordinator of the Environmental Toxicology Research Program within the School of Pharmacy at The University of Mississippi. He attended Florida Institute of Technology where he obtained a Bachelor of Science degree in Biology (Environmental and Ecological Biology Option) in 1976. Dr. Benson obtained his M.S. and Ph.D. in Toxicology from the University of Kentucky in 1980 and 1984, respectively. While in graduate school he was the first recipient of the Society of Environmental Toxicology and Chemistry Pre-Doctoral Fellowship sponsored by The Procter & Gamble Company.

Dr. Benson has published over 50 scientific publications and has received more than $1.8 million in grants and contracts dealing with environmental toxicology and chemistry. His research activities have focused on assessing the acute and chronic health effects of environmental contaminants. He has done extensive research in the areas of metal and pesticide bioavailability, reproductive and developmental effects in aquatic organisms, use of biomarkers in environmental monitoring, and residue health effects.

Professor Benson was a recipient of the Dow–Dow Corning Distinguished Lecture in Environmental Toxicology at Michigan State University, and has served on the Board of Directors of the Society of Environmental Toxicology and Chemistry. He is active in several other professional societies including the American Chemical Society and Society of Toxicology and participates on state and federal technical committees, as well as acting as a consultant.

Workshop Participants and Contributors

James J. Alberts, Ph.D.
University of Georgia Marine Institute
Sapelo Island, GA 31327

Herbert E. Allen, Ph.D.
University of Delaware
Department of Civil Engineering
Newark, DE 19716

William H. Benson, Ph.D.
School of Pharmacy
The University of Mississippi
University, MS 38677

Harold L. Bergman, Ph.D.
Department of Zoology and Physiology
University of Wyoming
Laramie, WY 82071

Terry D. Bills, Ph.D.
National Biological Survey
National Fisheries Research Center
2630 Fanta Reed Road
LaCrosse, WI 54602–0818

Wolfgang Calmano, Ph.D.
Technische Universität Hamburg-Harburg
Umweltschutztechnik
Eissendorfer Str. 40
21073 Hamburg, Germany

James R. Clark, Ph.D.
Environmental Toxicology Division
Exxon Biomedical Sciences, Inc.
Mettlers Road, CN 2350
East Millstone, NJ 08875–2350

Donald G. Crosby, Ph.D.
Department of Environmental Toxicology
University of California
Davis, CA 95616

William Davison, Ph.D.
Institute of Environmental and Biological
   Sciences
Lancaster University
Lancaster LA1 4YQ, U.K.

Kenneth L. Dickson, Ph.D.
Institute of Applied Sciences
University of North Texas
P.O. Box 13078, NT Station
Denton, TX 76203

Russell J. Erickson, Ph.D.
U.S. EPA
Environmental Research Laboratory
6201 Congdon Boulevard
Duluth, MN 55804

Ulrich Förstner, Ph.D.
Technische Universität Hamburg-Harburg
Postfach 90 10 2
21073 Hamburg, Germany

John P. Giesy, Ph.D.
Michigan State University
Department of Fisheries and Wildlife
#13 Natural Resources Building
East Lansing, MI 48824-1222

Frank A.P.C. Gobas, Ph.D.
School of Resource and Environmental
   Management
Simon Fraser University
Burnaby, British Columbia
V5A1S6 Canada

Jerry L. Hamelink, Ph.D.
Dow Corning Corporation
2200 West Salzburg Road
Midland, MI 48686–0994

David J. Hansen
U.S. EPA
Environmental Research Laboratory
27 Tarzwell Drive
Narragansett, RI 02882

William L. Hayton, Ph.D.
The Ohio State University
College of Pharmacy
500 West 12th Avenue
Columbus, OH 43210–1291

Jihua Hong, Ph.D.
Department of Chemical Engineering
Tulane University
New Orleans, LA 70118

George E. Howe
U.S. Fish and Wildlife Service
National Fisheries Research Center
2630 Fanta Reed Road
LaCrosse, WI 54602–0818

Carlton D. Hunt, Ph.D.
Battelle Ocean Sciences
397 Washington Street
Duxbury, MA 02332

Chad T. Jafvert, Ph.D.
Purdue University
School of Civil Engineering
1284 Civil Engineering Building
West Lafayette, IN 47907–1284

Charles H. Jagoe
Biogeochemical Ecology Division
Savannah River Ecology Laboratory
P.O. Drawer E
Aiken, SC 29803

John P. Knezovich, Ph.D.
Health and Ecological Assessment
   Division L-453
Lawrence Livermore National Laboratory
Livermore, CA 94550

Jussi Kukkonen, Ph.D.
University of Joensuu
Department of Biology
P.O. Box 111
FIN-80101 Joensuu, Finland

Peter F. Landrum, Ph.D.
Great Lakes Environmental Research Lab
2205 Commonwealth Boulevard
Ann Arbor, MI 48105–1593

Darrel J. Laurén, Ph.D.
Entrix, Inc.
555 University Avenue, Suite 224
Sacramento, CA 95825

Henry Lee II, Ph.D.
U.S. EPA
2111 S.E. Marine Science Drive
Newport, OR 97365–5258

Donald Mackay, Ph.D.
University of Toronto
Institute for Environmental Studies
Toronto, Ontario M5S 1A4
Canada

Leif L. Marking
U.S. Fish and Wildlife Service
National Fisheries Research Center
2630 Fanta Reed Road
LaCrosse, WI 54602–0818

Foster L. Mayer, Ph.D.
U.S. EPA
Environmental Research Laboratory
1 Sabine Island Drive
Gulf Breeze, FL 32561–5299

Lynn S. McCarty, Ph.D.
Scientific Research and Consulting
280 Glen Oak Drive
Oakville, Ontario L6K 2J2
Canada

Anne E. McElroy, Ph.D.
New York Sea Grant Institute
Dutchess Hall–Room 147
Stony Brook, NY 11794–5001

James M. McKim, Ph.D.
U.S. EPA
Environmental Research Laboratory
6201 Congdon Boulevard
Duluth, MN 55804

Joseph S. Meyer, Ph.D.
Department of Zoology and Physiology
University of Wyoming
Laramie, WY 82071

Michael C. Newman, Ph.D.
Savannah River Ecology Lab
Biogeochemical Ecology Division
P.O. Drawer E
Aiken, SC 29803

James A. Noblet
Department of Environmental Science
  and Engineering
UCLA School of Public Health
Los Angeles, CA 90024

James T. Oris, Ph.D.
Department of Zoology
The Miami University
Oxford, OH 45056

Rodney Parrish
SETAC/SETAC Foundation
1010 North 12th Avenue
Pensacola, FL 32501

Mark R. Servos, Ph.D.
Great Lakes Laboratory for Fisheries and
  Aquatic Sciences
Canada Dept. of Fisheries and Oceans
P.O. Box 5050/867 Lakeshore Road
Burlington, Ontario L7R 4A6
Canada

Anne Spacie, Ph.D.
Department of Forestry and Natural
  Resources
Purdue University
1159 Forestry Building
West Lafayette, IN 47907

I. H. Suffet, Ph.D.
Department of Environmental Science
  and Engineering
UCLA School of Public Health
Los Angeles, CA 90024

Bjorn Sundby, Dr. Philos.
INRS-Océanologie
University of Québec
310 Allée des Ursulines
Rimouski, Québec G5L 3A1
Canada

Lisa L. Williams
Michigan State University
Pesticide Research Center
Room 201
East Lansing, MI 48824

Lee Wolfe, Ph.D.
U.S. EPA
Environmental Research Laboratory
College Station Road
Athens, GA 30613–0801

Xin Zhang, Ph.D.
School of Resource and Environmental
  Management
Simon Fraser University
Burnaby, British Columbia V5A 1S6
Canada

Contents

SESSION 1

INTRODUCTION

Jerry L. Hamelink, Peter F. Landrum, Harold L. Bergman, and William H. Benson

Many current practices in environmental toxicology and chemistry, and in the resulting applications for risk assessment, are founded on empirical relationships; these relationships depend mainly on experience or observation. That does not mean that they are incorrect or inappropriate—just that they are limited because our understanding of the physical, chemical, and biological mechanisms underlying these empirical observations is often limited. The empirical approach has allowed us to address pressing issues and has facilitated regulation of toxic materials in the environment. However, as we work to produce the empirical data necessary to set allowable limits for chemicals in the environment, our understanding of the principles and processes controlling environmental exposure to, and effects of, toxic materials must advance beyond empirical relationships to a sound mechanistic framework.

The intent of this workshop, and the Proceedings volume published from it, was not to advocate abandoning, or in any way undermining, use of the important empirical relationships that currently underpin many of our environmental regulations for protection of aquatic life. These empirical relationships have often been used in setting water quality criteria and standards, and they have been used as the basis for extrapolating from constant exposure conditions in the laboratory to variable conditions in the field or for predicting long-term responses of organisms or communities from short-term laboratory test results. Examples of these important empirical relationships include the inverse relationship between water hardness and the toxicity of many metals to aquatic organisms,[1,2] the direct relationship between octanol:water partition coefficients and bioaccumulation for many organic chemicals,[3,4] and the relationship between early life-stage toxicity and full life-cycle toxicity for fish.[5] As stated above, these empirical relationships, and others, have been important tools in regulating water quality. Through their use we have established an approach for regulating on the basis of water quality criteria and standards, and by using this approach we have achieved considerable improvement in surface water quality. However, as laboratory and field research has progressed over the past several decades, it has become more and more apparent that each of these empirical relationships has exceptions or limitations.

The empirical relationship between water hardness and metal toxicity provides an excellent example of a variable water-quality standard with limitations. Without question, many heavy metals are less toxic to aquatic biota in hard water, and the relationship is sufficiently robust to allow adjustment of a water-quality standard to account for the effect of receiving water hardness on metal toxicity. However, it is now clear that the reported effects of hardness on metal toxicity were sometimes caused not by hardness, per se, but rather by pH or alkalinity.[2] The laboratory experiments that were used to establish the empirical relationship between metal toxicity and water hardness for some metals did not always control or even measure pH and alkalinity, and we now know that pH and alkalinity were confounding covariants in some of these experiments. It is also now clear that other water-quality factors that modify toxicity, such as dissolved organic matter, can have a very large effect on metal bioavailability and toxicity in many surface waters. As a consequence of these kinds of findings, the U.S. EPA has recently proposed revisions in methods for determining site-specific water-quality criteria to more fully account for the modifying effect of receiving water chemistry on bioavailability and toxicity of discharged metals.[6]

1-56670-086-8/94/$0.00+$.50
© 1994 by CRC Press, Inc.

Thus, though empirical relationships that can be used to predict environmental fate or effects of chemicals are useful and necessary, it is very clear that we also need to pursue other research strategies to better understand the physical, chemical, and biological mechanisms that control contaminant fate and effects in the aquatic environment. This kind of mechanistic knowledge can help us in several ways, including identifying and explaining the exceptions and limitations of existing or new empirical relationships and the application of these relationships to water-quality regulation; identifying and measuring the chemical species that are actually the bioavailable, toxic species and, therefore, the appropriate chemical species for measurement of exposure; predicting through extrapolation from the fate or effects of one chemical to another chemical in the same or a similar structural class; predicting through extrapolation from effects in one organism or community to effects in another organism or community; and predicting fate or effects of chemical(s) in the field under different and often more variable geochemical or physiological conditions than were used to measure fate or effects in the laboratory.

It should be evident, then, that we now need to go beyond and improve upon the empirical approach to water-quality research and regulation by achieving a mechanistic understanding of the many physical, chemical, and biological interactions that control contaminant bioavailability. Achieving a better understanding of mechanisms is not merely an esoteric, academic pursuit. Rather, better mechanistic understanding is required to predict accurately the fate and effects of environmental contaminants, and this knowledge may have substantial social and economic consequences. Underestimating the effect of chemicals may result in serious environmental contamination and adverse health effects, while overestimating potential hazards can result in social disruptions and undue economic burdens. Thus, a mechanistic understanding of exposure conditions, bioaccumulation processes, and chemical effects becomes imperative if we are to institute appropriate safeguards to maintain proper environmental health while maximizing economic activity.

This workshop was conceived to be the first of several possible workshops on the mechanisms controlling bioavailability of toxic contaminants in aquatic environments. The first workshop was to review key issues related to contaminant bioavailability as well as to focus on events external to the organism, while later workshops will emphasize events at the membrane-water interface and events within the organism. The workshop steering committee, consisting of Jerry L. Hamelink, William H. Benson, Harold Bergman, and Peter F. Landrum, first met in September, 1989, at the R. A. MacMullen Conference Center in Roscommon, Michigan, under support provided by the Dow Corning Corporation. After further planning and fund raising, coordinated through the SETAC Foundation for Environmental Education, the workshop was conducted August 17–22, 1992, at the University of Michigan Biological Station near Pellston, Michigan.

WORKSHOP OBJECTIVES AND AGENDA

The 1992 Pellston workshop, the 13th in the series, was titled: *Bioavailability: A Mechanistic Understanding*. The workshop was convened to address the following objectives:

- Provide a review of the state-of-the-art for developing a mechanistic understanding of processes affecting the bioavailability and bioaccumulation of toxic contaminants.
- Discuss limitations of the methods currently being used to assess the bioavailability of environmental contaminants.
- Identify research priorities for improving our understanding of bioavailability.
- Discuss the feasibility of incorporating environmental factors limiting bioavailability in appropriate regulations.

In addition to these overall objectives, the following questions were presented to the workshop participants for consideration during their deliberations:

- What important empirical observations/relationships are used to describe the behavior or predict the fate and effects of chemicals in the environment?
- Which of these observations or relationships are actually used directly in regulating chemicals?
- What are the known or suspected exceptions to these relationships?
- Do we know the fundamental physical, chemical, and biological mechanisms that underlie these relationships, and if so, what are they?
- What are the key hypotheses about suspected mechanisms that need to be tested?
- What problems may arise when any of the key empirical relationships are applied, and how will understanding underlying mechanisms help to resolve these problems?

Seven individual workshop sessions were conducted to address these objectives and questions. Each session was initiated by one or more well-defined discussion initiation papers which were prepared and presented by an invited author selected by the steering committee. These papers were presented during half-day plenary sessions of the workshop, followed by a discussion period open to all participants. Each discussion was lead by a session chairperson, who was also responsible for coordinating the activities of the respective consensus committees during the week. During the last session, each chairperson summarized the consensus conclusions of the participants regarding each individual session, and the chairperson of the overall workshop summary committee presented the consensus conclusions and recommendations for the entire workshop.

The workshop sessions and the roles of the participants are described below:

Session 1. Introduction

Jerry Hamelink first presented an overview of the goals and objectives for the workshop. He then introduced all of the session chairs and made team assignments. Harold Bergman followed with an informal, unpublished discussion initiation paper entitled *Improving on Empiricism.*

Session 2. Water Chemistry

Foster L. (Sonny) Mayer presented a discussion initiation paper that focused on freshwater, followed by John Knezowich with a paper that focused on saltwater. The discussion synopsis team for this session included Russ Erickson (chair), Jim Clark, Anne McElroy, Terry Bills, and Dave Hansen.

Session 3. Inorganic Toxicants

Mike Newman addressed some of the interactions known to occur with metals. By describing a ligand as an anion that forms a complex with metals, he was able to discuss some of the limitations to our current adsorption and speciation models. Both sediments and events at the gill membrane were drawn into the discussion session. The synopsis team consisted of Bill Benson (chair), Carlton Hunt, Herb Allen, and Jim Alberts.

Session 4. Organic Toxicants

Anne Spacie gave the first paper in this session. Because of illness, Frank Gobas was unable to present his discussion initiation paper, so Peter Landrum presented it with a set of overheads sent by Gobas. The discussion synopsis team consisted of I. H. (Mel) Suffet (chair), Lisa Williams, Chad Jafvert, Jussi Kukkonen, Mark Servos, Anne Spacie, and Jerry Hamelink.

Session 5. Dynamic Environmental Factors

Don Crosby opened the session with a paper concerning photochemistry, the generally forgotten but surprisingly active force in aqueous environments, even under cloudy conditions. Ulrich

Fostner then focused on pH-redox conditions, primarily as they affected the bioavailability of metals. Bjorn Sundby finished the session by teaching all of us to think of sediments as large, living, breathing, well-structured, dynamic organisms. The discussion synopsis team for this session consisted of Joe Meyer (chair), Bill Davison, Jisua Hong, Bjorn Sundby, Darrel Lauren, and Jim Oris.

Session 6. Kinetic Limitations or Dissolution

Don MacKay opened this session by discussing the kinetic limitations surrounding the dissolution of chemicals in aquatic environments. Jim McKim then described some of the limitations imposed by biological kinetics. Henry Lee capped this session off by briefly describing how different inhabitants of the same bed of sediments are subjected to very different exposures because of the differences in their life styles. The discussion session synopsis team consisted of Peter Landrum (chair), Bill Hayton, Henry Lee, and Lynn McCarty.

Session 7. Summary and Conclusions

The chairperson of each discussion session synopsis team first presented the consensus position derived by their team for each session. An overall workshop summary and conclusion paper followed these presentations before the floor was opened for discussion. The summary and conclusions team was chaired by Ken Dickson. The other team members were John Giesy, Lee Wolfe, and Rod Parrish.

REFERENCES

1. National Academy of Sciences, National Academy of Engineering, Water Quality Criteria, 1972, Ecological Research Series Report EPA-R-73-033, U.S. Environmental Protection Agency, Washington, D.C., 1973.
2. Sprague, J.B., Factors that affect toxicity, in Rand, G.M. and Petrocelli, S.R., Eds., *Fundamentals of Aquatic Toxicology*, Hemisphere, Washington, D.C., 1985, p. 124–163.
3. Veith, G.D., Macek, K.J., Petrocelli, S.R., and Carroll, J., An evaluation of using partition coefficients and water solubility to estimate bioconcentration factors for organic chemicals in fish, *Fed. Regist.*, 44, 15926–15981, 1979.
4. Spacie, A. and Hamelink, J.L., Bioaccumulation, in Rand, G.M., and Petrocelli, S.R., Eds., *Fundamentals of Aquatic Toxicology*, Hemisphere, Washington, D.C., 1985, p. 495–525.
5. McKim, J.M., Evaluation of tests with early life stages of fish for predicting long-term toxicity, *J. Fish. Res. Bd. Can.*, 34, 1148–1154, 1977.
6. U.S. Environmental Protection Agency, Interim guidance on the interpretation and implementation of aquatic life criteria for metals, *Fed. Regist.*, 57, 24041, 1992.

SESSION 2

WATER CHEMISTRY

Chapter 1

Physicochemical Factors Affecting Toxicity in Freshwater: Hardness, pH, and Temperature

Foster L. Mayer, Jr., Leif L. Marking, Terry D. Bills, and George E. Howe

INTRODUCTION

Knowledge of the way in which different variables affect interactions among organisms and pollutants in the water column is needed to ensure the most effective protection and management of freshwater ecosystems. These interactions are influenced by changes in environmental variables that can alter the physiological condition of the biota or the physicochemical characteristics of the system.[1] Several investigators have reported on the toxicity of chemicals to freshwater organisms in relation to hardness,[2-4] pH,[2-5] and temperature.[2-4,6]

Much of the existing toxicity data are from laboratory experiments conducted at optimal and constant conditions (dissolved oxygen, pH, salinity, temperature, etc.). Where factors affecting toxicity have been examined, it has been done with one factor at a time for the most part; a situation which does not occur in the real world. Yet, the role of environmental fluctuations and extremes in confounding biological responses to contaminant exposures must be characterized to facilitate extrapolation from the laboratory to the field. Hence, we will summarize reported effects of hardness, pH, and temperature on the toxicity of contaminants to freshwater organisms, and discuss whether environmental factors affect toxicity or bioavailability and exposure for different kinds of toxic contaminants. For the purposes of this paper, bioavailability is the external availability of a chemical to an aquatic organism,[7] and exposure is the concentration of a chemical in an aquatic organism.

PHYSICOCHEMICAL FACTORS

Although there is a considerable amount of literature describing the effects of hardness, pH, and temperature on toxicities of chemicals to aquatic organisms, the variety of methodologies and, often, incomplete method descriptions can make comparisons and conclusions difficult. Most research has been on acute toxicity and animal species with little effort being expended on chronic toxicity or plant species.

We conducted an on-line computer search of the literature through 1992. The data bases included *Aquatic Science Abstracts, Biological Abstracts, Chemical Abstracts*, NTIS, and the *Federal Register*. Abstracts for the years not included in the computer search were manually reviewed. Tables 1 through 5 represent a summary of extensive information developed by Nishiuchi from 1977–1982.[8-15] Although the experimental design was good, he unfortunately did not provide sufficient detail to enable a complete evaluation of his data. For example, he tested organisms at pHs of 5 to 10 in 1-unit increments, but furnished little or no detail on methods, procedures, or quality control. English translations of his reports primarily provided results for a wide array of formulated pesticides.

Hardness

Water hardness is known to affect chemical toxicity.[16,17] In general, the toxicities of most inorganic chemicals are reduced by increased water hardness. Conversely, Inglis and Davis[16] and Pickering and Henderson[18] found that total hardness had little effect on the toxicity of organic chemicals, and that any changes were probably due to differences in pH among the hardnesses tested. Decreases in the toxicities of lindane[19] and pentachlorophenol[20] have also been reported. Surfactant toxicity either increases or decreases, depending on the surfactant type,[21] but pH appears to play a major role.[22] Mayer and Ellersieck,[3] using analysis of covariance on 39 tests with 25 organic chemicals (herbicides, insecticides, and solvents) and fishes, found that the slopes of only two acute toxicity/hardness regressions—those for dimethylamine and dodecyl/tetradodecyl amine salts of 2,4-D—differed significantly from zero. An increase in hardness (from 12 to 250–300 mg/l as $CaCO_3$) caused a 2.8-fold increase in the LC50 of the dimethylamine salt (96-h LC50 range, 285 to 800 mg/l) and a 2.1-fold increase in that of the dodecyl/tetradodecyl amine salt (96-h LC50 range, 1.9 to 4.0 mg/l). These increases were probably due to the increase in pH (1.5 units) between the very soft and very hard test waters.[3] Because 2,4-D is a weak acid ($pK_a = 2.80$),[23] its toxicity would be predicted to decrease in alkaline water. Hardness-related toxicity changes of carbamate and organophosphate insecticides have been indicated by others,[24-26] with these types of chemicals, toxicity changes may arise which are due more to pH than hardness because hydrolysis products are formed more rapidly under alkaline pH.

Effects of total hardness on the toxicities of most organic chemicals have not been demonstrated. Most studies that show an effect of hardness on toxicity rarely include the pH at the various hardness levels tested. Marking and Olson[27] demonstrated an 8- to 10-fold decrease in acute toxicity of the lampricide, TFM (3-trifluoromethyl-4-nitrophenol), as water hardness increased from 10 to 300 mg/l (as $CaCO_3$). However, pH increased from 6.6 to 8.2 and alkalinity increased from 10 to 235 mg/l (as $CaCO_3$) as hardness increased. Conversely, Bills et al.[28] demonstrated no change in the acute toxicity of TFM when hardness and alkalinity were maintained at constant levels. However, as pH increased from 6.5 to 9.5, the toxicity of TFM decreased about 100-fold.

Sprague[4] and Wang[29] provided extensive overviews of hardness and other factors affecting metal toxicity and accumulation by aquatic organisms. Recent studies have indicated that hardness has little or no effect on the toxicity of aluminum to freshwater clams (*Anodonta anatina, Unio pictorum*)[30] or boron to daphnids (*Daphnia magna*) and midge larvae (*Chironomus decorus*).[31] However, increased hardness has been related to decreased toxicity in many species for cadmium,[32-35] cobalt,[36] copper,[37-40] zinc,[41-43] and a mixture of trace elements.[44] With copper, hardness has also been found to have no effect under constant alkalinity.[45,46] In addition, increased hardness has been reported to be related to decreased mercury and zinc residues in fishes.[47-49]

Sprague[4] summarized the effects of hardness on metals, and stated that heavy metals are an order of magnitude more lethal in very soft water than in very hard water. He observed that sublethal changes follow a similar pattern. He attributed this response to changes in gill permeability caused by calcium. He further generalized that the free or ionic metal is very toxic. Ionized hydroxides of metal may also be toxic, but nonionized carbonates are probably not. Other authors[41,50-55] have further substantiated or added to Sprague's observations.

Hydrogen Ion Concentration (pH)

Playle and Wood[56] have stated that any contaminant whose toxicity varies with pH may be more or less toxic than predicted from bulk water chemistry alone. Investigators have shown that many organic chemicals appear to penetrate cells by diffusion at rates correlated with their solubility in lipids, and there is evidence that the nonionized form of most organic chemicals is the form that penetrates cell membranes most readily.[57,58] For instance, laboratory studies showed that residues of the lampricide, TFM, in fish tissues are correlated with the dissociation of the molecule in water at different pHs.[5] When rainbow trout (Oncorhynchus mykiss) were exposed to a 1 mg/l solution of TFM for 12 h at pHs 6, 7, 8, or 9, the residues in muscle were 3.2, 1.5, 0.33, and 0.03 µg/l, respectively. The pK_a of TFM, a weak organic acid, is 6.1; hence, above pH 6.1 the molecule becomes increasingly ionized and is less readily transported across gill membrane. Laboratory toxicity tests indicated that TFM was 50-fold less acutely toxic for salmonids at pH 9.5 than at 6.5, and more than 20-fold less for warmwater fishes.[27] Thus, the results on toxicity supported the ionization theory of Hunn and Allen[5] and identified pH as the environmental factor having the greatest influence on TFM toxicity. The toxicity of ionizable organic chemicals (weak acids and bases) is greatly affected by pH—decreases in pH increase the toxicity of organic acids but decrease that of organic bases.

Of the factors tested that modify the toxicity of chemicals, pH causes the greatest change.[3] Covariance analysis was conducted on 49 chemicals in 100 pH tests with fishes and aquatic invertebrates, and slopes significantly different from zero were observed in 23 of the tests on 10 chemicals. The ratio of the highest to lowest LC50 within a pH test averaged 16 (range, 4.2 to 45) among the 10 chemicals identified. The toxicity of nonpolar organics was not affected by pH. Regression slopes were consistent among species for a given chemical, indicating chemical rather than biological differences.

Sprague,[4] in a review of the effects of pH, reported that undissociated molecules may be more toxic because they penetrate cell membranes more readily, and speculated that pH is the major factor in the penetration of membranes by toxics. He cited the toxicity of ammonia to aquatic biota as an example: ionized ammonia has little or no acute toxicity, whereas the nonionized form is highly toxic. Thurston et al.[59] investigated the effects of pH on toxicity of ammonia and the ammonium ion to rainbow trout and fathead minnows (Pimephales promelas) in acute flow-through exposures. With pH changes of 6.5 to 9.0, the 96-h LC50s for unionized NH_3 increased from 0.13 to 0.53 mg/l, respectively, for rainbow trout. The trend was similar for fathead minnows; toxicity to both fish species increased as pH increased. This relationship was further confirmed by Erickson[60] and has also been observed with shrimp (Penaeus monodon).[61] The authors concluded that NH_4^+ exerts some measure of toxicity and that increased H^+ concentration increased the toxicity of NH_3; however, the nonionized form of ammonia is the most toxic. Sprague[4] concluded that although the toxicity of ammonia is complex, it is primarily governed by pH-related ionization and, thus, is predictable. Most metals act in the opposite way. Generally, the free ionic form is more toxic than the undissociated compound. Hence, metals are generally more toxic in the acid than the alkaline pH range. In a recent review of metal bioavailability and toxicity to fish in low-alkalinity lakes (pH ≤ 6.0 to 6.5), Spry and Wiener[62] stated that fish often have higher body or tissue burdens of aluminum, cadmium, lead, and mercury than do fish in nearby lakes with higher pH. They attributed this to greater metal bioavailability at low pH and to low concentrations of aqueous calcium that increase biological membrane permeability to these metals.

In addition to being more toxic and accumulated more at low pHs,[63–65] aluminum is eliminated from fish tissues more rapidly under these conditions.[66] The toxicity of aluminum to freshwater diatoms appears to depend on the interaction of pH-dependent availability, protonation of cell-surface ligands, and chelator-mediated metal speciation.[67] The effects of cadmium are also more pronounced at low pH in the green alga, Chlorella vulgaris.[68] Belanger and Cherry[37] found that the cladoceran, Ceriodaphnia dubia, was most sensitive to copper and zinc in low pH/low hardness waters and was least sensitive in high pH/high hardness waters. Bluegreen algae (Aphanizomenon gracile, Oscillatoria redekei) are also more sensitive to copper under acidic conditions.[69] The bioavailability of both copper[70] and zinc[71] is greater in acidic pH.

Table 1.   Toxicity of Chemicals to Common Carp (*Cyprinus carpio*) at Different pHs

Common and chemical name	24-h LC50 (mg/l) at pH					
	5.0	6.0	7.0	8.0	9.0	10
Molinate: S-ethyl hexahydro-1H-azepine-1-carbothioate	34	36	36	36	36	40
NIP (Nitrophen): 2,4-dichlorophenyl p-nitrophenyl ether	0.85	0.85	0.63	0.88	1.1	0.86
DCPA (Propanil): 3',4'-dichloropropionanilide	15	15	15	14	14	14
Trifluralin: α,α,α-trifluoro-2,6-dinitro-N,N-dipropyl-p-toluidine	1.8	1.0	0.72	0.87	1.2	0.80
CNP: p-nitrophenyl 2,4,6-trichlorophenyl ether	6.9	7.5	8.0	10	>13	>13
PCP-sodium: sodium salt of pentachlorophenol	0.06	0.06	0.10	0.12	0.13	>0.42

Source: From Nishiuchi, Y., *Suisan Zoshoku*, 25(2), 75–78, 1977. With permission.

The effects of pH on toxicities of several organic chemicals have been demonstrated in some recent studies.[72–74] At lower pHs, substituted phenols were typically more toxic to invertebrates and fishes.[75,76] Schooling and spawning behavior of fathead minnows were also disrupted,[75] and bioaccumulation increased as the pH decreased.[76] Toxicity of aminocarb, a base, increased significantly with increased pH; toxicity of acidic 2,4-D increased with decreased pH; and that of neutral fenitrothion did not change significantly at pH values of 4.6, 5.6, 6.9, and 8.5.[77] Lindane was significantly more toxic at pH 6 than at pH 4 and 8.[78] The difference in toxicity was attributed to two factors. First, the uptake of radiolabeled lindane was significantly less at pH 4 than at other pHs. Second, although equal amounts of total radioactivity were present at equilibrium at pH 6 and 8, a significantly greater amount of parent compound was present at pH 6, indicating possible chemical hydrolysis at pH 8.

The extensive works of Nishiuchi[8–15] describing the effects of pH and temperature on acute toxicity are presented in Tables 1 through 5. Since only LC50s and no confidence limits were provided in the reports, the significance of the reported values is unknown. However, the data identify the range of toxicity and demonstrate the trends of effects produced by modifying pH and temperature. Toxicities to common carp (*Cyprinus carpio*) of formulated pesticides, other than CNP and pentachlorophenol, generally were unaffected by pH. For these two compounds, toxicity was greater at lower pHs (Table 1). Information is also provided for 58 formulated pesticides from tests on a single fish species, the medaka (*Oryzias latipes*). Eight of the listed formulated pesticides were more toxic at lower pHs by a factor of fivefold or more (Table 2)—none was more toxic in basic media—the rest were apparently unaffected by pH. Four of the eight compounds—DNOC, DCNP, DNBPA, and DNBP—are phenolic and weakly acidic. Correspondingly, toxicity was dramatically increased at the low pHs.[9] Toxicities of dichlofluanid and captan[13] were also more toxic at pH 5.0 (60-fold and 130-fold, respectively) than at pH 10. In tests with toad larvae (*Bufo bufo japonicus* and *Bufo vulgaris formosus*), toxicities of most formulated pesticides were not affected by pH. Diquat and paraquat were more toxic to the larvae at high pHs (Tables 3 and 4). Hence, these reports suggest that while the toxicities of some formulated pesticides are increased at low pHs, toxicities of most organic chemicals are generally not affected by pH.

In interactive studies on pH and temperature, changes in pH alone did not affect the efficacy of the biocide, monochloramine, on control of Asiatic clams (*Corbicula fluminea*), but mortality

Table 2.   Toxicity of Chemicals to Medaka (*Oryzias latipes*) at Different pHs

Source and common and chemical name	24-h LC50 (mg/l) at pH					
	5.0	6.0	7.0	8.0	9.0	10
From Nishiuchi, Y., *Suisan Zoshoku*, 25(2), 75, 1977.						
DNOC: DNOC-sodium; sodium-4,6-dinitro-*o*-tolyloxide	0.21	0.56	0.83	1.6	2.3	10
DCNP: Chloronitrophen; sodium-2,4-dichloro-6-nitrophenoxide	0.93	0.17	0.27	0.56	0.68	4.8
DNBPA: Dinoseb-acetate, Aretit; 6-*sec*-butyl-2,4-dinitrophenyl acetate	0.020	0.023	0.032	0.036	0.056	0.20
DNBP: Dinoseb, Dormant, and Premerge; 2-*sec*-butyl-4,6-dinitrophenol	0.023	0.083	0.24	0.28	0.42	0.63
From Nishiuchi, Y., *Suisan Zoshuku*, 25(4), 151, 1978.						
Oxine-copper: copper 8-quinolinolate	0.075	0.078	0.080	0.083	0.087	0.095
DBEDC: copper *bis* (ethylenediamine) *bis* (dodecylbenzenesulfonate)	8.3	>13	>13	>13	>13	>13
Zineb: zinc ethylenebis (dithiocarbamate)	>42	>42	>42	>42	>42	>42
Maneb: manganese ethylenebis (dithiocarbamate)	5.8	6.1	6.4	6.5	7.1	7.5
Amobam: ammonium ethylenebis (dithiocarbamate)	>42	>42	>42	>42	>42	32
Polycarbamate: dizinc *bis* (dimethyl dithiocarbamate) ethylenebis (dithiocarbamate)	>1.3	1.3	0.83	0.70	0.86	> 1.3
Propineb: polymeric zinc propylenebis (dithiocarbamate)	23	36	38	42	42	42
TPN: tetrachloroisophthalonitrile	0.090	0.11	0.12	0.18	>0.18	> 0.18
EDDP: *O*-ethyl diphenyl phosphorodithiolate	1.2	1.4	1.7	1.9	2.0	2.3
Polyoxins	35	>42	>42	>42	>42	>42
Streptomycin	>42	>42	>42	>42	>42	>42
Hymexazol: 3-hydroxy-5-methylisoxazole	>42	>42	>42	>42	>42	>42
Echlomezol: 5-ethoxy-3-trichloromethyl-1,2,4-thiadiazol	8.4	10	>13	>13	>13	>13
Fentin hydroxide: triphenyltin hydroxide	0.27	0.26	0.25	0.26	0.27	0.25
Captan: *N*-(trichloromethylthio)-4-cyclohexene-1,2-dicarboximide	0.82	0.85	1.1	>1.8	>1.8	>1.8
Captafol: *N*-(1,1,2,2-tetrachloroethylthio)-4-cyclohexene-1,2-dicarboximide	0.13	0.15	0.19	>0.23	>0.23	>0.23
Quinomethionate: *S,S*-6-methylquinoxaline-2,3-diyl dithiocarbonate	13	26	32	38	>42	>42
Binapacryl: 2-*sec*-butyl-4,6-dinitrophenyl 3-methylcrotonate	0.10	0.10	0.13	>0.18	>0.18	>0.18
Thiophanate-methyl: dimethyl 4,4'-*O*-phenylenebis (3-thioallophanate)	>23	>23	>23	>23	>23	>23
Dichlofluanid: *N*-(dichlorofluoromethylthio)-*N'*, *N'*-dimethyl-*N*-phenylsulfamide	>0.32	>0.32	>0.32	0.81	1.0	1.0
CECA: *N*-(2-cyanoethyl) chloroacetamide	>42	>42	>42	>42	>42	>42
Dimethylymol: 5-butyl-2-dimethylamino-4-hydroxy-6-methylpyrimidine	>42	>42	>42	>42	>42	>42
From Nishiuchi, Y., *Suisan Zoshoku*, 27(3), 185, 1979.						
PCP-calcium: calcium pentachlorophenoxide	0.13	0.24	0.27	0.48	0.65	2.4

Table 2.   (Continued)

Source and common and chemical name	24-h LC50 (mg/l) at pH					
	5.0	6.0	7.0	8.0	9.0	10
PCP-barium: barium pentachlorophenoxide	0.19	0.22	0.27	0.50	0.68	3.2
CPMC: o-chlorophenyl methylcarbamate	15	14	8.3	7.5	12	23
MIPC: o-cumenyl methylcarbamate	10	13	13	13	13	15
XMC: 3,5-xylyl methylcarbamate	>42	>42	>42	>42	>42	>42
Chlomethoxynil: 2,4-dichlorophenyl 3-methoxy-4-nitrophenyl ether	>42	>42	>42	>42	>42	>42
Diquat dibromide: 6,7- dihydrodipyrido [1,2-a:2′,1′c] pyrazinedium dibromide	>42	>42	>42	>42	>42	>42
Butachlor: 2-chloro-2′,6′ diethyl-N-(butoxymethyl)acetanilide	1.3	1.2	1.3	1.3	1.3	1.3
Benthazone: 3-isopropyl-2,1,3-benzothiadiazinone-(4)-2,2-dioxide	27	>42	>42	>42	>42	>42
ACN: 2-amino-3-chloro-1,4-naphthoquinone	2.5	2.5	2.4	2.4	2.5	2.5
CAT (simazine): 2-chloro-4,6-bis (ethylamino)-1,3,5-triazine	>42	>42	>42	>42	>42	>42
DCMU (diuron): 3-(3,4-dichlorophenyl)-1,1-dimethylurea	>42	>42	>42	>42	>42	>42
EPTC: S-ethyl dipropylthiocarbamate	>42	>42	>42	>42	>42	>42
MCPA-sodium: sodium [(4-chloro-o-tolyl)oxyl] acetate	>42	>42	>42	>42	>42	>42
Diphenamid: N,N-dimethyl-2,2-diphenylacetamide	>42	>42	>42	>42	>42	>42
Linuron: 3-(3,4-dichlorophenyl)-1-methoxy-1-methylurea	35	>42	>42	>42	>42	>42
Lenacil: 3-cyclohexyl-5,6-trimethyleneuracil	12	19	21	19	19	19
From Nishiuchi, Y., *Suisan Zoshoku,* 27(4), 232, 1980.						
Dichlofluanid: N-(dichlorofluoromethylthio)-N′, N′-dimethyl-N-phenylsulfamide	0.75	0.86	0.86	0.86	27	48
Fentiazon; Celdion: 3-benzylideneamino-4-phenylthiazoline-2-thione	1.0	1.0	1.0	1.3	1.5	1.9
CNA; dicloran, ditranil: 2,6-dichloro-4-nitroaniline	23	23	22	23	23	23
Phthalide: 4,5,6,7-tetrachlorophthalide	>40	>40	>40	>40	>40	>40
Phenazine oxide: phenazine 5-oxide	20	26	26	26	26	26
PCNB; quintozene: pentachloronitrobenzene	>40	>40	>40	>40	>40	>40
DDPP; pyridinitril: 2,6-dichloro-3,5-dicyano-4-phenyl pyridine	0.20	0.20	0.20	0.21	0.25	0.42
CPA: pentachlorophenyl acetate	>10	>10	>10	>10	>10	>10
DAPA: sodium p-dimethylaminobenzenediazo sulfonate	20	20	20	25	32	38
Triazine, anilazine, Dyrene: 2,4-dichloro-6-(o-chloroanilino)-1,3,5-triazine	0.36	0.36	0.36	0.36	0.45	0.75
Dithianon: 2,3-dicyano-1,4-dithio 1,4-dihydro anthraquinone	0.013	0.66	0.034	0.068	0.5	21.7
Folpet: N-(trichloromethylthio) phthalimide	0.60	0.66	0.86	0.86	0.86	0.86
Dichlon: 2,3-dichloro-1,4-naphthoquinone	0.15	0.15	0.23	1.0	1.1	1.1
Captan: N-(trichloromethylthio)-4 cyclohexane-1,2-dicarboximide	0.36	0.45	1.1	3.5	12	48
Nitralin: 4-(methylsulfonyl)-2,6-dinitro-N,N-dipropylaniline	>10	>10	>10	>10	>10	>10

Table 3.   Toxicity of Chemicals to Toad Larvae (*Bufo bufo japonicus*) at Different pHs

Common and chemical name	24-h LC50 (mg/l) at pH					
	5.0	6.0	7.0	8.0	9.0	10
Ametryn(e): 2-ethylamino-4-isopropylamino-6-methylthio-1,3,5-triazine	4.0	3.8	4.4	5.6	5.6	5.6
Alachlor: 2-chloro-2',6'-diethyl-N-(methoxymethyl) acetanilide	11	11	12	12	12	12
Orthobencarb: S-(2-chlorobenzyl)-N,N-diethylthiocarbamate	2.4	2.7	2.7	2.7	2.7	2.7
Diquat dibromide: 6,7-dihydrodipyrido [1,2-a:2',1'-c] pyrazinedium dibromide	340	330	320	300	280	140
Paraquat dichloride: 1,1'-dimethyl-4,4'-bipyridinium dichloride	27	13	9.0	5.6	3.2	2.5
Paraquat: 1,1'-dimethyl-4,4'-bipyridinium *bis* (methyl sulfate)	27	27	25	23	22	20
Benfluralin: N-butyl-N-ethyl-α,α,α-trifluoro-2,6-dinitro-p-toluidine	11	11	11	11	11	11
2,4-D-dimethylamine: dimethylamine 2,4-dichlorophenoxyacetate	>40	>40	>40	>40	>40	>40
2,4-D-sodium: sodium 2,4-dichlorophenoxyacetate	>40	>40	>40	>40	>40	>40
CDAA (allidochlor): N,N-diallyl-2-chloroacetamide	2.8	2.9	3.3	3.3	3.4	3.5
CMMP(pentanochlor): 3'-chloro-2-methyl-p-valeroto-luidide	8.0	8.4	8.6	8.6	8.6	8.6
CMPT: 5-chloro-4-methyl-2-propinamido-1,3-thiazole	18	23	27	35	35	35
DBN (Dichlobenil): 2,6-dichlorobenzonitrile	14	15	16	17	18	21
IPC (chloropropham): isopropyl m-chloraocarbanilate	8.6	8.6	8.6	8.6	8.6	8.6
MCPA-allyl: allyl(4-chloro-o-tolyl) oxy acetate	0.64	0.64	0.64	0.64	1.8	1.9
MPCA-potassium: potassium(4-chloro-o-tolyl)oxy acetate	4.9	>130	>130	>130	>130	>130
MCPB-sodium: sodium 4-[(4-chloro-o-tolyl)oxy] butyrate	1.1	4.8	32	36	38	42
MCPE: 2-[(4-chloro-o-tolyl)oxy] ethanol	21	21	21	21	21	22

Source: From Nishiuchi, Y., *Suisan Zoshoku,* 27(4), 232–237, 1980; and 28(2): 107–112, 1980. With permission.

increased significantly with increased temperature.[79] The influence of pH on carbaryl and para-thion toxicity to midge larvae (*Chironomus riparius*) was almost nonexistent, while temperature increased toxicity from 2- to 100-fold over a 20° C temperature range.[80,81]

Temperature

An early review by Cairns et al.[6] on the effects of temperature on toxicity indicated that relatively few studies had been published as of 1975. Testing the influence of temperature on the toxicity of chemicals may be complicated because temperature alone may induce stress outside an organism's thermal tolerance zone. Because aquatic organisms are ectothermic, their metabolism increases approximately twofold with every 10° C rise in temperature. This concept is commonly referred to as Q_{10}.[82] Cairns et al.[6] pointed out that the rate of temperature change (acclimation) also may influence the survival of an organism—perhaps as much as the total change in temperature. They suggested that the toxic effects of substances that act on cellular enzymes involved in energy metabolism and the rate of chemical uptake are likely to increase with temperature. Thus, at higher temperatures, organisms may be exposed to greater amounts of toxicant because of increased diffusion or more active uptake. Increased metabolism would also likely

Table 4.   Toxicity of Chemicals to Toad Larvae (*Bufo vulgaris formosus*) at Different pHs

Common and chemical name	24-h LC50 (mg/l) at pH					
	5.0	6.0	7.0	8.0	9.0	10
Isoxathion: diethyl-5-phenyl-1,3-isoxazolyl phosphorothionate	5.6	6.0	8.2	8.4	8.7	8.7
Chlorpyriphos: O,O-diethyl-0–3,5,6-trichloro-2-pyridine phosphorothionate	13	15	16	15	16	15
Salithion: 2-methoxy-4H-1,3,2-benzodioxaphosphorin-2-sulfide	6.0	6.4	7.0	8.3	8.8	8.8
Dimethoate: dimethyl S-(N-methylcarbamoylmethyl) phosphorothiolothionate	42	42	>42	>42	>42	>42
Diazion: diethyl 2-isopropyl-4-methyl-6-pyrimidinyl phosphorothionate	9.0	11	11	11	11	12
Thiometon: S-(2-ethylthioethyl) dimethyl phosphorothiolothionate	5.8	6.2	8.1	8.7	8.7	8.7
Vamidothion: dimethyl S-[2-(1-methylcarbamoylethylthio)ethyl] phosphorothiolate	>42	>42	>42	>42	>42	>42
Prothiophos: O-2,4-dichlorophenyl-O-ethyl-S-propyl phosphorodithioate	>42	>42	>42	>42	>42	>42
Formothion: S-(N-formyl-N-methylcarbamoylmethyl) dimethyl phosphorothiolothionate	16	19	20	22	23	25
Malathion: S-[1,2-bis (ethoxycarbonyl)ethyl] dimethyl phosphorothiolothionate	16	17	18	20	27	27
CYAP: p-cyanophenyl dimethyl phosphorothionate	13	15	15	15	15	15
DDVP: 2,2-dichlorovinyl dimethyl phosphate	25	25	28	27	34	34
DEP: dimethyl 2,2,2-trichloro-1-hydroxyethylphosphonate	>42	>42	>42	>42	>42	>42
DMTP: S-[5-methoxy-2-oxo-2,3-dihydro-1,3,4-thiadiazolyl-(3)-methyl]dimethyl phosphorothiolothionate	19	19	19	20	23	20
EPN: ethyl p-nitrophenyl phenylphosphonothionate	6.4	7.5	9.3	10	11	11
ESP: S-(2-ethylsulfinyl-1-methylethyl)dimethyl phosphorothiolate	38	42	>42	>42	>42	>42
MEP: dimethyl 4-nitro-m-tolyl phosphorothionate	7.8	7.8	8.2	8.7	8.7	8.7
MPP: dimethyl 4-methylthio-m-tolyl phosphorothionate	3.9	3.9	4.0	5.8	6.0	6.0
PAP: S-[α-(ethoxycarbonyl)benzyl dimethyl] phosphorothiolothionate	6.6	8.3	8.5	8.5	8.5	8.5
BPMC: O-sec-butylphenyl methylcarbamate	21	21	21	21	21	21
MPMC: 3,4-xlyl methylcarbamate	25	27	29	34	36	38
MTMCS: m-tolyl methylcarbamate	>42	>42	>42	>42	>42	>42
NAC: 1-naphthyl methylcarbamate	3.5	3.8	3.9	3.9	3.9	3.9
PHC: O-isopropoxyphenyl methylcarbamate	19	23	26	27	27	27

Source: From Nishiuchi, Y., *Suisan Zoshoku,* 25(3), 108–111, 1977. With permission.

parallel increased rates of movement of water and solutes across the gills or other cellular membranes. In summary, the various factors mentioned suggest that temperature increases are likely to increase toxicity. However, some studies do not support this relationship. The toxicities of DDD, DDT, methoxychlor,[83] pyrethroids,[84] and ammonia[85,86] are negatively correlated with temperature.

Sprague[4] reported that he could find no single pattern to explain the effects of temperature on the toxicity of pollutants to aquatic organisms. Changes in toxicity may reflect the type of test species and methods used to determine toxicity rather than indicate temperature effects per se. He further stated that temperature effects were highly unpredictable, citing tests on the toxicity

Table 5.   Toxicity of Chemicals to *Daphnia pulex* at Different Temperatures

Source and common and chemical name	3-h LC50 (mg/l) at temperature (°C)			
	10	17.5	25.0	32.5
Isoxathion: diethyl-5-phenyl-1,3-isoxazolyl phosphorothionate	0.018	0.004	0.001	0.001
Chloropyriphos: *O,O*-diethyl-*O*-3,5,6-trichloro-2-pyridine phosporothionate	0.015	0.013	0.007	0.007
Rotenone: Rotenone suspension	0.038	0.023	0.010	0.005
DNBD: Dinoseb; 2-*sec*-butyl-4,6-dinitrophenol	>40	23	10	3.2
DNOC: DNOC-sodium; sodium-4,6-dinitro-*o*-tolyloxide	28	20	10	7.8
MIPC: *o*-cumenyl methyl carbamate	0.11	0.085	0.070	0.043
PAP: *S*-[α-(ethoxycarbonyl)benzyl dimethyl] phosphorothiolothionate	0.13	0.015	0.001	0.0006
Dithianon: 2,3-dicyana-1,4-dithio-1,4-dihydro anthraquinone	>40	>40	>40	30
Orthobencarb: *S*-(2-chlorobenzyl)-*N*,*N* diethylthiocarbamate	2.3	2.2	1.0	0.85
Butachlor: 2-chloro-2',6'-diethyl-*N*-(butoxymethyl)acetanilide	>40	>40	21	2.5
PCP-sodium: sodium salt of pentachlorophenoxide	15	7.3	3.0	1.8

Source: From Nishiuchi, Y., *Suisan Zoshoku*, 30, 158–162, 1982. With permission.

of zinc to rainbow trout. In these tests, toxicity inexplicably increased, decreased, or produced no change as temperature changed within the temperature tolerance range of the fish. Hodson and Sprague[87] reported that zinc toxicity to Atlantic salmon (*Salmo salar*) was most severe at high temperatures within the first 24 h, but highest at low temperatures after 2 weeks of exposure. However, Atlantic salmon that died at 19° C had about 2.5 times more zinc in their gill tissue than did those that died at 3 and 11° C. The effects of temperature on the toxicity of heavy metals suggest that rising temperatures stimulate uptake in acute exposures, but this phenomenon often ceases or even reverses in long-term exposures.[87] In tests with four inorganics on five species of fish, toxicity due to temperature changes (5 to 30° C) increased only by a factor of 3, but differences as great as ten-fold were observed between different fish species at a given temperature.[88] The toxicities of aluminum,[89] arsenate,[90] and tributyltin[91] also increased slightly with temperature, but no changes were observed with arsenite.[90]

Sprague[4] also reported that numerous chemicals are more toxic in warm water than in cold water. The effect was noted for a variety of compounds and species, including endrin (5 species of fish), 14 pesticides (rainbow trout and bluegills, *Lepomis macrochirus*), 3 herbicides (cutthroat, *Oncorhynchus clarki*, and lake trout, *Salvelinus namaycush*), and pentachlorophenol (common carp). Chemicals reported to be more toxic at low temperatures included DDT and permethrin, a synthetic pyrethroid. Pesticides whose toxicities were not affected by temperature included mexacarbate, rotenone, and toxaphene. For this latter group, there was no effect on eventual lethality though survival times varied.

Likewise, Nishiuchi[15] found that the acute toxicities of 11 pesticides increased (1.3- to 217-fold) with temperature (10 to 32.5° C) in the daphnid, *Daphnia pulex*. The isopod, *Asellus aquaticus*, was more sensitive to lindane as temperature increased.[92] When isopods were exposed to phenol,[93] immobilization was more rapid at 10° C than at 20° C, but recovery was significantly faster at 20° C. Midge larvae (*Chironomus plumosus*) were slightly more sensitive (1.9-fold) to pentachlorophenol at 35° C than at 15° C.[94] Studies with daphnids (*D. magna*, *D. pulex*)[95] and rotifers (*Brachionus plicatus*),[96] including an array of inorganic and organic chemicals, also resulted in increased toxicity with increased temperature, but the differences were not always statistically significant. However, although the acute toxicity (96-h LC50) of endosulfan to European eels (*Anguilla anguilla*) tended to increase with increasing temperature (15° C = 38 μg/l, 29° C = 20 μg/l), that for lindane tended to decrease (15° C = 320 μg/l, 29° C = 450 μg/l).[97,98] The fish rohu (*Labeo rohita*), exposed to urea developed abnormal behavior at higher temperatures,[99] and although a trend towards increasing toxicity with increasing temperature occurred with blue-gills exposed to anthracene, no significant temperature effect was observed.[100]

Bluegills had lower uptake and elimination rates of benzo-(a)-pyrene at 13° C than at 23° C; an effect attributed to increased fish metabolism.[101] Edgren,[102] using Eurasian perch (*Perca fluviatilis*) for DDT uptake and elimination experiments, found twice as much DDT in fish tested at 15° C than in those tested at 5° C. No significant difference in elimination rates was found in fish at three temperatures (8, 12, and 16° C). Similar results were observed with two PCBs (2,3′,4′,5-tetrachlorobiphenyl and 2,2′,4,4′,5,5′-hexachlorobiphenyl) using the same methods: accumulation rates were positively related to temperature, whereas depuration was unaffected. The bioconcentration of di-2-ethylhexyl phthalate increased significantly with temperature in sheepshead minnows (*Cyprinodon variegatus*).[103] Black et al.[104] found that respiratory functions decreased as temperature was lowered, resulting in decreased oxygen and toxicant uptake efficiencies.

Mayer and Ellersieck[3] pointed out that toxicity increased with temperature for most chemicals. However, compounds such as DDT, methoxychlor, and some of the pyrethroids did not follow this pattern. Gammon et al.[105] addressed negative temperature coefficients (toxicity decreases as temperature increases) and reported that the neurophysiological effects of *d-trans*-allethrin on the cockroach (*Periplaneta americana*) are excitatory at both high and low temperatures; the temperature-dependent differences being in the types of nerves directly affected. Both peripheral and central nervous systems are affected at 32° C, but only the peripheral system at 15° C. They concluded that the negative temperature coefficient of allethrin toxicity could be a result of increases in sensitivity of peripheral nerves at low temperatures. Differences in toxicity due to temperature have also been attributed to differences in respiration rate,[106] chemical absorption,[107] and excretion and detoxification of chemicals.[17]

The effects of temperature on toxicity generally conform to the Q_{10} concept.[3] Rises in temperature caused the acute toxicity of most chemicals to increase by an average of 3.1 times per 10° C rise; the factor was higher (up to 5.1 times) for organophosphate insecticides.[3] Cairns et al.[6] stated that changes observed with organophosphates may be very complex, due to the change in acetylcholinesterase (AChE) activity from temperature alone. The results of Hogan[108] demonstrate the increase in AChE activity with increased temperature for bluegills by the regression log AChE = 2.4539 + 0.0165 (° C), or a factor of 1.5-fold increase per 10° C. The slopes for organophosphates are greater than those for other chemicals,[3] with the average slope being 0.7113 (95% CL, 0.5836–0.8390) or a factor of 5.1-fold for organophosphates alone, and 0.4956 (95% CL, 0.3775–0.6137) or a factor of 3.1-fold for other chemicals. An increase in temperature, thus accelerating AChE activity, may simultaneously increase the rate of AChE inhibition by organophosphates. The interaction appears to be slightly more than additive. Carbamates, also AChE inhibitors, do not seem to be affected as much as organophosphates, on the average, and this decreased effect may be due to differences in modes of action between carbamates and organophosphates in AChE inhibition.[109]

TOXICITY VS. BIOAVAILABILITY/EXPOSURE

In recent studies,[110-112] the roles of pH and temperature were assessed in relation to modifying toxicity, bioavailability, and exposure. Acute toxicity tests with amphipods (*Gammarus pseudolimnaeus*) and rainbow trout were conducted using four chemicals (4-nitrophenol, 2,4-dinitrophenol, terbufos, and trichlorfon) in a factorial design testing pH (6.5, 7.5, 8.5, 9.5) and temperature (7, 12, 17° C) combinations.[110] Chronic tests with rainbow trout were performed for temperature and the two nitrophenols.[111] The hypotheses were

- *Acute toxicity:* physicochemical factors do not affect toxicity per se, but do affect bioavailability. Toxicity under different conditions of pH and temperature may be predictable from octanol/water partition coefficients (K_{ow}).
- *Chronic toxicity:* physicochemical factors only alter the rate of intoxication and, hypothetically, the no-observed-effect concentration (NOEC) in chronic exposures may not change with variations in temperature—only the time required to attain the same or similar no-effect concentration.

Toxicities of all four chemicals were significantly affected by pH in all tests except that of *Gammarus* and terbufos.[110] Both nitrophenols are ionizable weak acids, and more toxic (25- to 50-fold) at low pH than at high pH. The toxicity of terbufos to *Gammarus* and rainbow trout was less at pH 7.5 than at higher or lower pH values. The toxicity of trichlorfon increased with pH. In alkaline conditions, trichlorfon rapidly hydrolyzes to dichlorvos,[109] estimated to be 2.6 to 350 times more toxic than trichlorfon.[3] Trichlorfon is one of the few organophosphate insecticides that hydrolyzes to a more toxic compound.[109] For most organophosphate insecticides, hydrolysis products are generally poor inhibitors of acetylcholinesterase,[109] and are therefore less toxic than the parent compound. The relatively minor effects of pH on terbufos toxicity were similar to those of a previous study.[3] Terbufos is a thionate (i.e., sulfur-containing), and toxicity is usually not affected by high pH. Thionates do not hydrolyze readily in alkaline conditions.[109] Temperature also significantly affected the toxicity of all four chemicals to both species; toxicity increased with temperature in all tests except for rainbow trout exposed to nitrophenols.

Preliminary K_{ow} values, determined under selected pH and temperature conditions used in the acute toxicity tests, were affected by pH and temperature and were inversely related to the 96-h LC50s (Table 6).[112] The correlation of K_{ow} to acute toxicity (Figure 1) was best with the two nitrophenols (r = 0.97 to 0.99), but poor for the two organophosphate insecticides (r = 0.22 to 0.75). Terbufos toxicity was affected the least by pH and temperature and relates to little change in hydrolysis. Bioconcentration was also affected by temperature and pH, and was directly related to toxicity in nitrophenol tests, but not for terbufos. Terbufos bioconcentration factors for both amphipods and rainbow trout were approximately ten times greater than those for nitrophenols. This would be expected, since terbufos, with a higher log K_{ow} and very low water solubility, is much more lipophilic than the very water-soluble nitrophenols and thus would bioconcentrate more.

Water temperature did not significantly affect NOEC values in chronic tests.[111] For 4-nitrophenol, time-independent NOEC values at 7, 12, and 17° C were 3.4, 3.4, and 2.2 mg/l, respectively, for survival, and 1.2, 1.2, and 1.2 mg/l, respectively, for growth. For 2,4-dinitrophenol, NOEC values at 7, 12, and 17° C were 1.3, 1.9, and 1.6 mg/l, respectively, for survival, and 1.1, 0.50, and 0.80 mg/l for growth. However, temperature did affect the rate at which NOEC values were reached, with the rates being higher at higher temperatures. Time-independent NOEC values were achieved after 14 days at 12 and 17° C and remained constant throughout the test, whereas NOEC values for tests conducted at 7° C were established only after 42 days exposure. These findings are similar to those of Sprague[4] in which temperature effects are evident in acute, but not in chronic studies. Thus, NOECs in chronic studies do not change with temperature; only the time required to attain the same NOEC is altered.

Table 6.   Relation of Physicochemical Factors, Acute Toxicity, and Whole-Body Residue
Accumulation in Amphipods (*Gammarus pseudolimnaeus*) and Rainbow Trout
(*Oncorhynchus mykiss*) Exposed to 4-Nitrophenol, 2,4-Dinitrophenol, Terbufos, and
Trichlorfon

Species	Temperature (°C)	pH	Log K_{ow}	96-h LC50 (mg/l)[a][110]	Residue[b] (µg/g)	Bioconcentration factor[c]
				4-Nitrophenol		
Rainbow trout	7	6.5	1.96	3.1	122	39
	17	6.5	1.96	2.9	162	56
	12	7.5	1.51	6.9	124	18
	7	9.5	0.637	77	140	1.8
	17	9.5	0.467	82	223	2.7
Amphipods	7	6.5	1.96	2.9	119	41
	17	6.5	1.96	3.6	209	58
	12	7.5	1.51	6.6	127	19
	7	9.5	0.637	49	87	1.8
	17	9.5	0.467	39	84	2.2
				2,4-Dinitrophenol		
Rainbow trout	7	6.5	−0.0894	0.31	8.6	28
	17	6.5	−0.0538	0.41	8.2	20
	12	7.5	−0.279	1.8	14	7.9
	7	9.5	−0.590	21	36	1.7
	17	9.5	−0.518	35	68	1.9
Amphipods	7	6.5	−0.0894	0.71	22	31
	17	6.5	−0.0538	0.48	39	81
	12	7.5	−0.279	3.6	115	32
	7	9.5	−0.590	22	106	4.8
	17	9.5	−0.518	19	84	4.4
				Terbufos		
Rainbow trout	7	6.5	2.84	12	4.7	403
	17	6.5	3.43	6.7	2.2	336
	12	7.5	3.31	16	4.2	268
	7	9.5	3.13	14	5.6	387
	17	9.5	3.16	10	3.8	380
Amphipods	7	6.5	2.84	0.40	0.08	195
	17	6.5	3.43	0.26	0.06	231
	12	7.5	3.31	0.28	0.12	428
	7	9.5	3.13	0.55	0.08	145
	17	9.5	3.16	0.08	<0.01	8
				Trichlorfon		
Rainbow trout	7	6.5	0.20	41	—[d]	
	17	6.5	0.55	2.5	—	
	12	7.5	0.33	1.9	—	
	7	9.5	0.61	0.52	—	
	17	9.5	0.34	0.33	—	
Amphipods	7	6.5	0.20	11	—	
	17	6.5	0.55	0.14	—	
	12	7.5	0.33	0.35	—	
	7	9.5	0.61	0.07	—	
	17	9.5	0.34	0.07	—	

[a]Terbufos = µg/l.
[b]Interpolated residue at the 96-h LC50 derived from Howe et al.[110]
[c]BCF = body concentration (µg compound/g tissue) ÷ exposure concentration (mg/l).
[d]Not determined.

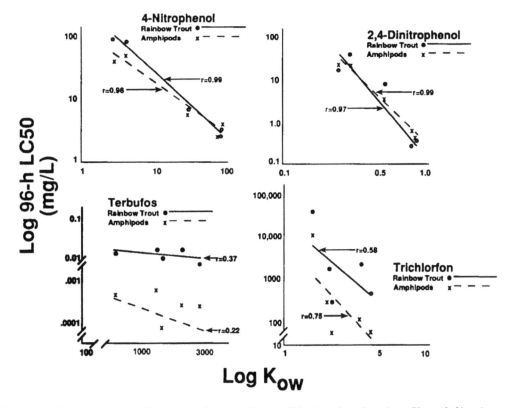

Figure 1. Relation of acute toxicity to octanol/water partition coefficients under selected conditions of pH and temperature.

SUMMARY

- Total hardness appears to have little if any effect on the acute toxicity of organic chemicals; the observed effects are probably due to pH differences in test solutions of differing hardnesses.
- Metals are generally more toxic in soft water than in hard water.
- pH appears to cause greater change in toxicity than other factors.
- Decreases in pH may increase the toxicity of acids, but decrease that of bases.
- Increased water temperature generally increases toxicity, but may decrease it with some chemicals.
- Temperature effects on toxicity often conform to the Q_{10} concept and may be predicted—organophosphates: log 96-h LC50 = $0.7113/10°$ C increase; other chemicals: log 96-h LC50 = $0.4956/10°$ C increase.
- Regression slopes of toxicity appear consistent among species within a chemical for both pH and temperature, indicating chemical rather than biological differences in toxicity.
- Physicochemical factors may not affect acute toxicity per se, but do affect bioavailability or chemical form and, therefore, exposure. Octanol/water partition coefficients, which may correlate well with acute toxicity, are altered by temperature and pH. Partition coefficients determined under physicochemical conditions of interest could replace some biological testing since the partition coefficient/acute toxicity relationship has been well established by other researchers (e.g., Veith et al.[113]).
- Temperature may only alter the rate of intoxication in chronic exposures—NOECs for fish exposed to nitrophenols are not affected by temperature; only the time required to attain the same NOEC varies.

ACKNOWLEDGMENTS

We thank Valerie Coseo, Computer Sciences Corporation, for typing the manuscript. This project was funded in part by the Ecological Risk Assessment Research Program, U.S. Environmental Protection Agency.

REFERENCES

1. Fales, R.R., The influence of temperature and salinity on the toxicity of hexavalent chromium to the grass shrimp *Palaemonetes pugio* (Holthuis), *Bull. Environ. Contam. Toxicol*, 20, 447–450, 1978.

2. Mayer, F.L., Jr. and Ellersieck, M.R., Manual of acute toxicity: Interpretation and data base for 410 chemicals and 66 species of freshwater animals, U.S. Fish. Wildl. Serv. Resour. Publ. No. 160, Department of the Interior, Washington, D.C., 1986.

3. Mayer, F.L., Jr. and Ellersieck, M.R., Experiences with single-species tests for acute toxic effects in freshwater animals, *Ambio*, 17, 367–375, 1988.

4. Sprague, J.B., Factors that modify toxicity, in Rand, G.M. and Petrocelli, S.R., Eds., *Fundamentals of Aquatic Toxicology*, Hemisphere, Washington, D.C., 1985, 123–163.

5. Hunn, J.B. and Allen, J.L., Movement of drugs across the gills of fishes, *Annu. Rev. Pharmacol.*, 14, 47–55, 1974.

6. Cairns, J., Jr., Heath, A.G., and Parker, B.C., The effects of temperature upon the toxicity of chemicals to aquatic organisms, *Hydrobiologia*, 47, 135–171, 1975.

7. Hamelink, J., Bioavailability of chemicals in aquatic environments, in Maki, A.W., Dickson, K.L., and Cairns, J., Jr., Eds., *Biotransformation and Fate of Chemicals in the Environment*, American Society for Microbiology, Washington, D.C., 1980, 56–62.

8. Nishiuchi, Y., Toxicity of formulated pesticides to some freshwater organisms. XXXXII, *Suisan Zoshoku*, 25(1), 27–35, 1977.

9. Nishiuchi, Y., Toxicity of formulated pesticides to some freshwater organisms. XXXXIII, *Suisan Zoshoku*, 25(2), 75–78, 1977.

10. Nishiuchi, Y., Toxicity of formulated pesticides to some freshwater organisms. XXXXXI, *Suisan Zoshoku*, 25(4), 151–155, 1978.

11. Nishiuchi, Y., Toxicity of formulated pesticides to some freshwater organisms. LXIV, *Suisan Zoshoku*, 27(3), 185–189, 1979.

12. Nishiuchi, Y., Toxicity of formulated pesticides to some freshwater organisms. LXIII, *Suisan Zoshoku*, 27(4), 232–237, 1980.

13. Nishiuchi, Y., Toxicity of formulated pesticides to some freshwater organisms. LXXIV, *Suisan Zoshoku*, 28(2), 107–112, 1980.

14. Nishiuchi, Y., Toxicity of formulated pesticides to some freshwater organisms. XXXII, *Suisan Zoshoku*, 25(3), 108–111, 1977.

15. Nishiuchi, Y., Toxicity of pesticides to freshwater organisms. LXXIX. Effects of water temperature on the sensitivity of *Daphnia pulex* to pesticides, *Suisan Zoshoku*, 30, 158–162, 1982.

16. Inglis, A. and Davis, E.L., Effects of water hardness on the toxicity of several organic and inorganic herbicides to fish, U.S. Fish. Wildl. Serv. Tech. Pap. No. 67, Department of the Interior, Washington, D.C., 1972.

17. Sprague, J.B., Measurement of pollutant toxicity to fish. II. Utilizing and applying bioassay results, *Water Res.*, 4, 3–32, 1970.

18. Pickering, Q.H. and Henderson, C., Acute toxicity of some important petrochemicals to fish, *J. Water Pollut. Control Fed.*, 35, 1419–1429, 1964.

19. Ferrando, M.D., Almar, M.M., and Andreu, E., Lethal toxicity of lindane on a teleost fish, *Anguilla anguilla*, from Albufera Lake (Spain). Hardness and temperature effects, *J. Environ. Sci. Health*, B23, 45–52, 1988.

20. Smith, P.D., Brockway, D.L., and Stancil, F.E., Jr., Effects of hardness, alkalinity and pH on the toxicity of pentachlorophenol to *Selenastrum capricornutum* (Printz), *Environ. Toxicol. Chem.*, 6, 891–900, 1987.

21. Eyanoer, H.F., Upatham, E.S., Duangsawasdi, M., and Tridech, S., Effect of water hardness and temperature on the toxicity of detergents to the freshwater fish *Puntius gonionotus*, *J. Sci. Soc. Thail.*, 11, 67–78, 1985.

22. Lewis, M.A., The effects of mixtures and other environmental modifying factors on the toxicities of surfactants to freshwater and marine life, *Water Res.*, 26, 1013–1023, 1992.

23. Weber, J.B., Interaction of organic pesticides with particulate matter in aquatic and soil systems, in Gould, R.F., Ed., *Fate of Organic Pesticides in the Aquatic Environment*, Adv. Chem. Ser. III, American Chemical Society, Washington, D.C., 1972, 55–129.

24. Schoettger, R.A. and Mauck, W.L., Toxicity of experimental forest insecticides to fish and aquatic invertebrates, in Swain, W.R. and Ivanikiw, N.K., Eds., Vol. II, Proc. Second USA-USSR Symp. Effects of Pollutants Upon Aquatic Ecosystems, Spec. Publ. No. EPA-600/3-78-076, U.S. Environmental Protection Agency, Duluth, MN, 1978, 11–27.

25. Woodward, D.F. and Mauck, W.L., Toxicity of five forest insecticides to cutthroat trout and two species of aquatic invertebrates, *Bull. Environ. Contam. Toxicol.*, 25, 846–853, 1980.

26. Neubert, J., On the acute toxicity of trichlorfon against selected aquatic organisms, *Acta Hydrochim. Hydrobiol.*, 14, 643–651, 1986.

27. Marking, L.L. and Olson, L.E., Toxicity of the lampricide 3-trifluoromethyl-4-nitrophenol (TFM) to non-target fish in static tests, U.S. Fish. Wildl. Serv. Invest. Fish Control 60, Department of the Interior, Washington, D.C., 1975.

28. Bills, T.D., Marking, L.L., Howe, G.E., and Rach, J.J., Relation of pH to toxicity of lampricide TFM in the laboratory, Great Lakes Fishery Comm. Tech. Rep. Ann Arbor, MI, No. 53, 9–17, 1988.

29. Wang, W., Factors affecting metal toxicity to (and accumulation by) aquatic organisms—Overview, *Environ. Int.*, 13, 437–457, 1987.

30. Pynnonen, K., Aluminum accumulation and distribution in the freshwater clams Unionidae, *Comp. Biochem. Physiol.*, C97, 111–118, 1990.

31. Maier, K.J. and Knight, A.W., The toxicity of waterborne boron to *Daphnia magna* and *Chironomus decorus* and the effects of water hardness and sulfate on boron toxicity, *Arch. Environ. Contam. Toxicol.*, 20, 282–287, 1991.

32. Michibata, H., Sahara, S., and Kojima, M.K., Effects of calcium and magnesium ions on the toxicity of cadmium to the egg of the teleost, *Oryzias latipes*, *Environ. Res.*, 40, 110–114, 1986.

33. Nakagawa, H. and Ishio, S., Effects of water hardness on the toxicity and accumulation of cadmium in eggs and larvae of medaka, *Oryzias latipes*, Nippon Suisan Gakkaishi, *Bull. Jpn. Soc. Sci. Fish.*, 55, 321–326, 1989.

34. Pascoe, D., Evans, S.A., and Woodworth, J., Heavy metal toxicity to fish and the influence of water hardness, *Arch. Environ. Contam. Toxicol.*, 15, 481–488, 1986.

35. Winner, R.W., Interactive effects of water hardness and humic acid on the chronic toxicity of cadmium to *Daphnia pulex*, *Aquat. Toxicol.*, 8, 281–293, 1986.

36. Diamond, J.M., Winchester, E.L., Mackler, D.G., Rasnake, W.J., Fanelli, J.K., and Gruber, D., Toxicity of cobalt to freshwater indicator species as a function of water hardness, *Aquat. Toxicol.*, 22, 163–179, 1992.

37. Belanger, S.E., and Cherry, D.S., Interacting effects of pH acclimation, pH, and heavy metals on acute and chronic toxicity to *Ceriodaphnia dubia* (Cladocera), *J. Crust. Biol.*, 10, 225–235, 1990.

38. Belanger, S.E., Farris, J.L., and Cherry, D.S., Effects of diet, water hardness, and population source on acute and chronic copper toxicity to *Ceriodaphnia dubia*, *Arch. Environ. Contam. Toxicol.*, 18, 601–611, 1989.

39. Belanger, S.E., Farris, J.L., Cherry, D.S., and Cairns, J., Jr., Validation of *Corbicula fluminea* growth reductions induced by copper in artificial streams and river systems, *Can. J. Fish. Aquat. Sci.*, 47, 904–914, 1990.

40. Clements, W.H., Farris, J.L., Cherry, D.S., and Cairns, J., Jr., The influence of water quality on macroinvertebrate community responses to copper in outdoor experimental streams, *Aquat. Toxicol.*, 14, 249–262, 1989.

41. Everall, N.C., Macfarlane, N.A.A., and Sedgwick, R.W., The interactions of water hardness and pH with the acute toxicity of zinc to the brown trout, *Salmo trutta* L, *J. Fish. Biol.*, 35, 27–36, 1989.

42. Moni, D. and Dhas, S.S.M., Effects of water hardness on the toxicity of zinc to *Sarotherodon mossambicus* Peters, *Uttar Pradesh J. Zool.*, 9, 263–270, 1989.

43. Paulauskis, J.D. and Winner, R.W., Effects of water hardness and humic acid on zinc toxicity to *Daphnia magna* Straus, *Aquat. Toxicol.*, 12, 273–290, 1988.

44. Dwyer, F.J., Burch, S.A., Ingersoll, C.G., and Hunn, J.B., Toxicity of trace element and salinity mixtures to striped bass *Morone saxatillis* and *Daphnia magna*, *Environ. Toxicol. Chem.*, 11, 513–520, 1992.

45. Jin, H., Zhang, Y., and Yang, R., Toxicity and distribution of copper in an aquatic microcosm under different alkalinity and hardness, *Chemosphere*, 22, 577–596, 1991.

46. Lauren, D.J. and McDonald, D.G., Influence of water hardness, pH, and alkalinity on the mechanisms of copper toxicity in juvenile rainbow trout, *Salmo gairdneri, Can. J. Fish. Aquat. Sci.*, 43, 1488–1496, 1986.

47. Everall, N.C., Macfarlane, N.A.A., and Sedgwick, R.W., The effects of water hardness upon the uptake, accumulation and excretion of zinc in the brown trout, *Salmo trutta* L., *J. Fish. Biol.*, 35, 881–892, 1989.

48. McMurtry, M.J., Wales, D.L., Scheider, W.A., Beggs, G.L., and Dimond, P.E., Relationship of mercury concentrations in lake trout (*Salvelinus namaycush*) and smallmouth bass (*Micropterus dolomieui*) to the physical and chemical characteristics of Ontario Canada lakes, *Can. J. Fish. Aquat. Sci.*, 46, 426–434, 1989.

49. Wren, C.D., Scheider, W.A., Wales, D.L., Muncaster, B.W., and Gray, I.M., Relation between mercury concentrations in walleye (*Stizostedion vitreum vitreum*) and northern pike (*Esox lucius*) in Ontario lakes and influence of environmental factors, *Can. J. Fish. Aquat. Sci.*, 48, 132–139, 1991.

50. Bradley, R.W. and Sprague, J.B., The influence of pH, water hardness, and alkalinity on the acute lethality of zinc to rainbow trout (*Salmo gairdneri*), *Can. J. Fish. Aquat. Sci.*, 42, 731–736, 1985.

51. Davies, P.H., Chemical Factors Affecting Bioavailability and Toxicity of Cadmium to Rainbow Trout, Ph.D. Dissertation, Colorado State University, Fort Collins, 1989.

52. Hutchinson, N.J. and Sprague, J.B., Lethality of trace metal mixtures to American flagfish in neutralized acid water, *Arch. Environ. Contam. Toxicol.*, 18, 249–254, 1989.

53. Playle, R.C., Gensemer, R.W., and Dixon, D.G., Copper accumulation on gills of fathead minnows: Influence of water hardness, complexation and pH of the gill micro-environment, *Environ. Toxicol. Chem.*, 11, 381–391, 1992.

54. Pratap, H.B., Fu, H., Lock, R.A.C., and Bonga, S.E.W., Effect of waterborne and dietary cadium on plasma ions of the teleost *Oreochromis mossambicus* in relation to water calcium levels, *Arch. Environ. Contam. Toxicol.*, 18, 568–575, 1989.

55. Winner, R.W. and Gauss, J.D., Relationship between chronic toxicity and bioaccumulation of copper, cadmium and zinc as affected by water hardness and humic acid, *Aquat. Toxicol.*, 8, 149–161, 1986.

56. Playle, R.C. and Wood, C.M., Water chemistry changes in the gill micro-environment of rainbow trout: Experimental observations and theory, *J. Comp. Physiol.*, B159, 527–538, 1989.

57. Buckler, D.R., Comparative Toxicity and Availability of Dissociable Compounds to Fishes as Affected by Ambient pH, Ph.D. Dissertation, Utah State University, Logan, 1987.

58. Goldstein, A., Aronow, L., and Kalman, S.M., Principles of drug action, in *The Basis of Pharmacology*, 2nd ed., John Wiley & Sons, New York, 1974.

59. Thurston, R.V., Russo, R.C., and Vinogradov, G.A., Ammonia toxicity to fishes. Effects of pH on the toxicity of the unionized ammonia species, *Environ. Sci. Technol.*, 15, 837–839, 1981.

60. Erickson, R.J., An evaluation of mathematical models for the effects of pH and temperature on ammonia toxicity to aquatic organisms, *Water Res.*, 19, 1047–1058, 1985.

61. Chen, J.-C. and Sheu, T.-S., Effect of ammonia at different pH levels on larval *Penaeus monodon*, *J. Fish Soc. Taiwan*, 17, 59–64, 1990.

62. Spry, D.J. and Weiner, J.G., Metal bioavailability and toxicity to fish in low-alkalinity lakes: A critical review, *Environ. Pollut.*, 71, 243–304, 1991.

63. Cleveland, L., Little, E.E., Ingersoll, C.G., Wiedmeyer, R.H., and Hunn, J.B., Sensitivity of brook trout to low pH, low calcium and elevated aluminum concentrations during laboratory pulse exposures, *Aquat. Toxicol.*, 19, 303–318, 1991.

64. Ingersoll, C.G., Mount, D.R., Gulley, D.D., La Point, T.W., and Bergman, H.L., Effects of pH, aluminum, and calcium on survival and growth of eggs and fry of brook trout (*Salvelinus fontinalis*), *Can. J. Fish. Aquat. Sci.*, 47, 1580–1592, 1990.

65. Marungi, J.I. and Robinson, J.W., Uptake and accumulation of aluminum by fish—The modifying effect of added ions, *J. Environ. Sci. Health*, A27, 713–719, 1992.

66. Cleveland, L., Buckler, D.R., and Brumbaugh, W.G., Residue dynamics and effects of aluminum on growth and mortality in brook trout, *Environ. Toxicol. Chem.*, 10, 243–248, 1991.

67. Gensemer, R.W., The effects of pH and aluminum on the growth of the acidophilic diatom *Asterionella ralfsii* var. *americana, Limnol. Oceanogr.*, 36, 123–131, 1991.

68. Rachlin, J.W. and Grosso, A., The effects of pH on the growth of *Chlorella vulgaris* and its interactions with cadmium toxicity, *Arch. Environ. Contam. Toxicol.*, 20, 505–508, 1991.

69. Luederitz, V. and Nicklisch, A., The effect of pH on copper toxicity to blue-green algae, *Int. Rev. Gesamt. Hydrobiol.*, 74, 283–291, 1989.
70. Meador, J.P., The interaction of pH, dissolved organic carbon and total copper in the determination of ionic copper and toxicity, *Aquat. Toxicol.*, 19, 13–32, 1991.
71. Bentley, P.J., Influx of zinc by channel catfish (*Ictalurus punctatus*): Uptake from external environmental solutions, *Comp. Biochem. Physiol.*, C101, 215–217, 1992.
72. Neilson, A.H., Allard, A.-S., Fisher, S., Malmberg, M., and Viktor, T., Incorporation of a subacute test with zebra fish into a hierarchical system for evaluating the effect of toxicants in the aquatic environment, *Ecotoxicol. Environ. Saf.*, 20, 82–97, 1990.
73. Stehyl, G.R. and Hayton, W.L., Effect of pH on the accumulation kinetics of pentachlorophenol in goldfish, *Arch. Environ. Contam. Toxicol.*, 19, 464–470, 1990.
74. Stephenson, G.L., Kaushik, N.K., and Solomon, K.R., Acute toxicity of pure pentachlorophenol and a technical formulation to three species of *Daphnia*, *Arch. Environ. Contam. Toxicol.*, 20, 73–80, 1991.
75. Holcombe, G.W., Fiandt, J.T., and Phipps, G.L., Effects of pH increases and sodium chloride additions on the acute toxicity of 2,4-dichlorophenol to the fathead minnow, *Water Res.*, 14, 1073–1077, 1980.
76. Spehar, R.L., Nelson, H.P., Swanson, M.J., and Renoos, J.W., Pentachlorophenol toxicity to amphipods and fathead minnows at different test pH values, *Environ. Toxicol. Chem.*, 4, 389–397, 1985.
77. Doe, K.G., Ernst, W.R., Parker, W.R., Julien, G.R.J., and Hennigar, P.A., Influence of pH on the acute lethality of fenitrothion, 2,4-D, and aminocarb and some pH-altered sublethal effects of aminocarb on rainbow trout (*Salmo gairdneri*), *Can. J. Fish. Aquat. Sci.*, 45, 287–293, 1988.
78. Fisher, S.W., Effects of pH on the toxicity and uptake of [^{14}C] Lindane in the midge, *Chironomus riparius*, *Ecotoxicol. Environ. Saf.*, 10, 202–208, 1985.
79. Cameron, G.N., Symons, J.M., Bushek, D., and Kulkarni, R., Effect of temperature and pH on the toxicity of monochloramine to the Asiatic clam, *Am. Water Works Assoc. J.*, 81, 62–71, 1989.
80. Lohner, T.W. and Fisher, S.W., Effects of pH and temperature on the acute toxicity and uptake of carbaryl in the midge, *Chironomus riparius*, *Aquat. Toxicol.*, 16, 335–354, 1990.
81. Lydy, M.J., Lohner, T.W., and Fisher, S.W., Influence of pH, temperature and sediment type on the toxicity, accumulation and degradation of parathion in aquatic systems, *Aquat. Toxicol.*, 17, 27–44, 1990.
82. Giese, A.C., *Cell Physiology*, W.B. Saunders, Philadelphia, 1968.
83. Macek, K.J., Hutchinson, C., and Cope, O.B., The effects of temperature on the susceptibility of bluegills and rainbow trout to selected pesticides, *Bull. Environ. Contam. Toxicol.*, 4, 174–183, 1969.
84. Mauck, W.L., Olson, L.E., and Marking, L. L., Toxicity of natural pyrethrins and five pyrethroids to fish, *Arch. Environ. Contam. Toxicol.*, 4, 18–29, 1976.
85. Knoph, M.B., Acute toxicity of ammonia to Atlantic salmon (*Salmo salar*) parr, *Comp. Biochem. Physiol.*, C101, 275–282, 1992.
86. Nimmo, D.W., Link, D., Parrish, L.P., Rodriguez, G.L., Wuerthele, W., and Davies, P.H., Comparison of on-site and laboratory toxicity tests: Derivation of site-specific criteria for un-ionized ammonia in a Colorado transitional stream, *Environ. Toxicol. Chem.*, 8, 1177–1189, 1989.
87. Hodson, P.V. and Sprague, J.B., Temperature induced changes in acute toxicity of zinc to Atlantic salmon (*Salmo salar*), *J. Fish. Res. Board Can.*, 32, 1–10, 1975.
88. Smith, M.J. and Heath, A.G., Acute toxicity of copper, chromate, zinc, and cyanide to freshwater fish: Effect of different temperatures, *Bull. Environ. Contam. Toxicol.*, 22, 113–119, 1979.
89. Poleo, A.B.S., Lydersen, E., and Muniz, I.P., The influence of temperature on aqueous aluminum chemistry and survival of Atlantic salmon (*Salmo salar* L.) fingerlings, *Aquat. Toxicol.*, 21, 267–277, 1991.
90. McGeachy, S.M. and Dixon, D.G., The impact of temperature on the acute toxicity of arsenate and arsenite to rainbow trout (*Salmo gairdneri*), *Ecotoxicol. Environ. Saf.*, 17, 86–93, 1989.
91. Fent, K., Embryotoxic effects of tributyltin on the minnow *Phoxinus phoxinus*, *Environ. Pollut.*, 76, 187–194, 1992.
92. Le Bras, S., Sensitivity of *Asellus aquaticus* L. (Crustacea, Isopoda) to lindane in relation to biotic size and metabolism and abiotic factors of concentration of insecticide and temperature, *Rev. Sci. Eau*, 3, 183–194, 1990.

93. McCahon, C.P., Barton, S.F., and Pascoe, D., The toxicity of phenol to the freshwater crustacean *Asellus aquaticus* (L.) during episodic exposure—Relationship between sub-lethal responses and body phenol concentrations, *Arch. Environ. Contam. Toxicol.*, 19, 926–929, 1990.

94. Fisher, S.W., Effects of temperature on the acute toxicity of PCP in the midge *Chironomus riparius* Meigen, *Bull. Environ. Contam. Toxicol.*, 36, 744–748, 1986.

95. Lewis, P.A. and Horning, W.B., II, Differences in acute toxicity test results of three reference toxicants on *Daphnia* at two temperatures, *Environ. Toxicol. Chem.*, 10, 1351–1358, 1991.

96. Snell, T.W., Moffat, B.D., Janssen, C., and Persoone, G., Acute toxicity tests using rotifers. III. Effects of temperature, strain and exposure time on the sensitivity of *Brachionus plicatilis*, *Environ. Toxicol. Water Qual.*, 6, 63–76, 1991.

97. Ferrando, M.D., Andreu-Moliner, E., Almar, M. M., Cebrian, C., and Nunez, A., Acute toxicity of organochlorine pesticides to the European eel, *Anguilla anguilla:* The dependency on exposure time and temperature, *Bull. Environ. Contam. Toxicol.*, 39, 365–369, 1987.

98. Ferrando, M.D. and Andreu-Moliner, E., Effects of temperature, exposure time and other water parameters on the acute toxicity of endosulfan to European eel, *Anguilla anguilla*, *J. Environ. Sci. Health*, B24, 219–224, 1989.

99. Sarkar, S.K., Effects of temperature on eggs, fry, and fingerlings of rohu (*Labeo rohita*) exposed to urea, *Prog. Fish-Cult.*, 53, 242–245, 1991.

100. McCloskey, J.T. and Oris, J.T., Effect of water temperature and dissolved oxygen concentration on the photo-induced toxicity of anthracene to juvenile bluegill sunfish (*Lepomis macrochirus*), *Aquat. Toxicol.*, 21, 145–156, 1991.

101. Jimenez, B.D., Cirmo, C.P., and McCarthy, J.F., Effect of feeding temperature on uptake, elimination and metabolism of benzo-(a)-pyrene in the bluegill sunfish (*Lepomis macrochirus*), *Aquat. Toxicol.*, 10, 41–57, 1987.

102. Edgren, M., Preliminary results on uptake and elimination at different temperatures of p,p'-DDT and two chlorobiphenyls in perch from brackish water, *Ambio*, 8, 270–272, 1979.

103. Karara, A.H. and Hayton, W.L., A pharmacokinetic analysis of the effect of temperature on the accumulation of di-2-ethylhexyl phthalate (DEHP) in sheepshead minnow, *Aquat. Toxicol.*, 15, 27–36, 1989.

104. Black, M.C., Milsap, D.S., and McCarthy, J.F., Effects of acute temperature change on respiration and toxicant uptake by rainbow trout (*Salmo gairdneri* Richardson), *Physiol. Zool.*, 64, 145–168, 1991.

105. Gammon, D.W., Brown, M.A., and Casida, J.E., Two classes of pyrethroid action in the cockroach, *Pestic. Biochem. Physiol.*, 15, 181–191, 1981.

106. Weiss, C.M. and Botts, J.L., Factors affecting the responses of fish to toxic materials, *Sewage Ind. Wastes*, 29, 810–818, 1957.

107. Wuhrmann, K., Concerning some principles of the toxicology of fish, *Bull. Centre Belge d'Etude et de Documentation des Eaux*, No. 15, Brussels, 1952.

108. Hogan, J.W., Water temperature as a source of variation in specific activity of brain acetylcholinesterase of bluegills, *Bull. Environ. Contam. Toxicol.*, 5, 347–353, 1970.

109. O'Brien, R.D., *Insecticides, Action and Metabolism*, Academic Press, New York, 1967.

110. Howe, G.E., Marking, L.L., Bills, T.D., Rach, J.J., and Mayer, F.L., Effects of water temperature and pH on toxicity of terbufos, trichlorfon, 4–nitrophenol, and 2,4–dinitrophenol to the amphipod *Gammarus pseudolimnaeus* and rainbow trout (*Oncorhynchus mykiss*), *Environ. Toxicol. Chem.*, 13, 51–66, 1994.

111. Howe, G.E., Marking, L.L., Bills, T.D., Boogard, M.A., and Mayer, F.L., Effects of water temperature on the toxicity of 4-nitrophenol and 2,4–dinitrophenol to developing rainbow trout (*Oncorhynchus mykiss*), *Environ. Toxicol. Chem.*, 13, 79–84, 1994.

112. Mayer, F.L., Marking, L.L., Howe, G.E., Brecken, J.A., Linton, T.K., and Bills, T.D., Physicochemical factors affecting toxicity: Relation to bioavailability and exposure duration, *Proc. Soc. Environ. Toxicol. Chem.*, 12, 141, 1991.

113. Veith, G.D., Call, D.J., and Brooke, L.T., Structure-toxicity relationships for the fathead minnow, *Pimephales promelas:* Narcotic industrial chemicals, *Can. J. Fish. Aquat. Sci.*, 40, 743–748, 1983.

Chapter 2

Chemical and Biological Factors Affecting Bioavailability of Contaminants in Seawater

John P. Knezovich

INTRODUCTION

Organic and inorganic contaminants enter seawater from a variety of sources. Storm water runoff, sewage treatment effluents, industrial discharges, and oil refining process waters contribute to the contamination of bays, estuaries, and ocean waters. Because these contaminants typically enter near-shore environments that have economic and aesthetic value, it is important to understand the extent to which they will interact with and impact marine species. While this discussion will focus on understanding the bioavailability of contaminants in seawater, it must be recognized that most contaminants that enter seawater are initially present in freshwater (e.g., as in the case of sewage treatment effluents). This situation creates a transitory environment in which a wide range of salinities must be considered.

As an introduction, it is worthwhile to consider some of the basic properties of seawater in order to contrast them with what is known about freshwater. In the simplest sense, the transition from freshwater to seawater encompasses the extreme range of water hardness that is found in the environment. Freshwater, which has a salt composition that varies as dictated by local geological conditions and is dominated by calcium, is considered soft at a salinity of approximately 0.05‰ and hard at a salinity of approximately 0.30‰. Oceanic seawater, on the other hand, has a salinity of 33 to 38‰ that is generally consistent in its composition and is dominated by sodium (see Table 1).[1] Furthermore, the elements that make up sea salt are conservative, that is, they do not change significantly due to biochemical or geochemical activities. Indeed, one of the greatest differences between seawater and freshwater is that the composition and chemical properties of seawater are much more consistent than freshwater. Chemical fate processes that are affected by salinity should therefore be applicable to most oceanic waters.

As a result of its consistency, the pH of seawater is maintained in a relatively narrow range (i.e., 8.1 to 8.5) that is largely determined by the concentrations of bicarbonate and borate.[1,2] This occurs because carbon dioxide, which occurs largely as bicarbonate in seawater (approximately 25 mg C/L), is present in excess of amounts required for plant growth. In estuarine environments, where salinity varies as a function of freshwater input, pH is typically lower than that of oceanic seawater. This occurs because river waters usually contain low concentrations of excess bases and often carry substantial concentrations of acidic humic material. Accordingly, the pH of water in estuarine systems varies over space and time and can be substantially lower than that found in oceanic seawater.

In contrast to freshwater, the bioavailability of contaminants that are present in oceanic seawater is not likely to be affected by the relatively narrow pH range of this medium. Information derived from studies of contaminant toxicities in freshwater indicate that water hardness may influence contaminant bioavailability more as a function of changes in pH than in hardness itself. By extrapolation, this observation would indicate that slight changes in the salinity of oceanic seawater, which would not be accompanied by significant changes in pH, would have a minimal influence on contaminant bioavailability. However, large depressions in the pH of seawater that result from its dilution with freshwater in estuaries would increase the bioavailability of heavy metals.

Table 1.   The Chemical Composition of Salt in Seawater

Ion	%	Ion	%
Na^+	30.61	Cl^-	55.04
Mg^{2+}	3.69	SO_4^{2-}	7.68
Ca^{2+}	1.16	HCO_3^-	0.41
K^+	1.10	H_3BO_3	0.07
Sr^{2+}	0.03	Br^-	0.19

Source: From Parsons, T. and Takahashi, M., *Biological Oceanographic Processes*, Pergamon Press, Elmsford, NY, 1978. With permission.

The majority of information that is available concerning contaminant bioavailability in seawater has been derived from measurements of toxicological effects observed at different salinities. Unless experiments were conducted with regard for the physiological state of the organism, however, the data obtained may not reflect the influence of salinity on the bioavailability of the contaminant but rather the effects of salinity-induced stress on the organism. Accordingly, the influence of salinity on contaminant toxicity may provide a false indication of chemical availability. Such data should therefore be considered carefully when they are used to make inferences about contaminant availability.

In the discussion that follows, the influence that salinity has on the bioavailability of the two largest classes of contaminants, trace metals and organic compounds, will be discussed. Although data on contaminant toxicity will be used to draw inferences about chemical availability, this discussion will focus on the properties that contaminants are likely to exhibit in waters of varying salinities. In addition, information on physiological changes that are affected by salinity will be used to illustrate how biological effects can alter the apparent availability of contaminants.

BIOAVAILABILITY OF CONTAMINANTS TO MARINE AND FRESHWATER ORGANISMS

One approach for evaluating the bioavailability of contaminants in seawater is to compare levels at which selected contaminants are toxic to marine and freshwater species. The underlying assumption to this approach is that similar organisms (e.g., fish, bivalves) have similar sensitivities to a specific toxicant. Therefore, by performing a statistical comparison of available toxicity data for all tested species, one might expect to see marine species being more or less susceptible to a given toxicant if there are inherent differences in chemical availability that result from differences in water salinity. Although this is a rather simplistic approach and does not address physiological differences between species, Klapow and Lewis[3] found that, in general, marine and freshwater species appear to have the same sensitivity to a variety of organic and inorganic contaminants. The results of their analyses indicate that there are no observable differences in the bioavailability of most contaminants as evidenced by acute toxicity (i.e., LC50 values). It is interesting to note that the only exception found was for cadmium, which was generally more toxic to freshwater organisms. This observation is consistent with the speciation of cadmium, which exists primarily as the ionic, bioavailable form in freshwater and as the complexed form (i.e., $CdCl_2$) in seawater.[4]

The results of the above analyses indicate that the bioavailability of most contaminants is not appreciably different in seawater than in freshwater. Subtle differences in contaminant bioavailability that result from changes in salinity are not addressed by this approach, however, and are likely to be more important due to the transition of most contaminants from freshwater to seawater matrices. In the sections that follow, the influence that different levels of salinity have on the physico-chemical form of inorganic and organic contaminants will be addressed and related to their bioavailability.

Influence of Salinity on Metal Availability

Metals commonly enter marine environments as components of freshwater discharges, which presents an interesting case for examining the influence of salinity on metal bioavailability. Metals contained in such effluents are usually sorbed to dissolved ligands and particulate matter, which reduces their availability to aquatic organisms. However, the high dilution ratios that accompany such discharges favor metal desorption because they shift the equilibrium toward the dissociation of the sorbed species. Metal desorption is also enhanced initially in seawater due to the presence of soluble anions (i.e., Cl^-, SO_4^{2-}, and HCO_3^-), which are abundant in seawater and compete for the sorption of metals to form soluble complexes. In addition, competitive exchange of inorganic cations (e.g., sodium, potassium, calcium, and magnesium) on the sorptive substrate may also enhance the desorption process.

Trace metals that are sorbed to dissolved and particulate matter will also be affected following their discharge into seawater. Copper, for example, exists in a variety of chemical forms in aquatic environments, including free ions, inorganic complexes, and metal adsorbed on or incorporated into particulate matter. Upon discharge into seawater, the dissociation of bound metal will largely depend upon the nature of the initial substrate to which it was sorbed. For example, the release of copper from digested-sludge particulates has been shown to be relatively low ($\leq 9\%$), while release of copper from other substrates can be much greater (e.g., 65 to 70%).[5] The metal-binding capacity (e.g., the extent to which dissolved ligands complex free metal) of sewage effluent has also been shown to decrease as a result of increases in salinity. This effect indicates that metals discharged into marine environments in a complexed form will be released into solution to a greater extent than from effluents discharged to freshwater.[6] Although these metals are likely to be rapidly re-adsorbed on particulate material, their bioavailability in the zone of initial dilution may be significant.

Metals that are desorbed from ligands upon entry into marine environments, or are initially present in their ionic form, will rapidly bind with ligands that are present in seawater. Such processes have been defined for relatively large, episodic discharges of ionic copper into marine systems. Based on laboratory determinations of rate and equilibrium constants for the sorption and desorption of copper complexes, Orlob et al.[7] developed a model to simulate the fate of discharges of ionic copper into marine environments. Using this approach, they predicted that a significant fraction of a relatively high discharge of ionic copper would be bound primarily to dissolved organic matter and secondarily to suspended solids, with apparent equilibria being reached within 5 h. While such an event would result in an immediate increase in metal availability within the dilution zone, the results of this model indicate that the effect would be transitory.

Many studies have been conducted to define the influence of salinity on the bioavailability and toxicity of trace metals.[8-15] In general, the toxicity of trace metals has been found to be inversely related to salinity.[8] For example, Coglianese[9] studied the effect that different salinities had on the toxicity of trace metals to oyster embryos and found that decreased salinities resulted in increased sensitivities to copper and silver, which could not be accounted for by osmotic stress alone. Similar results have been reported for cadmium toxicity in fish,[10,11] and chromium toxicity in amphipods, worms, and bivalves.[12] Coglianese[9] has postulated that such effects may be due to increased ion fluxes or unmasking of a carrier or transmembrane pore at low salinities.

The observed influences of salinity on trace metal tolerance in marine organisms appears to be linked to the disruption of the normal pattern of hyper/hypoosmoregulation.[14] Such effects make it difficult to resolve the influence of salinity on metal bioavailability in the water column from the effects on physiological processes in the organism. Arsenic is a notable exception,[8] primarily because it exists as an anion and is not in competition for uptake sites. Furthermore, not all marine species exhibit increased metal toxicity associated with decreased salinity. Forbes[13] has recently reported that a marine gastropod (*Hydrobia ventrosa*), which normally inhabits water at a salinity of 23‰, exhibited a greater reduction in growth in the presence of cadmium at a salinity of 33‰ than at a salinity of 13‰. The results of this study illustrate the difficulty in relating a biological response (e.g., growth or mortality) to bioavailability. Because the relationship between cadmium toxicity and salinity is believed to function as a result of the free ion concentration (i.e., Cd^{2+}), the decrease in growth due to cadmium exposure could not be resolved from effects caused by osmotic stress at the higher salinity. Therefore, metal-induced toxicity was more evident at a higher salinity even though there was a lower percentage of metal available for accumulation.

Due to the occurrence of physiologically-based osmotic effects, it is not advisable to generalize the influence of salinity on trace metal bioavailability from studies of metal toxicity. This is due, in part, to species-specific differences in osmoregulation that affect the uptake and toxicity of trace metals. For example, Voyer and McGovern[15] reported that cadmium toxicity to *Mysidopsis bahia*, as evidenced by reduced reproduction, was greater at low salinities (e.g., 13‰) than at high salinities (e.g., 32‰). These results contrast with those cited above.[13] They concluded that although metal speciation can be correlated with acute toxic responses of aquatic organisms, effects due to long-term exposures may involve differing and multiple routes of uptake and mechanisms of toxicity that are not totally explainable in terms of metal speciation. These conclusions are especially important when considering the hazard posed by the input of metals into estuarine systems, which have normal fluctuations in salinity that may subject organisms to osmotic stress.

The bioconcentration of trace metals by marine organisms will also be related to their overall availability within the ecosystem. As with the information presented above for toxicity, however, there is no evidence that marine species accumulate more or less metals than freshwater species. For example, the EPA has listed bioconcentration factors (BCFs) for copper for several aquatic species.[16] These BCFs cover a wide range of values and are not related to the trophic level or ecological niche of the organism. For example, values listed range from log BCFs of 3.3 to 3.6 for freshwater algae, from 4.2 to 4.3 for a bivalve mollusc, and from 0 to 2.5 for bluegill fish and fathead minnows, respectively. The BCFs listed for marine species are equally variable and range from log values of 1.8 to 2.8 for several species of algae, from 2.3 to 3.4 for polychaete worms, and from 1.9 to 4.4 for several species of bivalve molluscs.

The log BCF values listed above span four orders of magnitude (log BCFs from 0 to 4.4) for species that are consumed by humans (i.e., bivalve molluscs and fish). Bivalve molluscs represent the higher end of this range (log BCFs from 2.7 to 4.4) and fish the lower end (log BCFs from 0 to 2.5). Due to the very wide range of bioconcentration factors that occur between species and within a population of the same species, this index is a poor predictor of metal availability in aquatic environments. Furthermore, trace metal accumulation and toxicity may be mediated by inducible metal-binding proteins that affect how trace metals are accumulated and stored within the organism. The induction of such metal-binding proteins can increase the resistance of organisms to toxic effects, which can complicate the interpretation of toxicity studies. For example, Roesijadi et al.[17] have demonstrated that preexposure of marine mussels to mercury confers resistance to mercury toxicity in subsequent exposures. Such effects may occur independent of metal availability, and call into question the use of toxicity data to infer the bioavailability of trace metals present in the water column.

Influence of Salinity on Organic Compound Availability

Possibly the most important factor that differentiates the behavior of organic compounds in seawater from their behavior in freshwater is related to differences in solubility. Neutral organic compounds are less soluble in seawater than in freshwater because of what is referred to as a "squeezing out" phenomenon that results from the compression of seawater due to the presence of high concentrations of salts.[18] This effect results in measurably lower solubilities for a variety of neutral organic compounds (see Table 2). The extent of their solubility is predictable, based

Table 2.   The Solubilities of Organic Compounds Measured in Distilled
            Water and Seawater

Molecule	Solubility (mg/l)		Ref.
	Distilled water	Seawater	
Toluene	535	379	20
p-Xylene	156	110	20
Naphthalene	31	22	18
t-Butylbenzene	29	21	20
Biphenyl	7.4	4.7	18
Dodecane	3.7	2.9	21
Octadecane	2.1	0.8	21
Phenanthrene	1.0	0.7	18

on a knowledge of the salt content of the water, and follows an inverse relationship that is accurate for salinities encountered in estuarine and oceanic waters.[19] Compound solubilities are lowest in oceanic water, which has a consistent salinity, and increase in a linear fashion as the transition is made to brackish and fresh waters.[20] As with trace metals, the transition from freshwater to saltwater can have an immediate effect on the distribution of organic compounds in marine ecosystems. The most obvious effect results from the salting-out of contaminants that were present in freshwater at levels of saturation. Such compounds would be rapidly transformed from the soluble to the particulate phase upon entry into a marine system and would presumably be less available for accumulation by marine organisms.[21]

An interesting aspect of the salting-out phenomenon is its influence on organic contaminant partitioning and its relationship to bioconcentration. The partitioning of neutral organic compounds between aqueous and organic phases has been well established, and octanol/water partition coefficients (K_{ow}s) are commonly used to estimate their potential for bioaccumulation. Because K_{ow}s are determined in distilled water, however, they are likely to underestimate the partitioning of a compound from seawater to an organic phase (e.g., biota, sediment). This phenomenon will occur because seawater, due to its high ionic strength, is a poorer solvent for neutral organic compounds than pure distilled water. For this reason, salt is commonly added to aqueous solutions to increase the efficiency with which organic compounds are extracted by immiscible organic phases. Accordingly, if one considers the immiscible organic phase to be lipid, then the bioaccumulation of a neutral organic compound should be higher in a marine organism than in a freshwater organism exposed to equivalent dissolved concentrations of the chemical. This effect is particularly important for some ionized compounds, which in the presence of neutral salts can form ion pairs that are more soluble in organic phases and hence have measurably higher K_{ow}s.[22]

Because the relationships between chemical solubility, K_{ow}, and bioconcentration have been well established for some aquatic organisms,[23-25] it is possible to calculate BCFs for a given species based on solubility data (Table 3). The results of this exercise indicate, as expected, that the bioaccumulation of a given compound should be greater in seawater than in freshwater. Such predicted increases are not likely to be observed, however, due to the relatively small magnitude of the expected change and the high degree of variability associated with measurements of BCFs.

This point can be further illustrated by the direct comparison of BCF data acquired for freshwater and marine species. This was accomplished by using data derived from studies of chemical accumulation in a marine bivalve (bay mussel)[23] and a freshwater fish (fathead minnow)[24] that span a relatively wide range of K_{ow}s. By plotting the relationships between K_{ow}s and BCFs for these organisms on the same graph (see Figure 1) it is apparent that there are no discernible differences between the accumulation of neutral organic compounds by these species. Even though these are different species, it is apparent from the distribution of the BCFs for both data sets that the slight differences that may result from salinity-induced differences in solubility are not likely to be observed. For practical purposes, therefore, the bioavailability of neutral organic compounds in marine and freshwater ecosystems can be considered to be equivalent.

Table 3.   Bioconcentration Factors (BCF) Calculated for Marine Mussels From
Water Solubilities in Freshwater and Seawater[a]

Molecule	log BCF	
	Freshwater	Seawater
Toluene	3.1	3.2
p-Xylene	3.4	3.5
Naphthalene	3.9	4.0
t-Butylbenzene	3.9	4.0
Biphenyl	4.3	4.5
Dodecane	4.6	4.6
Octadecane	4.7	5.0
Phenanthrene	4.9	5.0

[a]Bioconcentration factors (BCF) calculated for *M. edulis* according to Geyer et al. (log BCF = $-0.682 \times$ log WS + 4.94). Solubility data are presented in Table 2.
Source: From Geyer, H., Sheehan, P., Kotzias, D., Freitag, D., and Korte, F., *Chemosphere*, 11, 1121–1134, 1982. With permission.

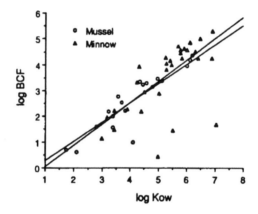

Figure 1. The relationship between K_{ow} and bioconcentration factors (BCF) determined for the mussel *Mytilus edulis* and the fathead minnow *Pimephales promelas* for a variety of organic compounds. Data were derived from studies conducted by Geyer et al.[23] and Veith et al.[24]

The influence of salinity on the toxicity of organic contaminants has been reported for several species.[26-29] As is the case with trace metals, however, it is not clear that such increases in toxicity are associated with increased contaminant availability in the water column. In fact, the effect of salinity on organic compound toxicity appears to be related to changes in metabolic capability as well as bioavailability. For example, Johnston and Corbett[26] reported that fenitrothion was more toxic to blue crabs at salinities of 34‰ than at 17‰. They postulated that toxic effects may have been related to changes in metabolic capabilities rather than chemical availability. Thomas and Rice[27] confirmed this assumption by demonstrating that euryhaline trout exposed to napthalene and toluene in seawater exhibited increased toxic effects as compared to fish exposed in freshwater. The increased toxicity of these compounds in seawater was related to the reduced ability of the seawater-exposed trout to metabolize and excrete the parent compounds. The end result of this effect was that these fish accumulated greater chemical residues due to their inability to metabolize and excrete the parent compounds.

The influence of salinity on metabolic capabilities may be common, but is difficult to generalize due to species-specific differences in physiological response. For example, Tedengren et al.[28] investigated the effects of salinity on the toxicity of diesel oil to two species of amphipods (*Gammarus* spp.) and reported that the species that was more tolerant of changes in salinity was also less sensitive to diesel oil. They postulated that organisms that are physiologically broad-niched are less likely to be affected by salinity-induced toxic effects. It is apparent from the results of these studies that toxicity data must be interpreted carefully when they are being used to evaluate salinity-induced changes in contaminant availability. This occurs because the availability of contaminants in the water column can not account for the effects of salinity on membrane permeability and osmoregulatory mechanisms, which will influence contaminant accumulation.

CONCLUSIONS

The influence of salinity on the bioavailability of contaminants is a function of the physiological state of the exposed organism as well as the physical and chemical properties of the contaminant. These factors are difficult to resolve because most studies of contaminant bioavailability in seawater, in actuality, are studies of the effects of salinity on contaminant toxicity. Salinity-induced changes in an organism's physiology may alter the availability of a contaminant by effecting changes at the site of uptake or in its metabolic capabilities. Such changes can affect the uptake and accumulation of contaminants, but are not necessarily related to the availability of the contaminant in the water column. Furthermore, many experiments are conducted at salinities that are at the extreme range of those normally encountered by the test organisms. Care must be taken, therefore, when attempting to generalize results obtained from toxicity studies that were conducted under different salinity regimes.

ACKNOWLEDGMENT

Work performed under the auspices of the U.S. Department of Energy by the Lawrence Livermore National Laboratory under Contract W-7405-Eng-48.

REFERENCES

1. Parsons, T. and Takahashi, M., *Biological Oceanographic Processes*, Pergamon Press, Elmsford, NY, 1973.
2. Reid, G.K., *Ecology of Inland Waters and Estuaries*, D Van Nostrand, New York, 1961.
3. Klapow, L.A. and Lewis, R.H., Analysis of toxicity data for California marine water quality standards, *J. Water Pollut. Control Fed.*, 51, 2054–2070, 1979.
4. Sunda, W.G., Engel, D.W., and Thuotte, R.M., Effect of chemical speciation on toxicity of cadmium to grass shrimp, *Palaemonetes pugio:* importance of free cadmium ion, *Environ. Sci. Technol.*, 12, 409–413, 1978.
5. Rohatgi, N. and Chen, K.Y., Transport of trace metals by suspended particulates on mixing with seawater, *J. Water Pollut. Control Fed.*, 47, 2298–2316, 1975.
6. Blustein, H. and Exton, M.J., Binding capacity of wastewater discharge to seawater, *Water Res.*, 17, 1505–1509, 1983.
7. Orlob, G.T., Hrovat, D., and Harrison, F.L., Mathematical model for simulation of the fate of copper in a marine environment, in *Particulates in Water;* Adv. Chem. Ser. No. 109, American Chemical Society, Washington, D.C., 1980.
8. McLusky, D.S., Bryant, V., and Campbell, R., The effects of temperature and salinity on the toxicity of heavy metals to marine and estuarine invertebrates, *Oceanogr. Mar. Biol. Annu. Rev.*, 24, 481–520, 1986.
9. Coglianese, M.P., The effects of salinity on copper and silver toxicity to embryos of the Pacific oyster, *Arch. Environ. Contam. Toxicol.*, 11, 297–303, 1982.
10. Voyer, R.A., Heltsche, J.F., and Kraus, R.A., Hatching success and larval mortality in an estuarine teleost, *Menidia menidia* (Linnaeus), exposed to cadmium in constant and fluctuating salinity regimes, *Bull. Environ. Contam. Toxicol.*, 23, 475–481, 1979.

11. Voyer, R.A., Cardin, J.A., Heltshe, J.F., and Hoffman, G.L., Viability of embryos of the winter flounder *Pseudopleuronectes americanus* exposed to mixtures of cadmium and silver in combination with selected fixed salinities, *Aquat. Toxicol.*, 2, 223–233, 1982.

12. Bryant, V., McLusky, D.S., Roddie, K., and Newbery, D.M., Effect of temperature and salinity on the toxicity of chromium to three estuarine invertebrates (*Corophium volutator, Macoma balthica, Nereis diversicolor*), *Mar. Ecol. Prog. Ser.*, 20, 137–149, 1984.

13. Forbes, V.E., Response of *Hydrobia ventrosa* (Montagu) to environmental stress: effects of salinity fluctuations and cadmium exposure on growth, *Functional Ecol.*, 5, 642–648, 1991.

14. McLusky, D.S. and Hagerman, L., The toxicity of chromium, nickel and zinc: effects of salinity and temperature, and the osmoregulatory consequences in the mysid *Praunus flexuosus.*, *Aquat. Toxicol.*, 10, 225–238, 1987.

15. Voyer, R.A. and McGovern, D.G., Influence of constant and fluctuating salinity on responses of *Mysidopsis bahia* exposed to cadmium in a life-cycle test, *Aquat. Toxicol.*, 19, 215–230, 1991.

16. EPA, Ambient Water Quality Criteria for Copper, EPA/440/5–84–031, Environmental Protection Agency, Washington, D.C., 1985.

17. Roesijadi, G., Drum, A.S., Thomas, J.T., and Fellingham, G.W., Enhanced mercury tolerance in marine mussels and relationship to low molecular weight, mercury-binding proteins, *Mar. Pollut. Bull.*, 13, 250–253, 1982.

18. Eganhouse, R.P. and Calder, J.A., The solubility of medium molecular weight aromatic hydrocarbons and the effects of hydrocarbon co-solutes and salinity, *Geochim. Cosmochim. Acta*, 40, 555–561, 1976.

19. Gordon, J.E. and Thorne, R.L., Salt effects on non-electrolyte activity coefficients in mixed aqueous electrolyte solutions. II. Artificial and natural seawaters, *Geochim. Cosmochim. Acta*, 31, 2433–2443, 1967.

20. Sutton, C. and Calder, J.A., Solubility of alkylbenzenes in distilled water and seawater at 25.0° C, *J. Chem. Eng. Data*, 20, 320–322, 1975.

21. Sutton, C. and Calder, J.A., Solubility of higher-molecular-weight *n*-paraffins in distilled water and seawater, *Environ. Sci. Technol.*, 8, 654–657, 1974.

22. Esser, H.O. and Moser, P., An appraisal of problems related to the measurement and evaluation of bioaccumulation, *Ecotoxicol. Environ. Saf.*, 6, 131–148, 1982.

23. Geyer, H., Sheehan, P., Kotzias, D., Freitag, D., and Korte, F., Prediction of ecotoxicological behaviour of chemicals: relationship between physico-chemical properties and bioaccumulation of organic chemicals in the mussel *Mytilus edulis*, *Chemosphere*, 11, 1121–1134, 1982.

24. Veith, G.D., DeFoe, D.L., and Bergstedt, B.V., Measuring and estimating the bioconcentration factor of chemicals in fish, *J. Fish. Res. Board Can.*, 36, 1040–1048, 1979.

25. Veith, G.D. and Kosian, P., Estimating bioconcentration potential from octanol/water partition coefficients, in Mackay, D., Patterson, S., Eisenreich, S.J., and Simmons, M.S., Eds., *Physical Behavior of PCBs in the Great Lakes*, Ann Arbor Science, Ann Arbor, MI, 1983, 269–282.

26. Johnston, J.J. and Corbett, M.D., The effects of temperature, salinity and a simulated tidal cycle on the toxicity of fenitrothion to *Callinectes sapidus*, *Comp. Biochem. Physiol.*, 80C, 145–149, 1985.

27. Thomas, R.E. and Rice, S.R., The effects of salinity on uptake and metabolism of toluene and napthalene by Dolly Varden, *Salvelinus malma*, *Mar. Environ. Res.*, 18, 203–214, 1986.

28. Tedengren, M., Arner, M., and Kautsky, Nils, Ecophysiology and stress response of marine and brackish water *Gammarus* species (Crustacea, Amphipoda) to changes in salinity and exposure to cadmium and diesel-oil, *Mar. Ecol. Prog. Ser.*, 47, 107–116, 1988.

29. Stickle, W.B., Sabourin, T.D., and Rice, S.D., Sensitivity and osmoregulation of coho salmon, *Oncorhynchus kisutch*, exposed to toluene and napthalene at different salinities, in Vernberg, W.B., Calabrese, A., Thurberg, F.P., and Vernberg, F.J., Eds., *Physiological Mechanisms of Marine Pollutant Toxicity*, Academic Press, New York, 1982, 331–348.

Chapter 3

Synopsis of Discussion Session on Physicochemical Factors Affecting Toxicity

Russell J. Erickson (Chair), Terry D. Bills, James R. Clark, David J. Hansen, John Knezovich, Foster L. Mayer, Jr., and Anne E. McElroy

INTRODUCTION

Concerns about the bioavailability of toxic chemicals to aquatic organisms originated from observations that concentrations associated with toxic endpoints (e.g., LC50s) are affected by various physicochemical factors of the test water such as hardness, pH, temperature, salinity, and the presence of other inorganic and organic constituents. Many of these factors can change the chemical speciation of the toxicant, which can alter the ability of the toxicant to be adsorbed or absorbed by the gills or skin of the organism. Other groups in this workshop considered the role of various inorganic and organic constituents whose principal action is to chemically associate with a toxicant, and thereby affect its bioavailability. However, there are other factors whose actions are not necessarily so straightforward, most notably hardness, pH, temperature, and salinity, which were addressed in the discussion initiation papers by Mayer et al.[1] and Knezovich[2] for this section of the workshop.

This paper documents the workshop discussion regarding the role of these factors in altering toxicity. For each factor, the nature, magnitude, and uncertainty of its empirical relation to the toxicity of various chemicals or chemical classes were discussed. Limitations in the empirical data base regarding the variety of species and endpoints tested were addressed. Possible mechanisms underlying the empirical relations were identified, and the confidence with which mechanisms are known was considered. Particular attention was paid to whether these factors actually affect bioavailability, or alter toxicity in some other way. Finally, research needed to better understand these effects was discussed.

HARDNESS

Hardness refers to the concentration of certain multivalent cations, principally calcium and magnesium, in water. Because hardness constituents are cations, they do not affect the speciation of cationic toxicants except to the extent that they compete for the same ligands. Hardness can alter the speciation of anionic toxicants, although the association constants must be high for this to be a significant factor in most waters. When interpreting the effects of hardness, it is important to account for its correlation with other variables such as alkalinity and pH. The effects of such variables should not be confused with an effect of hardness per se. Furthermore, it is sometimes necessary to consider the separate effects of the different constituents of hardness, rather than their combined effects.

Organic Toxicants

For most organic toxicants, effects of hardness on toxicity and bioaccumulation are small or nonexistent, or have not been separated from the effects of pH.[3-7] For organic surfactants, studies have indicated that hardness affects toxicity, but effects vary depending on the test species and chemical formulation and are often confounded by differences in pH.[8] Complexation with calcium and magnesium increases the toxicities of some anionic surfactants, but greatly decreases the toxicities of others.[9] This is possibly related to the extent the complexes remain dissolved and can interact with absorptive surfaces of test organisms. Calcium has been shown to affect the partitioning of weak organic acids onto solids and into organic phases, either by formation of ion pairs with the acid anions or by modification of adsorptive surfaces.[10,11] By such a mechanism, hardness could have an effect on the toxicity of these organic chemicals, but relevant toxicity information regarding this is not available.

Metals

Various studies have shown that the toxicities of many cationic metals to aquatic organisms decrease as hardness increases, whereas other studies have indicated either a negligible interaction or a complicated relation of toxicity and hardness (see reviews by Sprague[7] and Hunt[12]). Most studies have examined the effects of hardness on acute lethality, but increased hardness has also been reported to ameliorate copper effects on fish growth.[13] In contrast, the effect of hardness on chronic toxicity of copper to *Daphnia pulex* has been reported to be negligible.[14] Hardness was correlated with alkalinity and pH in many of these studies, so the importance traditionally attributed to hardness might be partly due to the influence of these other parameters on metal speciation and availability. The effects of hardness on metal toxicity also might be influenced by incomplete acclimation of test organisms to the hardness used for a toxicity test or by the test duration. Based on detailed measurements of ion exchanges, Lauren and McDonald[15] disputed the importance of hardness on acute copper toxicity to rainbow trout acclimated to the test hardness. In general, the effects of hardness vary among different metals, toxic endpoints, test conditions, and organisms. Thus, despite considerable study, the empirical effects of hardness on metal toxicity are rather uncertain.

Effects of hardness on metals toxicity to aquatic animals have been examined mostly for acute lethality of metals for which the mechanism of toxicity appears to involve disruption of gill function, especially osmoregulation. Because of the influence of calcium on membrane permeability and ion exchange, it has been suggested that effects of hardness might reflect competition between calcium and the toxic metal for binding sites on the gill.[17,18] Hardness thus might be properly considered to affect metal bioavailability because such competition affects the accumulation of the toxicant. However, such competitive interaction has not been demonstrated, and it is also possible that hardness independently affects the ability of an organism to osmoregulate and withstand the stress imposed by a toxic metal. If so, hardness cannot really be considered to be altering bioavailability. In any event, considerable uncertainty remains over how hardness affects the toxicity of metals.

Some investigators have examined the effects of hardness on bioaccumulation as well as toxicity. In addition to finding no effect of hardness on acute copper toxicity to rainbow trout, Lauren and McDonald[15] reported no effect of hardness on copper accumulation in gill tissue. While this might seem to be a meaningful correlation, they also found alkalinity to have no effect on accumulation, although alkalinity did have a large effect on toxicity. Winner[14] reported bioaccumulation of copper by *Daphnia pulex* to increase at low hardness, whereas LC50s were unaffected by hardness. Increased calcium and magnesium also have been shown to reduce cadmium uptake into the perfusate of isolated, perfused fish gills, but to have no effect on accumulation of cadmium in the gill tissue.[16] Thus, effects of hardness on bioavailability, as indicated by bioaccumulation, do not necessarily correlate well with the effects on toxicity, and different bioaccumulation measurements can even suggest different bioavailabilities. However, this does not necessarily mean that effects of hardness on toxicity do not involve changes in metal bioavailability, because these bioaccumulation measurements might not be sufficiently correlated with toxicant accumulation at the specific site of action.

pH

Systemic Organic Toxicants

For nonionizable organic toxicants, there is little evidence that pH, within normal physiological ranges, alters toxicity or bioaccumulation. For weak organic acids, toxicity and bioaccumulation decrease with increasing pH for pH > pK, and are roughly constant for pH < pK; whereas, for organic bases, toxicity and bioaccumulation are constant for pH > pK and decrease with pH < pK.[19-24] These trends indicate that toxicity and bioaccumulation decrease as the fraction of chemical in the neutral form decreases; however, this pH dependence does not follow the dissociation curve for the chemical. Rather, accumulation and toxicity are usually greater than expected based on the fraction of chemical in the neutral form at a particular pH, and this deviation from the dissociation curve tends to increase with increasing hydrophobicity.[25] For the lampricide, 3-trifluoromethyl-4-nitrophenol (TFM), the relation of toxicity to pH also is different between bony fishes and more primitive fishes.[26]

A good mechanistic explanation exists for the generally observed relations of the toxicities of weak acids and bases to pH. The neutral form of an organic electrolyte crosses lipid membranes more readily, and has a higher equilibrium partition coefficient into biota, than the charged form. Therefore, accumulation and toxicity are greatest at pHs where the neutral form predominates (pH < pK for acids, pH > pK for bases) and decrease as the percentage of the charged form increases (pH > pK for acids, pH < pK for bases). However, the ionized form is not entirely unavailable. It has been shown that organic anions can partition into organic phases to some extent, especially for chemicals whose neutral form is highly hydrophobic.[10] Therefore, the decrease in toxicity and accumulation is less marked than would be expected given the decrease in the percentage of the neutral form. The relation of toxicity to pH for aquatic animals can also be affected by the fact that the pH at gill surfaces can be markedly different from that in the bulk water. The relative amounts of neutral and ionized chemical at the site of absorption can therefore be quite different than expected based on the bulk water pH. This can significantly alter the relation of LC50s to pH and cause interspecies differences in this relation.[27]

The above model is well substantiated by reported data. For substituted phenols, water concentrations associated with a toxic endpoint can vary greatly with pH, but body burdens of the phenols associated with the endpoint are approximately constant, suggesting that the differences in water concentrations reflect differences in accumulation[19,20] Uptake and toxicity follow similar pH relations consistent with uptake of both neutral and ionic forms of a chemical, modified somewhat by the pH at the gill surface.[21-23,27,28] Major uncertainties in understanding these relations include the pH gradients in the microenvironment at exchange surfaces and the relative membrane permeabilities of neutral and charged forms of organic electrolytes.

Metals

For cationic metals, the "free", or hydrated, metal ion has been suggested to be the most, or sole, bioavailable form. As such, the toxicity of many metals, on the basis of total metal, would be expected to decrease (LC50s increase) as pH increases because of complexation by hydroxide, carbonate, and other ligands whose affinities for metals increase with pH. At alkaline pHs, toxicity based on total metal, in fact, is generally observed to decrease as pH increases for metals that form such complexes;[7,12] however, at acidic pHs, toxicity is sometimes observed to decrease as pH *decreases*.[27,28] On the basis of free copper ion, toxicity does not remain constant, but rather it increases as pH increases across a wide pH range.[29,30] In general, these relations vary among metals and test organisms.

This pH dependence of metal toxicity could involve several major factors. First, species other than the hydrated metal might be available to some degree.[31] Second, hydrogen ions could compete with metal ions for adsorption sites.[30,32,33] The decrease in toxicity at low pH might reflect such competition. Third, the pH at the gill surface can differ markedly from that in the bulk water.[34-36] Any adequate mechanistic explanation for the relation of metal toxicity to bulk water pH should consider such pH differences. Fourth, pH, especially at extreme values, can affect the nature of the gill epithelium and mucus layer, and thereby affect the adsorption or absorption of toxic metals.[37] Finally, pH would be expected to affect any metal toxicity that depends on active uptake of the metal in which hydrogen ion serves as a counter ion. Although all these factors are plausible, the mechanistic explanation for how pH affects metal toxicity is poorly established. Better understanding is hampered by uncertainties regarding the complex speciation of metals, the interaction of metals with cell membranes, the chemical environment at gill surfaces, and the details of toxic action.

Inorganic Nonmetals

The toxicity of weak inorganic acids and bases, such as hydrogen sulfide, cyanide, and ammonia, decrease with pH for acids and increase with pH for bases.[38,39] These relations seem to reflect different bioavailabilities of the neutral and charged forms of a chemical, as previously described. The pH in the gill microenvironment also has a role in determining this relation. Relations can vary among species and can deviate from this joint toxicity model at extreme pHs.

TEMPERATURE

Organic Toxicants

As reviewed by Mayer et al.,[1] the acute toxicity of most organic chemicals to aquatic animals increases as temperature increases. The increase in toxicity roughly follows a log-linear relation, with the average slope corresponding to about a 3-fold decrease in 96-hr LC50s per 10° C increase. However, for acetylcholinesterase (AChE) inhibitors (e.g., organophosphate insecticides), the effect of temperature is greater (about a 5-fold increase per 10° C increase). For some chemicals (e.g., DDT, methoxychlor, pyrethroids), the temperature dependence is the reverse. Under chronic conditions, the temperature effect is often reduced or absent.

These effects of temperature on toxicity can involve several mechanisms. First, increased temperature can change respiration rates and permeabilities and thereby lead to more rapid uptake of a toxicant and hasten the accumulation of a toxic dose. This mechanism is supported by the fact that for many chemicals the effect of temperature follows a log-linear/Q_{10}-type relation and that this effect decreases with increasing duration. Second, chemical equilibrium partition coefficients can change with temperature and result in a different net uptake for the same water concentrations. For example, Karara and Hayton[40] reported that the bioconcentration factor for di-2-ethylhexylphthalate in sheepshead minnows increases markedly with temperature. Third, temperature can alter the biochemistry or physiology involved in the toxic response. For AChE inhibitors, the greater temperature effect might reflect a temperature sensitivity of the AChE system, whereas for pyrethroids the opposite effect of temperature might be due to increased metabolic transformation of the chemicals to nontoxic forms at higher temperatures. If so, the temperature effects should not be considered to involve changes in bioavailability. The relative importance of these possible mechanisms is poorly resolved due to a limited toxicity data base with relatively little information on chemical accumulation rates, partition coefficients, and effects on metabolic processes.

Inorganic Toxicants

For inorganic chemicals, toxicity can increase, decrease, or remain constant with temperature, depending on the toxicant, species of organism, and duration of test.[1,7,41] These effects can be substantial (2- to 3-fold per 10°C). In some cases, increased toxicity with increased temperature has been observed to become less pronounced as test duration increases, so temperature effects might largely reflect accelerated uptake and response. However, in other cases, temperature effects do not appear to simply affect bioavailability because they are poorly correlated with speciation changes, kinetic relations, or accumulation of the toxicant.[7,39,42] In general, the mechanisms responsible for these temperature effects are poorly resolved because temperature can alter such a variety of metabolic and physiological processes involved in toxicity.

SALINITY

Organic Toxicants

Based on a limited data base, effects of salinity on the toxicity and bioaccumulation of organic chemicals in euryhaline organisms are variable and generally limited (usually within a factor of two to three over a wide range of salinity).[7,43,44] For some studies these effects might reflect osmotic stress at low salinities, and for some chemicals toxicity seems to be at a minimum at intermediate salinities which are isosmotic with the organism.[7] Effects of salinity might also be partly related to changes in activity coefficients and, for toxicants which are weak acids or bases, to changes in the degree of ionization, but the current data base is insufficient to resolve this. In estuaries, effects of salinity on the environmental partitioning of organic contaminants might be more important than actual effects on toxic action.[2]

Inorganic Toxicants

Acute toxicity of inorganic toxicants to euryhaline organisms generally decreases as salinity increases, the effects being moderate (two- to threefold) to great (over an order of magnitude), depending on the toxicant and organism[7,43,45–47] In some cases toxicity reaches a minimum at intermediate salinities, suggestive of some importance of osmoregulation in toxicity.[7] For chronic toxicity, salinity effects are less evident based on an even more limited data base.[46] Uncertainties regarding acclimation and the lack of tests with diel salinity fluctuations further limit an understanding of these effects. In freshwater, increasing sodium chloride from 1 to 10 mM (about 0.6‰) decreased copper toxicity to fathead minnows by about 3-fold,[42] but effects of higher salt concentrations and the influence of acclimation on this response are unknown.

The reasons for decreased acute toxicity with increased salinity are uncertain, but the presence of a more favorable osmotic gradient might necessitate increased levels of toxicant in order to sufficiently disrupt osmoregulation. In such a case, the mechanism does not really involve changes in toxicant bioavailability. Gossiaux et al.[47] reported that an increase in salinity decreased bioavailability of cadmium to *Diporeia* sp. (presumably due to complexation by chloride), but also increased tolerance on the basis of accumulated cadmium. Other possible mechanisms for the effects of salinity include changes in activity coefficients and alteration of the adsorption or absorption characteristics of the gill surface. However, limited data and the lack of detailed measurements of interactions at the gills currently precludes a thorough analysis of mechanisms.

RESEARCH NEEDS

Research needs could be identified individually for each of the factors and chemical classes cited above, but those considered by the discussion group generally fall into five common themes.

Microenvironment of Exchange Surface

A major problem in interpreting the effects of physicochemical factors is that these factors are often much different at the exchange surface (cell wall of phytoplankton, gills and skin of aquatic animals) than in the bulk water where they are manipulated and measured. Even factors which do not differ between the bulk water and exchange surface, such as temperature, can alter surface characteristics and exchange properties. Research on these surfaces is needed to determine how the properties that influence toxicant exchange (pH, permeabilities, adsorption characteristics) respond to changes in physicochemical variables in bulk water.

Chemical Speciation and Activity

Chemical speciation and activity are of central importance to bioavailability, but their dependence on physicochemical factors remains uncertain. Of particular interest are the binding properties of dissolved and colloidal humic material, the effects of temperature and salinity on speciation, and the kinetics of association and dissociation of chemical species. Such information must be applied to the bulk environment and the chemical interactions at the exchange surfaces to better understand effects on toxicity.

Mechanistic Processes at Exchange Surfaces

More study is needed to determine the actual processes involved in toxicity, and how these processes determine chemical bioavailability. For example, Lauren and McDonald[15] extensively studied the relation between copper exposure and ion fluxes at fish gills, and the effects of water chemistry on this relation. Similar detailed work needs to be conducted for other chemicals and organisms. Simple empirical correlations of toxicity to environmental factors are not particularly effective in helping to understand mechanisms of toxicity or resolving many questions regarding bioavailability.

Chemical Accumulation and Mode of Action

When appropriate, toxicity tests on a chemical should include measurements of chemical accumulation, at least in the whole-body and preferably at the site of toxic action. While not always definitive, such information would help to determine whether environmental factors affect toxicity by restricting the rate or extent of accumulation, by altering metabolism and chemical distribution, or by affecting sensitivity to accumulated chemical. An understanding of toxicokinetic relations and toxicological effects at the site of action are needed to further resolve how environmental factors affect toxicity. Understanding the relation of effects to accumulation also will improve abilities to interpret environmental monitoring of fish chemical accumulation.

Data Bases for Relations

For many chemicals and physicochemical variables, the empirical data base is inadequate to understand the effect of the variable on toxicity. This inadequacy involves several factors. First, the variable of interest was often not investigated over a sufficient range and number of levels, and was often confounded with other variables. Second, too few species of organisms were tested to make generalizations about the results. Third, where relevant, accumulation of the chemical was not monitored. Fourth, the effects of acclimation of test organisms to test conditions are largely unknown. Fifth, the effects of the physicochemical variables on sublethal endpoints and on chronic toxicity have not been widely tested. Sixth, the tests did not include adequate control and measurement of certain variables that could help to elucidate mechanisms. Improvement of this database does not require a large number of indiscriminate tests, but rather selective tests under the appropriate conditions with the measurements needed to improve understanding of the mechanism of toxicity.

REFERENCES

1. Mayer, F.L. Jr., Marking, L.L., Bills, T.D., and Howe, G.E., Physicochemical factors affecting toxicity in freshwater: hardness, pH, and temperature, this volume, Session 2, Chapter 1, 1994.

2. Knezovich, J.P., Physicochemical factors affecting bioavailability of contaminants in seawater, this volume, Session 2, Chapter 2, 1994.

3. Pickering, Q.H. and Henderson, C., Acute toxicity of some important petrochemicals to fish, *J. Water Pollut. Control Fed.*, 35, 1419–1429, 1964.

4. Inglis, A. and Davis, E.L., Effects of water hardness on the toxicity of several organic and inorganic herbicides to fish, U.S.FWS Tech. Pap. No. 67, U.S. Fish and Wildlife Service, Department of the Interior, Washington, D.C., 1972.

5. Mayer, F.L., Jr. and Ellersieck, M.R., Experiences with single-species tests for acute toxic effects in freshwater animals, *Ambio*, 17, 367–375, 1988.

6. Nimmo, D.R., Pesticides, in Rand, G.M and Petrocelli, S.R., Eds., *Fundamentals of Aquatic Toxicology*, Hemisphere, New York, 1984, 335–373.

7. Sprague, J.B., Factors that modify toxicity, in Rand, G.M. and Petrocelli, S.R., Eds., *Fundamentals of Aquatic Toxicology*, Hemisphere, New York, 1984, 124–163.

8. Lewis, M.A., The effects of mixtures and other environmental modifying factors on the toxicities of surfactants to freshwater and marine life, *Water Res.*, 26, 1013–1023, 1992.

9. Henderson, C., Pickering, Q.H., and Cohen, J.M., The toxicity of synthetic detergents and soaps to fish, *Sewage Ind. Wastes*, 31, 295–306, 1959.

10. Jafvert, C.T., Sorption of organic acid compounds to sediments: initial model development, *Environ. Toxicol. Chem.*, 9, 1259–1268, 1990.

11. Jafvert, C.T., Westall, J.C., Grieder, E., and Schwarzenbach, R.P., Distribution of hydrophobic ionogenic organic compounds between octanol and water: organic acids, *Environ. Sci. Technol.*, 24, 1795–1803, 1990.

12. Hunt, D.T.E., Trace metal speciation and toxicity to aquatic organisms—a review. Rep. TR 247, Water Research Centre, Marlow, U.K., 1987.

13. Waiwood, K.G. and Beamish, F.W.H., Effects of copper, hardness and pH on the growth of rainbow trout, *Salmo gairdneri, Water Res.*, 12, 611–619, 1978.

14. Winner, R.W., Bioaccumulation and toxicity of copper as affected by interactions between humic acid and water hardness, *Water Res.*, 19, 449–455, 1985.

15. Lauren, D.J. and McDonald, D.G., Influence of water hardness, pH, and alkalinity on the mechanisms of copper toxicity in juvenile rainbow trout, *Salmo gairdneri, Can. J. Fish. Aquat. Sci.*, 43, 1488–1496, 1986.

16. Part, P., Svanberg, O., and Kiessling, A., The availability of cadmium to perfused rainbow trout gills in different water qualities, *Water Res.*, 19, 427–434, 1985.

17. Zitko, V. and Carson, W.G., A mechanism of the effects of water hardness on the lethality of heavy metals to fish, *Chemosphere*, 5, 299–303, 1976.

18. Pagenkopf, G.K., Gill surface interaction model for trace-metal toxicity to fishes: role of complexation, pH, and water hardness, *Environ. Sci. Technol.*, 17, 342–347, 1983.

19. Kobayashi, K. and Kishino, T., Effect of pH on the toxicity and accumulation of pentachlorophenol in goldfish, *Bull. Jpn. Soc. Sci. Fish.*, 46, 167–170, 1980.

20. Spehar, R.L., Nelson, H.P., Swanson, M.J., and Renoos, J.W., Pentachlorophenol toxicity to amphipods and fathead minnows at different test pH values, *Environ. Toxicol. Chem.*, 4, 389–397, 1985.

21. Saarikoski, J. and Viluksela, M., Influence of pH on the toxicity of substituted phenols to fish, *Arch. Environ. Contam. Toxicol.*, 10, 747–753, 1981.

22. Lo, I.-H. and Hayton, W.L., Effects of pH on the accumulation of sulfonamides by fish, *J. Pharmacokinet. Biopharm.*, 9, 443–459, 1981.

23. Hayton, W.L. and Stehly, G.R., pH control of weak electrolyte toxicity to fish, *Environ. Toxicol. Chem.*, 2, 325–328, 1983.

24. Marking, L.L., Howe, G.E., and Bills, T.D., Temperature and pH effects on acute and chronic toxicity of four chemicals to amphipods (*Gammarus pseudolimnaeus*) and rainbow trout (*Oncorhynchus mykiss*). Rep. EPA/600/X-90/286, U.S. Environmental Protection Agency, Gulf Breeze, FL, 1991.

25. Saarikoski, J., Lindstrom, M., Tyynila, M., and Viluksela, M., Factors affecting the absorption of phenolics and carboxylic acids in the guppy (*Poecilia reticulata*), *Ecotoxicol. Environ. Saf.*, 11, 158–173, 1986.

26. Applegate, V.C., Howell, J.H., and Smith, M.A., Use of mononitrophenols containing halogens as selective sea lamprey larvicides, *Science*, 127, 336–338, 1958.

27. McKim, J.M. and Erickson, R.J., Environmental impacts on the physiological mechanisms controlling xenobiotic transfer across fish gills, *Physiol. Zool.*, 64, 39–67, 1991.

28. Konemann, H. and Musch., A., Quantitative structure-activity relationships in fish toxicity studies. II. The influence of pH on the QSAR of chlorophenols, *Toxicology*, 19, 223–228, 1981.

29. Cusimano, R.G., Brakke, D.F., and Chapman, G.A., Effects of pH on the toxicities of cadmium, copper, and zinc to steelhead trout (*Salmo gairdneri*), *Can. J. Fish. Aquat. Sci.*, 43, 1497–1503, 1986.

30. Peterson, H.G., Healey, F.P., and Wagemann, R., Metal toxicity to algae: a highly pH dependent phenomenon, *Can. J. Fish. Aquat. Sci.*, 41, 974–978, 1984.

31. Newman, M.C. and Jagoe, C.H., Ligands and the bioavailability of metals in aquatic environments. This volume, Session 3, Chapter 1, 1994.

32. Borgman, U., Metal speciation and toxicity of free metal ions to aquatic biota, in Nriagu, J.O., Ed., *Aquatic Toxicology*, John Wiley & Sons, New York, 1983, 47–72.

33. French, P. and Hunt, D.T.E., Appendix A. The effects of inorganic complexing upon the toxicity of copper to aquatic organisms (principally fish), in Hunt, D.T.E., Ed., Trace Metal Speciation and Toxicity to Aquatic Organisms—a Review, rep. TR 247, Water Research Centre, Marlow, U.K., 1987, 43–51.

34. Wright, P., Heming, T., and Randall, D., Downstream pH changes in water flowing over the gills of rainbow trout, *J. Exp. Biol.*, 126, 499–512, 1986.

35. Randall, D.J., Lin, H., and Wright, P.A., Gill water flow and the chemistry of the boundary layer, *Physiol. Zool.*, 64, 26–38, 1991.

36. Playle, R.C. and Wood, C.M., Water chemistry changes in the gill microenvironment of rainbow trout: experimental observations and theory, *J. Comp. Physiol. B*, 159, 527–537, 1989.

37. Miller, T.G. and Mackay, W.C., Relationship of secreted mucus to copper and acid toxicity in rainbow trout, *Bull. Environ. Contam. Toxicol.*, 28, 68–74, 1982.

38. Broderius, S.J., Smith, L.L., Jr., and Lind, D.T., Relative toxicity of free cyanide and dissolved sulfide forms to the fathead minnow (*Pimephales promelas*), *J. Fish. Res. Board Can.*, 34, 2323–2332, 1977.

39. Erickson, R.J., An evaluation of mathematical models for the effects of pH and temperature on ammonia toxicity to aquatic organisms, *Water Res.*, 19, 1047–1058, 1985.

40. Karara, A.H. and Hayton., W.L., A pharmacokinetic analysis of the effect of temperature on the accumulation of di-2-ethylhexyl phthalate (DEHP) in sheepshead minnow, *Aquat. Toxicol.*, 15, 27–36, 1989.

41. Cairns, J., Jr., Buikema, A.L., Jr., Heath, A.G., and Parker, B.C., Effects of Temperature on Aquatic Organism Sensitivity to Selected Chemicals., Bull. 106, Virginia Water Resources Research Center, Blacksburg, 1978.

42. Erickson, R.J., Benoit, D.A., and Mattson, V.R., A Prototype Toxicity Factors Model for Site-Specific Copper Water Quality Criteria. Rep. U.S. Environmental Protection Agency, Duluth, MN, 1987.

43. Mayer, F.L., Jr., Marking, L.L., Pedigo, L.E., and Brecken, J.A., Physicochemical Factors Affecting Toxicity: pH, Salinity, and Temperature. I. Literature Review. Rep. EPA/600/X-89/033, U.S. Environmental Protection Agency, Gulf Breeze, FL, 1989.

44. Brecken, J.A. and D'Asaro, C.N., Acute Toxicity of Four Chemicals to Marine Organisms as Affected By Temperature and Salinity. Rep. U.S. Environmental Protection Agency, Gulf Breeze, FL, 1990.

45. Voyer, R.A. and McGovern, D.G., Influence of salinity and temperature on acute toxicity of cadmium to *Mysidopsis bahia* Molenock, *Arch. Environ. Contam. Toxicol.*, 19, 124–131, 1990.

46. Voyer, R.A. and McGovern, D.G., Influence of constant and fluctuating salinity on responses of *Mysidopsis bahia* exposed to cadmium in a life-cycle test, *Aquat. Toxicol.*, 19, 215–230, 1991.

47. Gossiaux, D.C., Landrum, P.F., and Tsymbal, V.N., Response of the amphipod *Diporeia* spp. to various stressors: cadmium, salinity, and temperature, *J. Great Lakes Res.*, 18, 364–371, 1992.

SESSION 3

INORGANIC TOXICANTS

Chapter 1

Ligands and the Bioavailability of Metals in Aquatic Environments

Michael C. Newman and Charles H. Jagoe

INTRODUCTION

Metal bioavailability* is influenced by physical, chemical, and biological factors in aquatic environments. Physical factors include temperature, phase association (solid, liquid, or gas), physical adsorption, sequestration by occlusion within a solid phase, or depositional regime as dictated by water movement. Chemical factors include those influencing speciation at thermodynamic equilibrium, complexation kinetics, lipid solubility, and phase transitions such as those associated with precipitation, coprecipitation, or chemical adsorption. Both organic and inorganic species contribute to these phenomena. A myriad of biological factors can also modify bioavailability including trophic interactions, biochemical or physiological adaptation, microhabitat utilization, animal size and age, and particular species characteristics. Physical and chemical factors can also interact with these biological factors. For example, temperature, pH, or Cl^- can modify gill function and, consequently, uptake of dissolved metals.

A major class of chemicals that modify bioavailability are the ligands. Ligands are anions or molecules that form coordination compounds or complexes with metals.[1] Ligand influence may be direct, e.g., sequestering the metal by complexation, or indirect, e.g., influencing gill function. Most studies of ligand effects on metal bioavailability have been descriptive and the mechanism for their influence often remains speculative. Our intention here is to describe the various roles that ligands may play in determining metal bioavailability. Turner and co-workers' classification of cation-ligand interactions[2] will be used as a central theme in our description. Ligand effects on biological processes will also be highlighted.

As suggested by the editors of this volume, emphasis will be placed on inorganic ligands. This does not imply that organic ligands are unimportant relative to metal bioavailability. Organic ligands can play a major role in many aquatic systems.[1] Metals bind to dissolved organic ligands primarily at carboxylic and phenolic functional groups. Association constants for these functional groups correlate with the stability constants for hydroxo and carbonato complexes,[2] as discussed later in this chapter. Further, steric configurations of some algal exudates may impart a degree of

---

*Bioavailability is the degree to which a contaminant in a potential source is free for uptake (movement into or onto an organism). Some definitions of bioavailability further imply that the contaminant must affect the organism ("a metal is considered to be in a biologically available state when it is taken up by the organism and can react with its metabolic machinery").[97] This is consistent with several pharmacological definitions, e.g., "the bioavailability of an active ingredient(s) [is] the rate and extent to which the ingredient is absorbed and *becomes available for the site of action*".[126]

specificity to metal binding.[1] Organic ligands compete with inorganic ligands and functional groups on biological surfaces for cation binding, and consequently influence metal bioavailability.

INORGANIC LIGANDS

The ligand-metal complex (ML) may involve electrostatic, covalent, or some intermediate binding. Electron pairs are shared between the metal (M) and ligand (L). The ligand retains most of the influence over the electron pair, with the degree of control varying between complexes.[3] The ligand is monodentate when one pair of electrons is involved in the complex. If more than one pair are shared, the ligand is multidentate, e.g., bi-, tri-, or tetradentate ligands. Multidentate ligands are also referred to as chelates,[3] e.g., many dissolved organic ligands. Inorganic ligands may be simple (Cl^- in $CdCl^+$) or complex (HCO_3^- in $Mg(HCO_3)_2^0$). In complex ligands, the atom responsible for the nucleophilic (electron contributing) qualities of the ligand is termed the donor or ligand atom, e.g., O in OH^-.[1]

Pankow[3] gives the important yet often overlooked example of H_2O as a ligand. Metal ions complex with H_2O to form hydrated ions. Although there are potentially two electron pairs on the donor oxygen atom available for complexation with metal coordination sites, their proximity to each other results in H_2O functioning as a monodentate ligand. The water molecules associated with the hydrated metal (often six water molecules in the first hydration shell[3,4]) give it qualities distinct from the nonhydrated ion. As ions may have second or third hydration shells, qualities such as hydrated ion radius may be difficult to measure precisely.[5]

Both charge and diameter of the metal ion influence the size of the hydration shell. Assuming equal charge, smaller diameter ions have larger hydration shells than larger diameter ions. The hydration shell increases the effective diameter of the metal ion and, consequently, influences its passage through membrane channels with steric constraints.[5] The resistance to removal of water molecules in the hydration shell during interactions with channel binding sites also influences metal passage through membranes.[5]

Major Inorganic Ligands and Complex Formation

Five inorganic ligands of primary importance in natural waters are F^-, Cl^-, SO_4^{2-}, OH^-, and HCO_3^-.[2] Also important are CO_3^{2-}, HPO_4^{2-}, NH_3 and, in anoxic waters, HS^- and S^{2-}.[6] Nitrate (NO_3^-) doesn't form strong complexes with metals.[1]

Several qualities of cations define their abilities to form complexes with these ligands.[1-3,6] Turner et al.[2] developed a very effective, bivariate scheme involving the cation polarizing power (estimated with the ion charge squared divided by the ion radius, Z^2/r) and ability to form covalent bonds (estimated by the difference between the logarithms of the metal's binding strengths in the fluoro and chloro complexes, $\Delta\beta = \log\beta_{MF} - \log\beta_{MCl}$). Differences linked to polarizing power imply electrostatic bonding, with Z^2/r regarded as a measure of the metal's ability to attract electrons in the hydrolysis scheme:

$$M^{n+} + H_2O \xrightarrow[\text{tion}]{\text{hydra-}} [M-OH_2]^{n+} \xrightarrow[\text{lysis}]{\text{hydro-}} [M-OH]^{(n-1)+} + H^+$$

The tendency of the reaction to shift to the right, and the strength of the M-O interaction, increase as the polarizing power of the cation increases. The cations are fully hydrolyzed beyond a Z^2/r of approximately 23: they are very weakly hydrolyzed below 11.[2]

Figure 1 categorizes cations into four groups based on these two variables. According to Turner et al.,[2] stability constants ($K = [ML]/[M][L]$ for $M + L \rightarrow ML$) for complexes with hard ligands (SO_4^{2-} and F^-) will increase from the top to the bottom of this diagram. Movement across

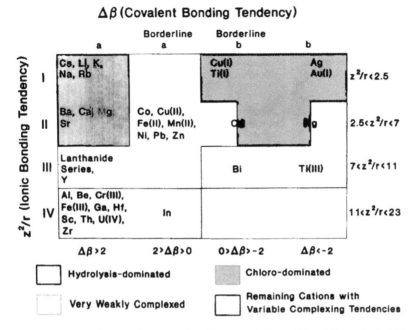

Figure 1. Complexation field diagram for cations from Turner et al. Figure 12[2] and Brezonik et al. Figure 6.[6]

the diagram does not affect complex stability for hard ligands. Stability of complexes with soft ligands (Cl^-, HS^-, and S^{2-}) will increase rapidly from left to right across the diagram. Stability of soft ligand complexes will increase only slowly down the complexation field diagram. Bisulfide and S^{2-} are much softer ligands than Cl^-. Complex stability for intermediate ligands (OH^- and CO_3^{2-}) will increase diagonally (top left to bottom right) across the diagram. These general trends are summarized in Figure 2. Hard ligands tend to bind primarily by electrostatic interactions, while soft ligands tend toward covalent bonding. As Turner et al.[2] found a strong correlation between hydroxide and carbonate complex stability and metal-organic matter complexation, carboxylic and phenolic groups were estimated to follow the trend noted for intermediate ligands.[2,6] The sulfhydryl group will react as a very soft ligand.

Based solely on the a- and b-metal characterization like that described above using $\Delta\beta$, Stumm and Morgan[1] give the following general trends for complex stabilities for ligands or ligand atoms: a-metal cations—$F > O > N = Cl > Br > I > S$, $OH^- > RO^- > RCO_2^-$, $CO_3^{2-} >> NO_3^-$, $PO_4^{3-} >> SO_4^{2-} >> ClO_4^-$, and b-metal cations—$S > I > Br > Cl = N > O > F$. Using both Z^2/r and $\Delta\beta$ for classification of the cations, cations forming weak complexes are those in groups Ia and IIa (Figure 1). Chloro-dominated cations are the Ib and I borderline b metals. The IVa, IV borderline a, III borderline b, and IIIb metals have hydrolysis-dominated interactions. Cadmium is intermediate between chloro-dominated cations and cations with variable complexing tendencies. Mercury is intermediate between the hydrolysis- and chloro-dominated groups. Both Cd and Hg are chloro-dominated in seawater. In freshwater, Hg shifts toward hydrolysis-dominated and Cd shifts to more complicated speciation.

Unfortunately, the remaining cations have variable complexing tendencies depending on their electron configurations. Transition metals have 0 to 10 d electrons and their complex stabilities are not easily predicted from the system outlined above.[2,6]* The Irving-Williams series ($Mn^{2+} <$

*Class-a metals have an inert gas configuration (d^0) and class-b metals have filled d orbitals with 10 to 12 electrons. Hence, class-a metals have electric fields that are very resistant to deformation by an adjacent ion's influence and class-b metals are more readily deformed. In this context, class-a and -b metals have "hard" or "soft" (easily polarizable) spheres, respectively.

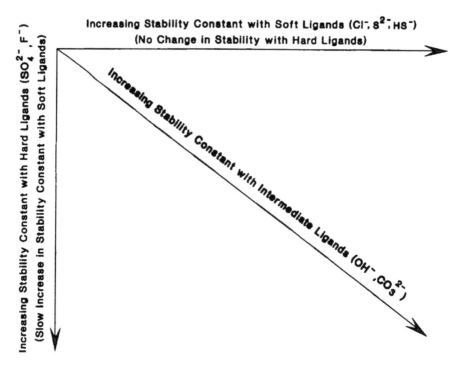

Figure 2. General trends within the Turner et al.[2] complexation field diagram shown in Figure 1.

$Fe^{2+} < Co^{2+} < Ni^{2+} < Cu^{2+} > Zn^{2+}$) can be used to predict the general order of complex stability for these metals, however.

Assessment of Speciation

Estimation of metal speciation in the presence of complexing ligands is necessary because various metal species differ in bioavailability and toxicity. The most widely used approach is to assume the system is at thermodynamic equilibrium and to calculate the equilibrium species distribution of the metal of interest in the presence of ligands at specified activities.[1,7] Typically, this is done by calculating the change in free energy associated with formation of the metal-ligand (ML) complex (ΔG_r), the Gibbs free energy of formation of the product (ML) minus the Gibbs free energy of formation of the reactants (M and L). The reaction equilibrium constant (K) is then derived with the relationship $\Delta G_r = -RT \ln K$, where R is the gas constant and T is the absolute temperature.

Tables of the Gibbs free energies of formation for a large number of inorganic substances are available,[1,7,8] as are tables of K values for reactions involving inorganic species.[8,9] Usually, by using the K for formation of a particular ML complex, boundary conditions are calculated by solving the equilibrium for L with ML and M present at equal activities. At ligand activities above this boundary, the ML complex is the dominant species, while M dominates at ligand activities below the boundary.

In natural systems, a number of metals and ligands are present together. Modeling such systems requires the simultaneous solution of a large number of equations, or iterative calculation. Consequently, a number of computer programs have been written to estimate equilibrium species distributions. Some are capable of predicting speciation of several metals (MINEQL[10]), while others focus on specific metals such as Al (ALCHEMI[11]).

While equilibrium thermodynamics models have been used successfully to assess metal speciation, several implicit assumptions can limit this approach. Many tabulated values assume conditions of 25° C, 1 atm pressure, and zero ionic strength, for example. These assumptions are reasonable in most cases and allow approximation of many equilibrium species distributions, but they are not valid in all situations. An extreme example would occur in cold seawater in a deep region of the ocean, where the actual metal-ligand complex distribution could be very different from that predicted using tabulated values.

A key assumption is that the reactions have reached equilibrium; kinetic constraints on metal speciation in natural waters are often overlooked. Some thermodynamically favored reactions occur too slowly to be relevant in certain situations. For example, Mn (II) occurs in anoxic waters, but it is oxidized to Mn (IV) in oxygenated water. Equilibrium calculation shows that virtually all of the Mn in a well-buffered, oxygenated stream is Mn (IV). Pankow and Morgan[12] calculated the oxidation of Mn (II) would require hours or days in neutral or basic waters. A significant fraction of the Mn present in a stream could therefore be in reduced form in water downstream from a groundwater seep or an input of water from an anoxic hypolimnion. Another rate-influencing factor is competition between cations for ligands that hinder ML complex formation. Hering and Morel[13] demonstrated the binding of Cu^{2+} to EDTA was slowed by Ca^{2+} at concentrations found in seawater. Their results suggested that the kinetics of exchange of Cr^{3+}, Fe^{3+}, and Ni^{2+} with the Ca-EDTA complex may be even slower than that of Cu^{2+}.

Another assumption often made in thermodynamic modeling is that bulk water chemistry accurately reflects the chemical conditions in the environment where ML interactions occur. In fact, the chemical conditions found in microenvironments and bulk water may be quite different. Examples include the boundary layers associated with sediments or surface coatings, and microenvironments created by biological activities. For instance, water pumped through burrows by polychaetes such as *Arenicola* sp. may generate oxygenated regions in oxygen-depleted sediments.

The water layer near the surfaces of gills is another important microenvironment that may be markedly different from the surrounding water, thereby affecting speciation. Both NH_3 and NH_4^+ are excreted at the gill.[14] The pK of the reaction, $NH_3 + H^+ \leftrightarrow NH_4^+$ is approximately 9.5. If the pH < 9.5, excreted NH_3 absorbs a hydrogen ion, thus raising the pH of water passing over the gill. Also, CO_2 and HCO_3^- move across the gill epithelium,[15] and can alter the pH of the water layer nearest the gill. Playle and Wood[16] calculated the base equivalents released by the gill, and showed that the pH of acidic, soft water increased as it passed over the gill. Wright et al.,[17] Randall and Wright,[18] and Lin and Randall[19] also found that pH changed as water passed over the gill, with the magnitude and direction of the change dependent on the initial pH of the water.

For example, aluminum speciation and solubility are affected by the change in pH occurring near the gill surface. As low pH, Al-enriched water passes over the gills, pH is raised, and $Al(OH)_2^+$ and $Al(OH)_3$ form. Aluminum may then precipitate and accumulate on epithelial surfaces. Playle and Wood[20] demonstrated that the respeciation of Al occurred rapidly enough (<2 s) to allow precipitation as water passed over the gills. In experiments, Al was extracted from water by fish gills and accumulated in gill tissue.[21] Temperature also affects Al speciation kinetics, and low temperatures may improve fish survival by slowing Al polymerization rates.[22] These studies clearly illustrate the importance of considering both kinetics and microlayer chemistry in metal bioavailability studies.

With consideration of the caveats discussed above, the equilibrium thermodynamics approach in bulk water yields a valid approximation of conditions in most natural systems. Hoffmann[23] found that rate constants for most reactions were rapid in comparison to residence time, and concluded that equilibrium modeling is generally valid even in dynamic systems. Tabulated values of K or Gibbs free energies of formation for many metal-ligand complexes are available, but considerably less data exist on the kinetics of formation of these complexes. Widespread incorporation of kinetic considerations into models is likely to be hampered by the scarcity of tabulated values for many half-lives or rate constants for reactions.

Lipid Solubility of Metal-Inorganic Ligand Complexes

Simkiss[24] noted that although most studies of metal bioavailability stress the hydrophilic nature of metal ions, many class-b metals such as Hg and Cd can form complexes that cross membranes based on their lipid solubility. Although metal-organic complexes are occasionally discussed in this regard,[25] inorganic complexes, especially those formed in marine systems, have been given inadequate attention. Simkiss[24] suggests that inorganic ligands can facilitate metal penetration of membranes by forming relatively uncharged complexes. For example, Cd and Hg in seawater can enter as chloro complexes, while Cu and Zn enter as hydroxo and carbonato complexes. Simkiss' experiments with metal partitioning between olive oil and sea water support these suppositions for Cu, Zn, Cd, and Hg. Consistent with the proposed mechanisms, two class-a metals (Na and Sr) showed no significant movement into the lipid layer. (Remember that class I and II-a metals are very weakly complexed.) Further, Simkiss and Taylor[5] estimate that $HgCl_2^0$ penetrates lipid membranes 10^6 times faster than its charged complexes. Although incorporated into data interpretation of a few studies,[26,27] this role played by inorganic ligands in facilitating passage through membranes is generally ignored. Its importance remains unclear at this time.

BIOAVAILABILITY FROM WATER

The aquo ion is thought to be the most available species,[28] although other complexes may also be taken up.[29] (More correctly, availability is proportional to the aquo ion activity.) Consequently, the effect of ligands on metal bioavailability is often deemed a direct result of dissolved ligand competition with binding sites on the gill or gut surface for the free metal ion. For example, Cu bioavailability as affected by CO_3^{2-}, HCO_3^-, PO_4^{3-}, and $P_2O_7^{4-}$ [30] has been linked to their influence on Cu^{2+} and $CuOH^+$ activities. Andrew[31] suggested that the influence of alkalinity components and other ligands on Cu toxicity could be explained by the observation that these ligands decreased the aquo complex activity. Similarly, Bingham et al.[32] correlated the effect of Cl^- on Cd bioavailability to its influence on Cd^{2+} activity.

Although considerable insight is gained by this approach, ligands affect biological processes and structures as well. For example, the effect of salinity on Cu bioavailability to marine bivalves could not be explained by the effect of Cl^- on the Cu^{2+} ion activity alone.[33] The combined impact of inorganic ligands on metal interactions with biological surfaces, membrane function, gill function, and the individual's general physiology must be considered. Furthermore, some ligands may have less obvious yet equally important roles in influencing bioavailability. For example, buffering associated with alkalinity components modulates pH shifts. As pH has a significant influence on metal bioavailability[26,34,35] such buffering would play an indirect but important role.

Adsorption Models of Uptake

As mentioned above, a major approach to predicting the influence of inorganic ligands on bioavailability is to simply treat the organism as one of many competing ligands, e.g., see Brezonik et al.[6] The ligands on the cell include those possessing O ligand atoms with high affinities for class-a metals and tendencies for ionic bond formation, e.g., carboxyl or phenolic groups. The ligand atom may also be N or S, e.g., amino or sulfhydryl groups. Such ligands have high affinities to class-b metals with covalent bond formation. Surface ligands compete for metal ions with ligands dissolved in the media or associated with seston.

An excellent example of this approach is Fisher's[36] work with metal bioaccumulation in cultured phytoplankton. The volume concentration factor (VCF = amount of metal per unit cell volume/amount per unit seawater volume) was measured for phytoplankton exposed to metals ranging from weakly complexed to hydrolysis-dominated. Figure 3a shows the clear correlation

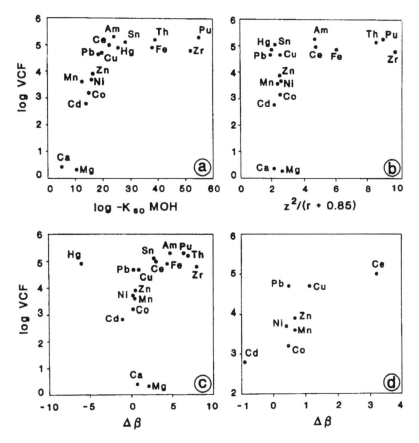

Figure 3. Phytoplankton bioaccumulation of metals based on metal-ligand complexation trends.[36] Panel a is Fisher's Figure 1a relating the volume concentration factor (VCF) to the log of the solubility products for the pertinent metal hydroxides. Panels b and c show the less obvious effects of Z^2/r and $\Delta\beta$ on the VCF. For metals intermediate between hydrolysis-dominated and weakly complexed metals ($11 < \log -K_{so}$ MOH < 23) (Panel d), $\Delta\beta$ has a clear influence, but the log $-K_{so}$ MOH still fits best.

between these VCF values and the log of the solubility products for the corresponding metal hydroxides. (A significant relationship was also derived using equilibrium constants with acetic acid, CH_3COOH.) There is a gradual increase in VCF with a plateau occurring for hydrolysis-dominated metals. Metals with the lowest VCFs were the weakly complexed metals from class Ia and IIa (Ca and Mg). Fisher attributed this strong relationship to the dominance of phenolic and carboxyl groups on phytoplankton and "organic films which coat particle surfaces in seawater." Plots of these metals (or those metals with intermediate tendencies for hydrolysis, i.e., $11 < \log -K_{os}$ MOH < 23, Figure 3d) against either Z^2/r or $\Delta\beta$ showed no marked improvement in predicting metal VCF, suggesting that neither characteristic alone dominated the correlation. However, our nonparametric analyses of Fisher's data suggest that both were significantly correlated ($\alpha = 0.05$) with the VCF. This is consistent with the Turner et al.[2] classification of phenolic and carboxyl groups as intermediate ligands, i.e., possessing complex stabilities highly correlated with those for OH^- (Figure 2).

Adsorption models are now common for describing metal-binding to algae,[37-42] bacteria,[43] and metazoan surfaces.[44-46] Pagenkopf[44] assumed that competition between gill binding sites and dissolved ligands explained the effect of Cu speciation on acute toxicity. Andrew[31] suggested that the toxic action of the aquo Cu complex resulted from its binding to sulfhydryl groups (soft ligand) of proteins. Protonation of surface sulfhydryl groups, and consequent lessening of Cu association with these groups, was put forward as the mechanism of pH effect on toxicity (see comment by Andrew in Reference 47). Reid and McDonald[46] related such gill surface binding of metals to a scheme similar to the complexation field diagram described above. Xue et al.[41] suggested that functional groups on algal cells compete for metals with ligands in solution during the initial, rapid adsorption phase of metal uptake. Crist et al.[40] related the relative strength of such adsorption to metal propensity for covalent and ionic bond formation.

Unfortunately, application of this parsimonious approach has its limitations. For example, Fujita and Hashizume[37] found that only 50% of Hg^{2+} taken up by algae was associated with passive adsorption. The other 50% entered the cell by some active mechanism. It is important to keep in mind that even unicellular algae are much more than simple collections of adsorption sites.

Several methodological limitations that could compromise results can arise during adsorption studies. Most studies consider only one adsorption isotherm type (L-curve isotherms as described by simple Langmuir or Freundlich equations) although four types of curves are possible.[48] Some use traditional, statistical fits to transformations that have undesirable qualities (see Kinniburgh, Reference 49). For example, Eadie-Hofstee or Scatchard plots have the undesirable condition of the measured variable (subject to significant error) in both the independent and dependent variables. The nonlinear regression or Langmuir reciprocal transform methods are preferable to Lineweaver-Burk, Eadie-Hofstee, or Scatchard transformations.

Finally, it is occasionally unclear whether adsorption ("the process through which a net accumulation occurs at a common boundary of two contiguous phases"[48]) or precipitation ("the accumulation of a substance to form a new bulk solid phase"[48]) is occurring. Although conformation of data to adsorption isotherms has been used to suggest adsorption, e.g., Les and Walker,[43] such conformity is not usually sufficient to ascribe removal to adsorption.[48] The term sorption is more appropriate when such ambiguity exists.

Inorganic Ligands and Membrane Transport

Figure 4 depicts the variety of mechanisms facilitating inorganic ligand modification of membrane transport. The eight mechanisms shown are not exclusive, e.g., adsorption followed by carrier-mediated flux may be one phase in the ion pump mechanism. Rather, the mechanisms shown are intended to illustrate the variety of ways that ligands may affect uptake.

Obviously, the hydration sphere surrounding the metal ion will influence metal passage through ion channels. As described above, adsorption and carrier-mediated metal flux are influenced by the presence of dissolved ligands. Precipitation of metal-ligand complexes, as discussed above for Al, also affects bioavailability and toxicity. Metal transport may also be carrier-mediated, with passage through ion channels with strong influence by the metal hydration sphere. Endocytosis occurring at clathrin-coated pits may involve engulfment of dissolved (pinocytosis) or particulate metal. Solute uptake through channels can be enhanced by solvent drag, e.g., the movement of the metal along the flow of water. Lipid solubilities of metals including those of metals complexed with ligands also modify bioavailability. Lipid permeation may be enhanced by hydrophilic sites on phospholipids within the membrane and by metal complexation with a dissolved ligand to form an uncharged complex.

Although not apparent from our treatment of biological surfaces to this point, inorganic ligands also influence movement across membranes of cells via the remaining mechanisms (ion pump, pinocytosis, and solvent drag). These effects result from ligand modification of ion and osmoregulation, metabolism, and excretion.

Gill Structure and Function

The gills of aquatic animals are the principal sites for exchange of dissolved substances, including metals. The remainder of the integument is relatively impermeable in most animals.

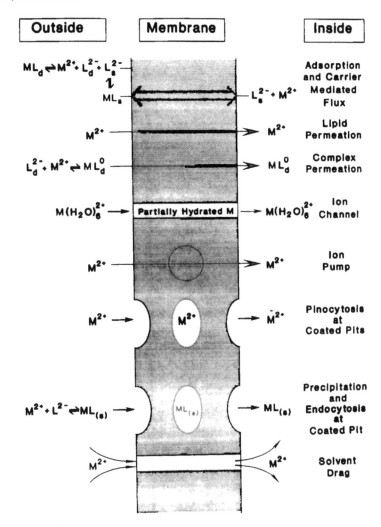

Figure 4. Possible mechanisms of ligand influence on metal passage across membranes. This figure is a composite of those of Simkiss and Taylor (Figure 6)[5] and Brezonik et al. (Figure 7)[6] with the addition of solvent drag from Gordon.[125]

However, exchanges of substances do occur by other routes. Fish in seawater drink to compensate for osmotic water losses, and the intestine functions as an ion-transporting epithelium.[50] In larvae, gas exchange may occur over the body surface rather than through gill ventilation.[51] This suggests that other exchanges across the general body surface are possible as well. In eggs, metals such as Cd can bind to functional groups present in the egg capsule.[52]

In many studies, metals are considered to bind to a gill "membrane," a hydrophobic layer dividing external and internal compartments, or to surface ligands on this membrane. While such conceptual simplification may be useful, a better understanding of metal accumulation and toxicity would result from consideration of actual gill structure and function.

Figure 5 shows the principal structural and chemical features of the fish gill surface. The three major cell types (respiratory or pavement, chloride and mucous) are attached by cell junctions. A layer of mucus (not to scale) covers the epithelium. Movement of substances across the gill maintains a boundary layer, whose chemistry may be different than that of the bulk water. The arrows represent the directions of major net fluxes of substances in freshwater (FW), seawater (SW), or both (if unlabeled). The plasma membrane of a cell is composed of a lipid bilayer (inset). The hydrophobic regions of the phospholipids are oriented towards the center of the bilayer. The

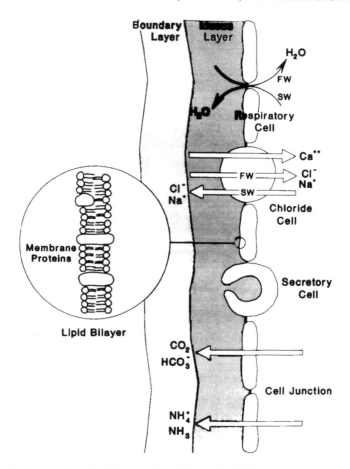

Figure 5. Basic structural and functional features of the teleost gill epithelium.

hydrophilic ends, representing potential metal binding sites, are oriented towards the aqueous environment and cytoplasm. Proteins float in this bilayer, and may extend partially or completely through it. Membrane proteins are involved in ion transport or other metabolic activities, and also represent potential metal binding sites.

The complex nature of the gill provides a variety of binding sites for dissolved metals. Cell membranes are composed of hydrophilic outer layers surrounding a hydrophobic layer. The lipids which form this structure include phospholipids, cholesterol, and glycolipids. The hydrophilic ends of these compounds are often nitrogen-containing bases, and may function as ligands for cation attachment. Embedded in the membrane are proteins functioning as ion channels and transport enzymes which may also serve as ligands.

Gills of vertebrates and invertebrates share a number of common features. These include short diffusion distances between water and circulatory fluid, and large surface areas to facilitate gas exchange. Due to these physical characteristics, ionic and osmotic fluxes across the gills are usually high. Gill surfaces are composed of single or multiple layers of epithelial cells, attached and sealed by cell junctions. Often, specialized cells (sensory, ion-transporting, mucus-secreting) are present.

In fish gills, primary lamellae (filaments) support two rows of secondary lamellae. The thin epithelium covering the secondary lamellae is the major site of gas exchange.[53] Ion-transporting (chloride) cells commonly occur along the primary lamellae, but spread to the secondary lamellae under conditions including low ionic strength media or exposure to metals.[54] These cells normally transport Na^+, Cl^-, and Ca^{2+}, but they may also accumulate toxic metals.[55] Metabolic waste products (NH_4^+, NH_3, HCO_3^-, CO_2) emitted here also modify the chemistry of the boundary layer, as discussed previously.

Gill epithelia are usually covered with an extracellular matrix. Cells of arthropod gills are covered with chitin, a polysaccharide. Specialized cells in the gills of many animals secrete mucus. In fish, mucus contains a variety of glycoproteins; the exact composition varies with species and environmental conditions.[56-58] The mucus layer is a polyanionic matrix functioning as an ion exchange system, with different affinities for different metals.[59] Mucus likely concentrates metals near gill surfaces in close proximity to membrane transport sites.[60] It may also serve to accumulate and slough off potentially toxic ions.[61]

Gill structure and function can be significantly modified by the presence of many ligands and metals. For example, in freshwater fish, Cl^- is exchanged for HCO_3^- to offset diffusional losses of Cl^- to the environment.[62] As Na^+ and Cl^- concentrations increase, euryhaline fish develop accessory cells that share a specialized cell junction with adjacent chloride cells. With increased salinity, water permeability decreases while permeability to ions increases.[63] Increased salinity also results in increased gill ATPase activity, reflecting increased active ion transport.[64] Activation of ion uptake mechanisms may increase uptake of nonessential metals as well.

There is evidence that H^+ and some metals compete for ligands on the gill at sites normally occupied by Ca^{2+}.[65,66] Calcium is essential to the maintenance of the tight junctions which seal the gill epithelia of freshwater fish. Removal of Ca^{2+} causes increased permeability to metals and water by opening paracellular pathways.[67] However, some metals may attach to binding sites in place of Ca^{2+} and maintain low paracellular permeability.[68]

Waste products expired from the fish gill, such as NH_3 and HCO_3^-, can themselves function as weak ligands. Further, they alter the pH near the gill and affect the charge state of titratable functional groups at the gill surface. Ammonia can also bind to some functional groups, potentially influencing metal binding.

An Illustration using Aluminum

Aluminum, an important toxicant in many acidified waters, provides an important example of ligand effects on metal bioavailability. The hardest trivalent ion commonly found in the environment is Al^{3+}. It has a very small ionic radius (0.51 Å) and a strong tendency to form hydrolytic species (see Reference 69 and Figure 1). Because hydrolysis reactions involve the H^+ ion, pH affects the species distribution of Al hydrolysis products (Figure 6a). Polymeric species of $Al(OH)_x$ have also been described, but the formation of these is kinetically much slower than reactions involving monomeric ligand complexes.[70] In dilute Al solutions without added base, polymeric species are relatively unimportant as hydrolysis products.[71]

Aluminum also forms complexes with F^-.[69,72] In a manner analogous to the effect of pH on the distribution of hydrolytic Al species, the equilibrium distribution of AlF_x species is controlled by F^- ion activity (Figure 6b). Sulfate can also complex with Al, but it is a weak ligand[72] and unlikely to modify bioavailability at reasonable concentrations in the presence of competing ligands. Silicic acid, $Si(OH)_4$, forms complexes with Al, and the toxicity of Al is greatly diminished when Si is present at higher activities than Al.[73] A variety of organic ligands complex Al,[69,74] but this topic is beyond the scope of this review except to note that organically complexed Al is generally biologically unavailable.

From consideration of the interactions of Al with inorganic ligands, the Al species contributing to bioavailability and toxic effects are Al^{3+}, $AlOH^{2+}$ and $AlOH_2^+$, and AlF_x, depending on the pH and F^- activity. In seawater, insoluble $AlOH_3$ is a dominant species, and Al concentrations measured in seawater are very low (about 1 μg/l).[75] In freshwater, concentrations more than two orders of magnitude or greater occur. Freshwater fish exposed to elevated levels of Al exhibit osmoregulatory disturbances due to increased passive efflux of physiologically important ions (Na^+, Cl^-), and inhibition of active ionic uptake mechanisms.[66,76] Fish exposed to elevated Al in freshwater also accumulate the metal, especially in gill tissue.[55,77]

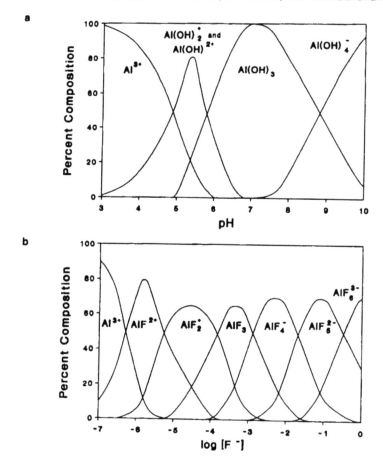

Figure 6. Aluminum speciation as affected by pH and F⁻ activity. a. Relative concentrations of hydrolytic Al species as a function of pH. b. Relative concentrations of aluminum-fluoride species as a function of F⁻ ion activity. (Panel a was redrawn from Figure 1 of Martell and Motekaltis,[69] and Panel b from Figure 1 of Hem.[72])

Several studies have shown that Al at pH ≤ 4.5 actually enhances the survival of adult fish, fry, and eggs. This may be caused by Al^{3+} substituting for Ca^{2+} at membrane binding sites, ameliorating the toxic effects of H^+.[78,79] The speciation of Al in water is pH dependent, so Al^{3+} cannot be the dominant species except when the pH is below approximately 4.9. Thus it is not possible to assess the effects of Al^{3+} except at low pH, when toxic effects due to H^+ are also likely occurring. However, accumulation of Al is reduced at low pH.[68,80] This could be due to competition between H^+ and Al^{3+} for gill surface binding sites based on the model of Pagenkopf,[44] who treated the metal-binding sites at the gill surface essentially as dilute ligands in solution. In his model, other ligands such as OH^- and F^- compete with the gill surface for Al^{3+}. Also, other cations such as Ca^{2+} and H^+ would compete with Al^{3+} for binding sites at the gill surface. To explain the accumulation and toxic effects of Al even in the presence of competing ligands such as OH^-, Neville and Campbell[80] proposed a modification of this model. They suggested mixed-ligand Al complexes, e.g., gill-L-Al-OH would be formed, as well as the gill-L-Al complexes predicted by the free ion model of Pagenkopf. The results of Wilkinson et al.[81] support this modification. In their study, Al bioaccumulation and toxicity persisted even in the presence of an excess of the competing ligand, F^-. The presence of F^- did ameliorate toxicity and bioaccumulation, but not to the extent predicted by the free ion model. This suggests that gill-L-Al-F complexes were indeed present under elevated F^- conditions.

Several authors have also proposed that direct precipitation of Al-OH or Al-OH polymers could occur at the gill surface, due to the higher pH present at the gill boundary layer.[16,20,80] Exeley et al.[82] suggested that Al binding to gill surface functional groups such as phosphatidyl-choline and phosphatidylserine would be facilitated in the higher pH conditions present at the gill boundary layer, so that actual precipitation of Al-OH species is unnecessary. Besides epithelial membranes, mucus containing a variety of polyanionic ligands[57] is present at the gill surface and may represent a significant Al binding site.[83]

BIOAVAILABILITY FROM SOLID-ASSOCIATED PHASES

Benthic Sediments

Bioavailability of sediment-associated metals is a major concern, as sediments are an important sink for metals and significant biotic communities are associated with sediments. Despite their importance, Luoma[84] recently acknowledged the paucity of information regarding concepts and methods critical to assessment of bioavailability of sediment-associated metals. Areas requiring study include:

"1) improved computational or analytical methods for analyzing distribution of metals among components of the sediments; 2) improved computational methods for assessing the influences of metal form in sediments on sediment-water metal exchange; and 3) a better understanding of the processes controlling bioaccumulation of metals from solution and food by metazoan species directly from sediments."

Important to each of these areas is the role of metal-ligand interactions in solid and dissolved phases.

Perhaps the most serious impediment has been interpretation of results from extraction techniques. These techniques are designed to identify various chemical fractions with which metals are associated. For example, sequential extraction procedures developed by Tessier et al.[85] provide approximate distributions of metals between forms such as readily exchangeable, carbonate bound, Fe and Mn oxide bound, organic matter bound, and "residual" metal. Unfortunately, the potential for inaccuracy in these techniques, and the associated tendencies to either overinterpret or categorically reject associated results, has led to much confusion. (See References 86 to 90 for examples.)

Despite the controversy and the critical need for more work in this area, several important themes have emerged from studies using these procedures. Work with surficial or oxic sediments suggests that several procedurally defined or laboratory-fabricated sediment fractions can reduce the bioavailability of metals. For some situations, it remains ambiguous whether the role of such fractions is direct (the metal associated with that specific ligand fraction is unavailable for uptake when ingested) or indirect (the role of the sediment form is to shift the equilibrium away from dissolved species in the interstitial waters).[91] In other situations, metal concentrations and complexes in both interstitial water and particulates are clearly important.[84] Regardless, several underlying themes emerge from studies of oxic sediments. First, easily extracted Fe (1 N HCl or an equivalent extractant) or hydrous Fe oxides tend to decrease the bioavailability of sediment-associated Ag, As, Co, Cu, Pb, and Zn.[92–98] Young and Harvey[98] suggest that Mn oxides may also decrease metal bioavailability. Secondly, increases in sediment organic carbon content may decrease Al, Co, Cu, Fe, Hg, Mn, Pb, and Zn availability.[97,98] However, the role of organic matter in lessening metal bioavailability varies between species.[99] Lastly, metals in the most readily extracted fractions in the extraction series tend to be most bioavailable.[96,98]

It is quite reasonable to assume that the reduction of metal bioavailability results from the avid adsorption of metals to hydrous Fe oxides, as Fe oxides are a major sink for metals in oxic sediments. This is readily illustrated by Davies-Colley et al.,[100] who found that 80% of Cd in estuarine sediments was associated with Fe oxides. In the simplest form, metal interactions can be envisioned as direct binding to negatively charged sites on the oxide surface. However, as shown below, more complicated interactions involving dissolved ligands are probable. Hydrous oxide adsorption sites may also bind dissolved complexes, e.g., the carbonato complex of Cu may adsorb to Mn oxides that have a net negative surface charge at pertinent pH conditions. A dissolved ligand may also adsorb to an oxide surface first, and act as a bridge for metal binding. For a more comprehensive discussion see Reference 48, pages 132 and 133.

Examples of Metal Cation Adsorption:

$$S - O_{(s)}^- + M^{2+} \rightleftharpoons S - OM_{(s)}^+$$

$$S - O_{(s)}^- + M(OH)^+ \rightleftharpoons S - OM(OH)_{(s)}$$

Examples of Adsorption of A Metal Present as An Oxyanion, e.g., CrO_4^{2-}:

$$S - OH_{(s)} + H^+ \rightleftharpoons S - OH_{2(s)}^+$$

$$S - OH_{2(s)}^+ + L^{n-} \rightleftharpoons S - L_{(s)}^{1-n} + H_2O$$

Example of Adsorption with Ligand Bridging:

$$S - L_{(s)}^- + M^{2+} \rightleftharpoons S - LM_{(s)}^+$$

It must also be remembered that hydrous oxides interact with metals in other ways that influence bioavailability. For example, Mn oxides can facilitate the oxidation of Cr(III) to Cr(VI).[101]

Consequently, total metal concentration in sediments is no longer considered a reliable indicator of bioavailable metal.[102] Instead, various normalization schemes have arisen to replace total metal for this purpose. For example, Ray et al.[103] recommended that metal associated with an EDTA extract is a good measure of available metal. Rule and Alden[104] suggested that the sum of the amount of metal found in exchangeable and easily reducible fractions can be used to indicate Cd bioavailability to the blue mussel, *Mytilus edulis*. The most common normalization method takes advantage of the negative influence of hydrous Fe oxides on bioavailability. The metal concentration of interest is divided by the concentration of Fe rendered soluble by an acid extractant such as 1 N HCl.[91,93,95] Although generally adequate for normalization purposes, the wide variation in the relationship between Fe normalized metal concentrations and extractable Fe suggests low predictive power of these techniques as presently employed. Further refinement is critical to enhance predictive use of associated results.

Under anoxic conditions, S^{2-} plays the major role in determining the bioavailability and toxicity of sediment-bound metals.[105-107] Under anoxic conditions, metals form predominantly sulfides[105] with very low solubility.* DiToro et al.[108] suggest that the predominance of metal sulfides in anoxic sediments greatly reduces the metal concentrations in interstitial waters and, in doing so, reduces metal bioavailability. Bioavailability of ingested metal sulfides or other solid forms in anoxic sediments has not been considered in this approach to date.

DiToro et al.[108] explain that the following equilibrium equations are pertinent for the interactions between metals and the predominant form of metal sulfide in anoxic sediments, FeS. Iron (Fe^{2+}) is at equilibrium with dissolved S^{2-} in the interstitial waters (log K_{sp} = −22.39 for M^{2+} + S^{2-} $\rightleftharpoons$ $MS_{(s)}$ and log K_{sp} = −3.64 for M^{2+} + HS^- $\rightleftharpoons$ $MS_{(s)}$ + H^+).

---

*Note from our discussion of ligand binding that the stability constants for complexes with S, a very soft ligand atom, will tend to increase with increasing "b-ness" of metals.

$$FeS_{(s)} \rightleftharpoons Fe^{2+} + S^{2-}$$

Added metal (Cd^{2+}) shifts the equilibrium.

$$Cd^{2+} + FeS_{(s)} \rightleftharpoons Cd^{2+} + Fe^{2+} + S^{2-}$$

Above a certain concentration of Cd^{2+}, CdS precipitates.

$$Cd^{2+} + FeS_{(s)} \rightarrow CdS_{(s)} + Fe^{2+}$$

The log K_{sp} values for Cd, Cu, Hg, Ni, Pb, and Zn range from approximately -57 to -28 (M^{2+} + $S^{2-} \rightleftharpoons MS_{(s)}$). Those estimated for M^{2+} + $HS^- \rightleftharpoons MS_{(s)}$ + H^+ range from -38 to -9. Consequently, under conditions controlled by HS^- and S^{2-}, interstitial water concentrations of these metals will be very low.

Because of the importance of these ligands in determining metal bioavailability in anoxic sediments, normalization of metal concentration to the amount of acid volatile sulfide (AVS)* in sediments is presently under development. This approach was used successfully in laboratory bioassays involving marine[108] and freshwater[109,110] sediments. DiToro et al.[108] speculated that the AVS normalization may be valid down to 1 μmol AVS per gram of sediment or lower.

To summarize, thus far, ML interactions control metal availability in oxic and anoxic sediments. Avid binding of metals to Fe (and Mn?) hydrous oxides and organic matter tends to lessen metal bioavailability under oxic conditions. However, considerable variation exists in the intensity of these effects. The dominant role of ligands under anoxic conditions involves the extreme "softness" of the HS^- and S^{2-} ligands.

Technique development for prediction of bioavailable metals in sediments requires further refinement and application to a wider range of conditions. For example, consensus on the best extraction procedure(s) for oxic sediments should be achieved and associated limitations of the procedures clearly defined. The use of Fe or organic matter normalization methods should be assessed over a wider range of phyla, as the majority of its present applications have involved molluscs. Also, the lower limit of the AVS normalization approach and its accuracy over a wide range of situations should reinforce its usefulness. The relative importance of ingestion vs. uptake from interstitial waters in anoxic sediments should be resolved and clearly stated.

Other Solid Phases
Submerged Surface Coatings

Materials coating submerged surfaces accumulate very high levels of metals.[111] Such materials, defined as periphyton or aufwuchs by applied biologists, have been used to monitor metal bioavailability in contaminated environments.[112-114] Central in each of these studies was concern about trophic transfer from this contaminated source.

In truth, these materials are a complex mixture of abiotic and biotic materials that are consumed by many species of scrappers and grazers.[111] When present on the surfaces of plants[115] or animals,[116] they may also contribute significantly to measured concentrations "in" those plants or animals. These facts leave open the question of metal bioavailability from this trophically significant yet poorly defined component of aquatic systems.

In two studies, Newman et al.[117,118] found that most of the As, Cd, Cr, Cu, Pb, and Zn in these materials was associated with an abiotic Fe- and Mn-rich matrix surrounding the microflora. The accumulation of these elements was attributed to adsorption and coprecipitation with hydrous metal oxides. Lead associated with these materials was accumulated by the grazing snail, *Physa integra*, at a rate much slower than that associated with plant tissue.[119]

---

*According to Carlson et al.[109] the AVS are procedurally defined as solid-phase sulfides soluble in cold HCl, and are thought to be primarily Fe and Mn sulfides.

Despite questions arising from these preliminary studies, the general role of hydrous metal oxides on the bioavailability of metals associated with submerged surface coatings remains poorly defined. Consequently, the plausible application of such normalization procedures as those used for assessing bioavailability in oxic sediments remains untested.[111] Considerable work remains to be done regarding metal bioavailability from submerged surface coatings.

Seston Including Phytoplankton

The bioavailability of metals in suspended materials (seston) has also received less attention than deserved. For example, the possible application of oxic sediment normalization procedures to abiotic components of the seston has not been reported, but recent studies suggest means of making these comparisons. Reinfelder and Fisher[120] examined the bioavailability of diatom-associated metals to zooplankton and related availability to the proportion of each metal in the diatom cytoplasm. Luoma et al.[121] recently completed examination of metalloid bioavailability in sediments (elementary Se) relative to that in diatoms. This approach has potential for estimating the effects of seston-associated ligands on metal bioavailability. Decho and Luoma[122] identified the role of specific stages of digestion in bivalves in Cr uptake. Such studies are badly needed for seston-associated metals as well as sediment-associated metals. Their extension to a wide range of metals with differing ligand chemistries would greatly enhance the generality of associated conclusions. Incorporation of species from other phyla and molluscan classes would also foster the emergence of general themes.

CONCLUSIONS

General Conclusions

Remarkable progress has been made in identifying the diverse roles played by inorganic ligands on metal bioavailability. General trends pertinent to both dissolved and solid phases are emerging. However, more attention is needed to further clarify major themes and express ligand roles in quantitative terms. Despite progress, it is difficult to imagine sufficient advancement in the near future to allow reliable and accurate predictions. Given the level at which mechanisms must be understood and quantified, the systems under scrutiny are too complex and too much information is lacking. Consequently, our immediate focus should be on refining present techniques and enriching our present understanding of underlying themes.

Several factors could impede progress in these areas. First, metals are often selected for study based solely on their regulatory importance, not their ligand complexing qualities. Beyond a certain point, this pragmatic selection process stymies the development of central themes. Second, technical advocacy of the understandable yet premature pressure for standardization can hinder technique refinement by imposing artificial constraints on investigative methodologies. Finally, our "preoccupation with description of the diversity of phenomena"[123] rather than the organization and classification of knowledge on the basis of explanatory principles (the goal of science as defined by Nagel[124]) continues to slow progress.

Specific Conclusions

1. Ligand influence on metal bioavailability may be direct, e.g., sequestering the metal by complexation or indirect, e.g., influencing gill function.
2. Five major ligands of primary importance in natural waters are F^-, Cl^-, SO_4^{2-}, OH^-, and HCO_3^-. Also important are CO_3^{2-}, HPO_4^{2-}, NH_3, HS^-, and S^{2-}.
3. The coordination chemistry of cations define their interactions with ligands. Turner et al.[2] developed an effective scheme involving cation polarizing power and cation ability to form covalent bonds.

4. Although thermodynamic equilibrium models are generally applicable to studies of bioavailability, several implicit assumptions must be kept in mind. Some problems that may limit the value of this otherwise extremely useful approach include significant deviation from the conditions associated with tabulated constants, the possibility of slow kinetics, and ignoring microenvironment chemistry.
5. The role of metal lipid solubility has not been given adequate attention in uptake studies, especially with regard to the solubility of metal-inorganic ligand complexes.
6. The bioavailabilities of metals are proportional to the aquo ion activity. The aquo ion is assumed to be the most bioavailable form although other complexes may also be available. The effect of ligands on metal bioavailability is often seen as a direct consequence of dissolved ligand competition with binding sites on gills or gut surface for the free metal ion. Although considerable insight is gained by this approach, it would be a mistake to forget that ligands affect biological processes and structures, too.
7. Use of adsorption models can be compromised by several assumptions and procedures. The assumption that adsorption controls uptake completely may be inappropriate. The use of a statistically compromised transformation of the data or an overly simplistic adsorption model may also compromise the results.
8. Dissolved ligands may modify metal transport across membranes by several mechanisms including competition with surface ligand sites, modification of lipid solubility, precipitation of complexes, and modifying biological processes. Modifications of biological processes include those associated with ion and osmoregulation, respiration, and excretion.
9. In many studies, metals are considered to bind to the gill "membrane." While such conceptual simplification is useful, a better understanding of metal accumulation would result from consideration of the actual gill structure and function. Such an approach could be used in conjunction with physiologically based, pharmacokinetic models.
10. Despite the importance of predicting the bioavailability of sediment-bound metals, the processes controlling bioavailability of ingested, sediment-bound metals still need to be addressed.
11. The controversy and uncertainty associated with interpreting results from sediment extraction techniques hampers progress in prediction of bioavailability of sediment-associated metals.
12. In oxic sediments, hydrous Fe oxides and, often, organic matter decrease metal bioavailability. The more readily extractable metals are generally more bioavailable than the more resistant metal fractions. In anoxic sediments, S^{2-} plays a major role in determining bioavailability. The predominance of metal sulfides in anoxic sediments greatly reduces the metal concentrations in interstitial waters and reduces metal availability.
13. As discussed in the text, the association of metals with those materials coating submerged surfaces and the abiotic components of seston have been generally ignored. Hence, more research is needed in this area.

ACKNOWLEDGMENTS

Financial support was obtained from Contract DE-AC09–76SR00819 between the U.S. Department of Energy and the University of Georgia. Additional support was obtained from the Society of Environmental Chemistry and Toxicology Foundation. The authors gratefully acknowledge the editorial and technical efforts of Ms. Rose Jagoe during all phases of manuscript preparation. Dr. Carl Strojan provided excellent comments on early versions of the manuscript.

REFERENCES

1. Stumm, W. and Morgan, J.J., *Aquatic Chemistry. An Introduction Emphasizing Chemical Equilibria in Natural Waters*, John Wiley & Sons, New York, 1981.
2. Turner, D.R., Whitfield, M., and Dickson, A.G., The equilibrium speciation of dissolved components in freshwater and seawater at 25° C and 1 atm pressure, *Geochim. Cosmochim. Acta*, 45, 855–881, 1981.

3. Pankow, J.F., *Aquatic Chemistry Concepts,* Lewis Publishers, Chelsea, MI, 1991.
4. Porter, W.W., *Concepts of Chemistry,* W.W. Norton, New York, 1972.
5. Simkiss, K. and Taylor, M.G., Metal fluxes across the membranes of aquatic organisms, *Rev. Aquat. Sci.,* 1, 173–188, 1989.
6. Brezonik, P.L., King, S.O., and Mach, C.E., The influence of water chemistry on trace metal bioavailability and toxicity to aquatic organisms, in Newman, M.C. and McIntosh, A.W., Eds., *Metal Ecotoxicology: Concepts and Applications,* Lewis Publishers, Chelsea, MI, 1991, 1–31.
7. Garrels, R.M. and Christ, C.L., *Solutions, Minerals and Equilibria,* W.H. Freeman, San Francisco, 1965.
8. Smith, R.M. and Martell, A.E., *Critical Stability Constants,* Vol. 4: Inorganic complexes, Plenum Press, New York, 1976.
9. Högfeldt, E., *Stability Constants of Metal-Ion Complexes Part A: Inorganic Ligands,* International Union of Pure and Applied Chem. Chem. Data Ser. No. 21, Pergamon Press, New York, 1982.
10. Westall, J.C., Zachary, J.L., and Morel, F.M.M., MINEQL: A Computer Program for the Calculation of Chemical Equilibrium Composition of Aqueous Systems, Tech. Note 18, Dept. of Civil Eng., MIT Press Cambrige, 1976.
11. Schecher, W.D. and Driscoll, C.T., An evaluation of uncertainty associated with aluminum equilibrium calculations, *Water Resour. Res.,* 23, 525–534, 1987.
12. Pankow, J.F. and Morgan, J.J., Kinetics for the aquatic environment, *Environ. Sci. Technol.,* 15, 1306–1313, 1981.
13. Hering, J.G. and Morel, F.M.M., Kinetics of trace metal complexation. Role of alkaline-earth metals, *Environ. Sci. Technol.,* 22 1469–1478, 1988.
14. Wright, P.A. and Wood, C.M., An analysis of branchial ammonia excretion in the freshwater rainbow trout: effects of environmental pH change and sodium uptake blockade, *J. Exp. Biol.,* 114, 329–353, 1985.
15. Randall, D.J. and Daxboeck, C., Oxygen and carbon dioxide transfer across fish gills, in Hoar, W.S. and Randall, D.J., Eds., *Fish Physiology,* Vol. 10A, Academic Press, New York, 1984, 283–314.
16. Playle, R.C. and Wood, C.M., Water chemistry changes in the gill microenvironment of rainbow trout: experimental observations and theory, *J. Comp. Physiol.,* 159B, 527–537, 1989.
17. Wright, P.A., Heming, T., and Randall, D.J., Downstream pH changes in water flowing over the gills of rainbow trout, *J. Exp. Biol.,* 126, 499–512, 1986.
18. Randall, D.J. and Wright, P.A., The interaction between carbon dioxide and ammonia excretion and water pH in fish, *Can. J. Zool.,* 67, 2936–2942, 1989.
19. Lin, H. and Randall, D.J., The effect of varying water pH on the acidification of expired water in rainbow trout, *J. Exp. Biol.,* 149, 149–160, 1990.
20. Playle, R.C. and Wood, C.M., Is precipitation of aluminum fast enough to explain aluminum deposition on fish gills? *Can. J. Fish. Aquat. Sci.,* 47, 1558–1561, 1990.
21. Playle, R.C. and Wood, C.M., Mechanisms of aluminum extraction and accumulation at the gills of rainbow trout, *Oncorhynchus mykiss* (Walbaum), in acidic soft water, *J. Fish Biol.,* 38, 791–805, 1991.
22. Poléo, A.B.S., Lydersen, E., and Muniz, I.P., The influence of temperature on aqueous aluminum chemistry and survival of Atlantic salmon (*Salmo salar* L.) fingerlings, *Aquat. Toxicol.,* 21, 267–278, 1991.
23. Hoffmann, M.R., Thermodynamic, kinetic and extrathermodynamic considerations in the development of equilibrium models for aquatic systems, *Environ. Sci. Technol.,* 15, 345–353, 1981.
24. Simkiss, K., Lipid solubility of heavy metals in saline solutions, *J. Mar. Biol. Assoc. U.K.,* 63, 1–7, 1983.
25. Blust, R., Verheyen, E., Doumen, C., and Decleir, W., Effect of complexation by organic ligands on the bioavailability of copper to the brine shrimp, *Artemia* sp., *Aquat. Toxicol.,* 8, 211–221, 1986.
26. Blust, R., Van der Linden, A., Verheyen, E., and Decleir, W., Effect of pH on the biological availability of copper to the brine shrimp *Artemia franciscana, Mar. Biol.,* 98, 31–38, 1988.
27. Holwerda, D.A., de Knecht, J.A., Hemelraad, J., and Veehof, P.R., Cadmium kinetics in freshwater clams, uptake of cadmium by the excised gill of *Anodonta anatina, Bull. Environ. Contam. Toxicol.,* 42, 382–388, 1989.
28. Morel, F., *Principles of Aquatic Chemistry,* John Wiley & Sons, New York, 1984.

29. Leland, H.V. and Kuwabara, J.S., Trace metals, in Rand, G.M. and Petrocelli, S.R., Eds., *Fundamentals of Aquatic Toxicology*, Hemisphere, New York, 1985, 374–415.

30. Andrew, R.W., Biesinger, K.E., and Glass, G.E., Effects of inorganic complexing on the toxicity of copper to *Daphnia magna, Water Res.,* 11, 309–315, 1977.

31. Andrew, R.W., Toxicity relationships to copper forms in natural waters, in Andrew, R.W., Hodson, P.V., and Konasewich, D.E., Eds., *Toxicity to Biota of Metal Forms in Natural Water,* Int. Joint Commission Research Advisory Board, Windsor, Ontario, 1976.

32. Bingham, F.T., Sposito, G., and Strong, J.E., The effect of chloride on the availability of cadmium, *J. Environ. Qual.,* 13, 71–74, 1984.

33. Wright, P.A. and Zamuda, C.D., Copper accumulation by two bivalve molluscs. Salinity effect is independent of cupric ion activity, *Mar. Environ. Res.,* 23, 1–14, 1987.

34. Merlini, M. and Pozzi, G., Lead and freshwater fishes. I. Lead accumulation and water pH, *Environ. Pollut.,* 12, 167–172, 1977.

35. Babich, H. and Stotzky, G., Influence of chemical speciation on the toxicity of heavy metals to the microbiota, in Nriagu, J.O., Ed., *Aquatic Toxicology*, John Wiley & Sons, New York, 1983, 1–46.

36. Fisher, N.S., On the reactivity of metals for marine phytoplankton, *Limnol. Oceanogr.,* 31, 443–449, 1986.

37. Fujita, M. and Hashizume, K., Status of uptake of mercury by the fresh water diatom, *Synedra ulna, Water Res.,* 9, 889–894, 1975.

38. Stary, J. and Kratzer, K., The cumulation of toxic metals on alga, *J. Environ. Anal. Chem.,* 12, 65–71, 1982.

39. Wang, H.-K. and Wood, J.M., Bioaccumulation of nickel by algae, *Environ. Sci. Technol.,* 18, 106–109, 1984.

40. Crist, R.H., Oberhoiser, K., Schwartz, D., Marzoff, J., Ryder, D., and Crist, D.R., Interactions of metals and protons with algae, *Environ. Sci. Technol.,* 22, 755–760, 1988.

41. Xue, H.-B., Stumm, W., and Sigg, L., The binding of heavy metals to algal surfaces, *Water Res.,* 22, 917–926, 1988.

42. Crist, R.H., Oberhoiser, K., McGarrity, J., Crist, D.R., Johnson, J.K., and Brittsan, J.M., Interaction of metals and protons with algae. III. Marine algae, with emphasis on lead and aluminum, *Environ. Sci. Technol.,* 26, 496–502, 1992.

43. Les, A. and Walker, R.W., Toxicity and binding of copper, zinc, and cadmium by the blue-green alga, *Chroococcus paris, Water Air Soil Pollut.,* 23, 129–139, 1984.

44. Pagenkopf, G.K., Gill surface interaction model for trace-metal toxicity to fishes. Role of complexation, pH, and water hardness, *Environ. Sci. Technol.,* 17, 342–347, 1983.

45. Krantzberg, G. and Stokes, P.M., The importance of surface adsorption and pH in metal accumulation by chironomids, *Environ. Toxicol. Chem.,* 7, 653–670, 1988.

46. Reid, S.D. and McDonald, D.G., Metal binding activity of the gills of rainbow trout (*Oncorhychus mykiss*), *Can. J. Fish. Aquat. Sci.,* 48, 1061–1068, 1991.

47. Andrew, R.W., Hodson, P.V., and Konasewich, D.E., *Toxicity to Biota of Metal Forms in Natural Water,* Int. Joint Commission Research Advisory Board, Windsor, Ontario, 1976.

48. Sposito, G., *The Surface Chemistry of Soils,* Oxford University Press, New York, 1984.

49. Kinniburgh, D.G., General adsorption isotherms, *Environ. Sci. Technol.,* 20, 895–904, 1986.

50. Simonneaux, V., Humbert, W., and Kirsch, R., Mucus and intestinal ion exchanges in the sea-water adapted eel, (*Anguilla anguilla* L.), *J. Comp. Physiol.,* 157B, 295–306, 1987.

51. El-Fiky N., Hinterleitner, S., and Wieser, W., Differentiation of swimming muscles and gills, and development of anaerobic power in the larvae of cyprinid fish (Pisces, teleostei), *Zoomorphology,* 107, 126–132, 1987.

52. Beattie, J. H. and Pascoe, D., Cadmium uptake by rainbow trout (*Salmo gairdneri*) eggs and alevins, *J. Fish Biol.,* 13, 631–637, 1978.

53. Laurent, P., Gill internal morphology, in Hoar, W. S. and Randall, D. J., Eds., *Fish Physiology,* Vol 10A, Academic Press, New York, 1984, 73–184.

54. Mallat, J., Fish gill structural changes induced by toxicants and other irritants. A statistical review, *Can. J. Fish. Aquat. Sci.,* 42, 630–648, 1985.

55. Karlsson-Norrgren, L., Dickson, W., Ljungberg, O., and Runn, P., Acid water and aluminum exposure; gill lesions and aluminum accumulation in farmed brown trout, *Salmo trutta* L., *J. Fish Dis.*, 9, 1–9, 1986.

56. Arillo, A., Margiocco, C., and Melodia, F., The gill sialic acid content as an index of environmental stress in rainbow trout, *Salmo gairdneri*, Richardson, *J. Fish Biol.*, 15, 405–410, 1979.

57. Gona, O., Mucous glycoproteins of teleostean fish: a comparative histochemical study, *Histochem. J.*, 11, 709–718, 1979.

58. Zuchelkowski, E. M., Pinkstaff, C.A., and Hinton, D.E., Mucosubstance histochemistry in control and acid-stresses epidermis of brown bullhead catfish, *Icatalurus nebulosus* (LeSueur), *Anat. Rec.*, 212, 327–335, 1985.

59. Part, P. and Lock, R.A.C., Diffusion of calcium, cadmium and mercury in a mucous solution from rainbow trout, *Comp. Biochem. Physiol.*, 75C, 259–265, 1983.

60. Kirchner, L. B., External charged layer and Na$^+$ regulation, in Karker-Jorgensen, C., Skadhauge, E., and Hess-Thaysen, J., Eds., *Osmotic and Volume Regulation*, Academic Press, New York, 1987, 310–332.

61. Varanasi, U. and Markey, D., Uptake and release of lead and cadmium in skin and mucus of coho salmon (*Onchorhynchus kisutch*), *Comp. Biochem. Physiol.*, 60C, 187–194, 1978.

62. Maetz, J. and Garcia-Romeu, F., The mechanism of sodium and chloride uptake by the gills of a freshwater fish *Carassius auratus*. II. Evidence for NH4$^+$/Na$^+$ and HCO$_3$$^-$/Cl$^-$ exchanges, *J. Gen. Physiol.*, 47, 1209–1227, 1964.

63. Evans, D. H., Studies on the permeability to water of selected marine, freshwater and euryhaline teleosts, *J. Exp. Biol.*, 50, 689–703, 1969.

64. Kamiya, M. and Utida, S., Changes in activity of sodium-potassium-activated adenosine triphosphatase in gills during adaptation of the Japanese eel to seawater, *Comp. Biochem. Physiol.*, 26, 675–685, 1968.

65. McWilliams, P.G., An investigation of the loss of bound calcium from the gills of brown trout, *Salmo trutta*, in acid media, *Comp. Biochem. Physiol.*, 74A, 107–116, 1983.

66. Booth, C. E., McDonald, D.G., Simons, B. P., and Wood, C.M., Effects of aluminum and low pH on net ion fluxes and ion balance in the brook trout (*Salvelinus fontinalis*), *Can. J. Fish. Aquat. Sci.*, 45, 1563–1574, 1988.

67. Cuthbert, A. W. and Maetz, J., The effects of calcium and magnesium on sodium fluxes through the gills of *Carassius auratus* L., *J. Physiol.*, 221, 633–643, 1972.

68. Wood, C.M., McDonald, D.G., Ingersoll, C.G., Mount, D.R., Johannsson, O.E., Landsberger, S., and Bergman, H.L., Whole body ions of brook trout (*Salvelinus fontinalis*) alevins: responses of yoke-sac and swim-up stages to water acidity, calcium, and aluminum, and recovery effects, *Can. J. Fish. Aquat. Sci.*, 47, 1604–1615, 1990.

69. Martell, A.E. and Motekaitis, R.J., Coordination chemistry and speciation of Al (III) in aqueous solution, in Lewis, T. E., Ed., *Environmental Chemistry and Toxicology of Aluminum*, Lewis Publishers, Chelsea, MI, 1989, 3–17.

70. Hem, J.D. and Roberson, C.E., Form and stability of aluminum hydroxide complexes in dilute solution, in Chemistry of Aluminum in Natural Water, U.S. Geological Survey Water-Supply Pap. 1827A:A1–A55, Department of the Interior, Washington, D.C., 1967.

71. Bertsch, P., Aqueous polynuclear aluminum species, in Sposito, G., Ed., *The Environmental Chemistry of Aluminum*, CRC Press, Boca Raton, FL, 1989, 87–115.

72. Hem, J. D., Graphical methods for studies of aqueous aluminum hydroxide, fluoride, and sulfate complexes, Chemistry of Aluminum in Natural Water, U.S. Geological Survey Water-Supply Pap. 1827B:B1–B33, Department of the Interior, Washington, D.C., 1968.

73. Birchall, J.D., Exeley, C., Chappell, J.S., and Phillips, M.J., Acute toxicity of aluminum to fish eliminated in silicon-rich acid waters, *Nature*, 338, 146–148, 1989.

74. Driscoll, C.T., Baker, J.P., Bisogni, J.J., and Schofield, C.L., Effect of aluminum speciation on fish in dilute acidified waters, *Nature*, 284, 161–164, 1980.

75. Sackett, W. and Arrhenius, G., Distribution of aluminum species in the hydrosphere. I. Aluminum in the ocean, *Geochim. Cosmochim. Acta*, 26, 955–968, 1962.

76. Witters, H.E., Acute acid exposure of rainbow trout, *Salmo gairdneri* Richardson: effects of aluminum and calcium on ion balance and hematology, *Aquat. Toxicol.*, 8, 197–210, 1986.

77. Buergel, P.M. and Soltero, R.A., The distribution and accumulation of aluminum in rainbow trout following a whole lake alum treatment, *J. Freshwater Ecol.*, 2, 37–44, 1983.

78. Baker, J.P. and Schofield, C.L., Aluminum toxicity to fish in acidic water, *Water Air Soil Pollut.*, 18, 289–309, 1982.

79. Hunn, J.B., Cleveland, L., and Little, E.E., Influence of pH and aluminum on developing brook trout in a low calcium water, *Environ. Pollut.*, 43, 63–73, 1987.

80. Neville, C.M. and Campbell, P.G.C., Possible mechanisms of aluminum toxicity in a dilute, acidic environment to fingerlings and older life stages of salmonids, *Water Air Soil Pollut.*, 42, 311–327, 1988.

81. Wilkinson, K.J., Campbell, P.G.C., and Couture, P., Effect of fluoride complexation on aluminum toxicity towards juvenile Atlantic salmon (*Salmo salar*), *Can. J. Fish. Aquat. Sci.*, 47, 1446–1452, 1990.

82. Exeley, C., Chappell, J.S., and Birchall, J.D., A mechanism for acute aluminum toxicity in fish, *J. Theor. Biol.*, 151, 417–428, 1991.

83. McCahon, C. P., Pascoe, D., and McKavanagh, C., Historical observations on the salmonids (*Salmo salar* L.) and (*Salmo trutta* L.) and the ephemeropterans (*Baetis rhodani*)(Pict.) and (*Ecdyonurus venosus*)(Fabr.) following a simulated episode of activity in an upland stream, *Hydrobiologia*, 153, 3–12, 1987.

84. Luoma, S.N., Can we determine the biological availability of sediment-bound trace elements?, *Hydrobiologia*, 176/177, 379–396, 1989.

85. Tessier, A., Campbell, P.G.C., and Bisson, M., Sequential extraction procedure for the speciation of particulate trace metals, *Anal. Chem.*, 51, 844–851, 1979.

86. Rendell, P.S., Batley, G.E., and Cameron, A.J., Adsorption as a control of metal concentrations in sediment extracts, *Environ. Sci. Technol.*, 14, 314–318, 1980.

87. Tipping, E., Hetherington, N.B., Hilton, J., Thompson, D.W., Bowles, E., and Hamilton-Taylor, J., Artifacts in the use of selective chemical extraction to determine distributions of metals between oxides of manganese and iron, *Anal. Chem.*, 57, 1944–1946, 1985.

88. Rapin, F., Tessier, A., Campbell, P.G.C., and Caignan, R., Potential artifacts in the determination of metal partitioning in sediments by a sequential extraction procedure, *Environ. Sci. Technol.*, 20, 836–840, 1986.

89. Kheboian, C. and Bauer, C.F., Accuracy of selective extraction procedures for metal speciation in model aquatic sediments, *Anal. Chem.*, 59, 1417–1413, 1987.

90. Belzile, N., Lecomte, P., and Tessier, A., Testing readsorption of trace elements during partial chemical extractions of bottom sediments, *Environ. Sci. Technol.*, 23, 1015–1020, 1989.

91. Jenne, E.A. and Luoma, S.N., Forms of trace elements in soils, sediments, and associated waters. An overview of their determination and biological availability, in Wildung, R.E. and Drucker, H., Eds., *Biological Implications of Metals in the Environment: CONF-750929*, NTIS, Springfield, VA, 1977, 110–143.

92. Luoma, S.N. and Jenne, E.A., The availability of sediment-bound cobalt, silver, and zinc to a deposit-feeding clam, in Drucker, H. and Wildung, R.E., Eds., *Biological Implications of Metals in the Environment*, NTIS, Springfield, VA, 1977, 213–230.

93. Luoma, S.N. and Bryan, G.W., Factors controlling the availability of sediment-bound lead to the estuarine bivalve *Scrobicularia plana*, *J. Mar. Biol. Assoc. U.K.*, 58, 793–802, 1978.

94. Cooke, M., Nickless, G., Lawn, R.E., and Roberts, D.J., Biological availability of sediment-bound cadmium to the edible cockle, *Cerastoderma edule*, *Bull. Environ. Contam. Toxicol.*, 23, 381–386, 1979.

95. Langston, W.J., Arsenic in U.K. estuarine sediments and its availability to benthic organisms, *J. Mar. Biol. Assoc. U.K.*, 60, 869–881, 1980.

96. Tessier, A. Campbell, P.G.C., Auclair, J.C., and Bisson, M., Relationships between the partitioning of trace metals in sediments and their accumulation in the tissues of the freshwater mollusc *Elliptio complanata* in a mining area, *Can. J. Fish. Aquat. Sci.*, 41, 1463–1472, 1984.

97. Campbell, P.G.C., Lewis, A.G., Chapman, P.M., Crowder, A.A., Fletcher, W.K., Imber, B., Luoma, S.N., Stokes, P.M., and Winfrey, M., Biologically Available Metals in Sediments (NRCC No. 27694), National Research Council Canada, Ottawa, 1988.

98. Young, L.B. and Harvey, H.H., Metal concentrations in chironomids in relation to the geochemical characteristics of surficial sediments, *Arch. Environ. Contam. Toxicol.*, 21, 202–211, 1991.

99. Crecelius, E.A., Hardy, J.T., Bobson, C.I., Schmidt, R.L., Apts, C.W., Gurtisen, J.M., and Joyce, S.P., Copper bioavailability to marine bivalves and shrimp. Relationship to cupric ion activity, *Mar. Environ. Res.*, 6, 13–26, 1982.

100. Davies-Colley, R.J., Nelson, P.O., and Williamson, K.J., Copper and cadmium uptake by estuarine sedimentary phases, *Environ. Sci. Technol.*, 18, 491–499, 1984.

101. Richard, F.C. and Bourg, A.C.M., Aqueous geochemistry of chromium. A review, *Water Res.*, 25, 807–816, 1991.

102. Thomson, E.A., Luoma, S.N., Johansson, C.E., and Cain, D.J., Comparison of sediments and organisms in identifying sources of biologically available trace metal contamination, *Water Res.*, 18, 755–765, 1984.

103. Ray, S., McLeese, D.W., and Peterson, M.R., Accumulation of copper, zinc, cadmium and lead from two contaminated sediments by three marine invertebrates. A laboratory study, *Bull. Environ. Contam. Toxicol.*, 26, 315–322, 1981.

104. Rule, J.H. and Alden, R.W., III, Cadmium bioavailability to three estuarine animals in relationship to geochemical fractions in sediments, *Arch. Environ. Contam. Toxicol.*, 19, 878–885, 1990.

105. Patrick, W.H., Jr., Gambrell, R.P., and Khalid, R.A., Physiochemical factors regulating solubility and bioavailability of toxic heavy metals in contaminated dredged sediment, *J. Environ. Sci. Health*, A12, 475–492, 1977.

106. Bryan, G.W., Bioavailability and effects of heavy metals in marine deposits, in Ketchum, B., Capuzzo, J., Burt, W., Duedall, I., Park, P., and Kester, D., Eds., *Wastes in The Sea*, Vol. 6, John Wiley & Sons, New York, 1985, 41–79.

107. Bjornberg, A., Hakanson, L., and Lundbergh, K., A theory of the mechanisms regulating bioavailability of mercury in natural waters, *Environ. Pollut.*, 49, 53–61, 1988.

108. DiToro, D.M., Mahony, J.D., Hansen, D.J., Scott, K.J., Hicks, M.B., Mayr, S.M., and Redmond, M.S., Toxicity of cadmium in sediments. The role of acid volatile sulfide, *Environ. Toxicol. Chem.*, 9, 1487–1502, 1990.

109. Carlson, A.R., Phipps, G.L., Mattson, V.R., Kosian, P.A., and Cotter, A.M., The role of acid-volatile sulfide in determining cadmium bioavailability and toxicity in freshwater sediments, *Environ. Toxicol. Chem.*, 10, 1309–1319, 1991.

110. Ankley, G.T., Phipps, G.L., Leonard, E.N., Benoit, D.A., Mattson, V.R., Kosian, P.A., Cotter, A.M., Dierkes, J.R., Hansen, D.J., and Mahony, J.D., Acid-volatile sulfide as a factor mediating cadmium and nickel bioavailability in contaminated sediments, *Environ. Toxicol. Chem.*, 10, 1299–1307, 1991.

111. Newman, M.C. and McIntosh, A.W., Appropriateness of Aufwuchs as a monitor of bioaccumulation, *Environ. Pollut.*, 60, 83–100, 1989.

112. Johnson, G.D., McIntosh, A.W., and Atchison, G.J., The use of periphyton as a monitor of trace metals in two contaminated Indiana lakes, *Bull. Environ. Contam. Toxicol.*, 19, 733–740, 1978.

113. Friant, S.L. and Koerner, H., Use of an *in situ* artificial substrate for biological accumulation and monitoring of aqueous trace metals. A preliminary field investigation, *Water Res.*, 15, 161–167, 1981.

114. Ramelow, G.L., Maples, R.S., Thompson, R.L., Mueller, C.S., Webre, C., and Beck, J.N., Periphyton as monitors for heavy metal pollution in the Calcasieu River estuary, *Environ. Pollut.*, 43, 247–261, 1987.

115. Patrick, F.H. and Loutit, M.W., The uptake of heavy metals by epiphytic bacteria on *Alisma plantago-aquatica*, *Water Res.*, 11, 699–703, 1977.

116. Johnson, I., Flower, N., and Loutit, M.W., Contribution of periphytic bacteria to the concentration of chromium in the crab *Helice crassa*, *Microb. Ecol.*, 7, 245–252, 1981.

117. Newman, M.C., McIntosh, A.W., and Greenhut, V.A., Geochemical factors complicating the use of aufwuchs as a biomonitor of lead levels in two New Jersey reservoirs, *Water Res.*, 17, 625–630, 1983.

118. Newman, M.C., Alberts, J.J., and Greenhut, V.A., Geochemical factors complicating the use of aufwuchs to monitor bioaccumulation of arsenic, cadmium, chromium, copper and zinc, *Water Res.*, 19, 1157–1165, 1985.

119. Newman, M.C. and McIntosh, A.W., Slow accumulation of lead from contaminated food sources by the freshwater gastropods, *Physa integra* and *Campeloma decisum*, *Arch. Environ. Contam. Toxicol.*, 12, 685–692, 1983.

120. Reinfelder, J.R. and Fisher, N.S., The assimilation of elements ingested by marine copepods, *Science*, 251, 794–796, 1991.

121. Luoma, S.N., Johns, C., Fisher, N.S., Steinberg, N.A., Oremland, R.S., and Reinfelder, J.R., Determination of selenium bioavailability to a benthic bivalve from particulate and solute pathways, *Environ. Sci. Technol.*, 26, 485–491, 1992.

122. Decho, A.W. and Luoma, S.N., Time-courses in the retention of food material in the bivalves *Potamocorbula amurensis* and *Macoma balthica*. Significance to the adsorption of carbon and chromium, *Mar. Ecol. Prog. Ser.*, 78, 303–314, 1991.

123. Rapport, D.J., Regier, H.A., and Hutchinson, T.C., Ecosystem behavior under stress, *Am. Nat.*, 125, 617–640, 1985.

124. Nagel, E., *The Structure of Science. Problems in the Logic of Scientific Explanation*, Harcourt, Brace and World, New York, 1961.

125. Gordon, M.S., *Animal Physiology: Principles and Adaptations*, 2nd ed., Macmillan, New York, 1972.

126. Wagner, J.G., *Fundamentals of Clinical Pharmacokinetics*, 2nd ed., Drug Intelligence Publications, Inc., Hamilton, IL, 1979.

Chapter 2

Synopsis of Discussion Session on the Bioavailability of Inorganic Contaminants

William H. Benson (Chair), James J. Alberts, Herbert E. Allen, Carlton D. Hunt, and Michael C. Newman

INTRODUCTION

Bioavailability is a widely accepted concept based on the implicit knowledge that before an organism may accumulate or show a biological response to a chemical, that element or compound must be available to the organism. While the concept of bioavailability is widely accepted, the processes that control it are poorly understood. This latter situation results from an incomplete understanding of basic processes in aquatic systems, which is further exacerbated by the numerous and sometimes conflicting qualitative definitions of bioavailability used by investigators from different scientific disciplines.

Bioavailability is defined here as the degree to which a chemical is able to move into or onto an organism. This definition is given here only to provide a framework in which to discuss limitations in the understanding of bioavailability as it pertains to inorganic species in aquatic systems—not because we feel that this should be the only definition. We focus on the bulk chemistry of the aquatic phase as it affects speciation and, presumably, availability of inorganic contaminants. It is assumed that transfer mechanisms operate within the diffusion layer existing between the bulk phase and membrane(s) to move the chemical into the organism. We further assume that transfer must occur within the solution phase regardless of the membrane type or location within the organism (e.g., gill, gut, or dermis). Thus, inorganic species in phases that are not dissolved within the bulk solution must involve dissolution for the chemical to become bioavailable.

Inorganic species are considered to be bioavailable based on laboratory bioassays that define "target values" and field observations of water column concentrations. By comparing the values derived from these measurements, decisions are made about the risk that an element poses to the environment. Thus, these laboratory measurements are the bases for assessment and regulation, and have served well in the past 15 to 20 years. However, scientists have always recognized that certain aspects of the solution phases of aquatic systems were too ill-defined to be adequately addressed by these techniques. The toxicological effects of metals on aquatic organisms have also been known for many years, and have been effectively used to address metal loading to the environment. This is evident in the fact that since the late 1970s water column metal concentrations have declined world-wide. In large part, this decline is a function of improved "clean" analytical techniques.[1] It also reflects the influence of point source controls in reducing the loads of metals to the environment and reducing the concentrations of metals in receiving waters. However, the degree to which pollution control measures have contributed to this decline is moot, as any "real" decreases in water column metal concentrations are usually masked by analytical problems that resulted in compilation of few reliable data sets prior to 1980. Unfortunately, generation of unreliable metals data continues even today.[2] These unreliable data can result in "false" high metals concentrations. If these "false" high values are associated with ambient water quality determinations in the permitting process, the number of National Pollution Discharge Elimination System (NPDES) permit violations becomes suspect.

Original bioassay studies included known "total" exposure concentrations that were well above ambient concentrations in receiving waters, and were conducted in relatively well-defined chemical compositions of the test solutions. This occurred because bioassay test concentrations were so high (e.g., in the milligram per liter range) that the dissolved phase dominated the partitioning of the chemicals between the true solution phase and other nondissolved phases in the water. As a result, extrapolation of the test results to aquatic environments of varying inorganic (e.g., H^+, Cl^-, Ca^{2+}, Mg^{2+}), organic (e.g., humic substances), and suspended solids content were made without adequate knowledge of the potential factors affecting availability. Some weaknesses of these assays have been improved by moving from the use of single to multiple test species, examining chronic end points rather than restricting testing to acute exposures, and varying solids contents as part of the test conditions.

Well-defined chemical compositions make it easier to estimate the effect of water quality on the response of an organism to an element. An outgrowth of these studies has been the perception and reliance on the concept that total soluble metal concentration is responsible for the toxicological response observed in the organism. However, a growing body of literature indicates that this perception is not true for all elements and that the "free" metal ion activity, not the total soluble metal concentration, is the fraction (species) that controls the bioavailability of many inorganic elements.[3-7] Unfortunately, we have often relied on empirical relationships and experimental design to provide estimates of the inorganic species that are bioavailable. Several analytical tools, developed in the 1970s and 1980s, enable measurement of some inorganic species; these methods need to be applied in experiments designed to evaluate the bioavailabilty of inorganic contaminants.

While advances have been made in our understanding of the physicochemical nature of the aquatic environment, these same advances continue to reinforce the need to extend our understanding of the fundamental thermodynamic and kinetic factors that control inorganic contaminant speciation. Such understanding must be advanced before a quantitative prediction of chemical speciation can be made. Once chemical speciation of inorganic contaminants can be predicted quantitatively, presumably, bioavailability may also be predicted. This hypothesis requires testing.

A parallel critical issue is understanding all of the fundamental interactions that control the bioavailabilty of inorganic contaminants to organisms. Furthermore, we need to understand which chemical species are involved in uptake processes in order to design the appropriate tests to estimate bioavailability and, thus, toxicity.

FUNDAMENTAL AND EMPIRICAL RELATIONSHIPS

As indicated above, toxicity of metals in aquatic bioassays has been and continues to be related to total dissolved metal concentration. This assumes that toxicity is directly related to bioavailability of the sum of the metal species in the aquatic phase. However, there is a body of literature which implies that free metal ion controls bioavailability, while showing that soluble metal complexes are not bioavailable.[3-7] The exceptions to this are the hydroxy complexes of some metals,[8,9] organometallic compounds (e.g., butyltin, methyl mercury), and oxyanions (e.g., chromate, arsenate). More disturbing are limited observations that the bioaccumulation[10] and toxicity[11,12] of low concentrations of metal are enhanced by humic substances. Thus, the dogma that free metal ion is the only reactive species involved in bioaccumulation/toxicity may be incorrect.

Aquatic Phases

In the most general sense, the bulk chemistry of aquatic systems can be described as a water containing truly dissolved inorganic species and phases that are not dissolved. In practice, particulate matter is separated from the bulk solution by operational means, usually filtration or centrifugation. The lower limit of size for particulate matter varies in these operational definitions,

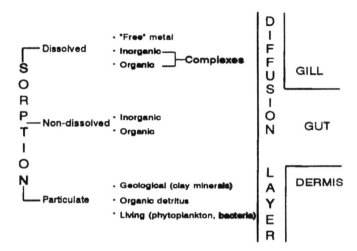

Figure 1. Simplified schematic of the types of materials interacting with inorganic species in aquatic environments and the diffusion layer of organisms.

but is typically between 0.2 and 0.5μm in diameter. While convenient, a significant drawback of operationally defining these particles is the implicit assumption that all material which is smaller than the operational particulate diameter is dissolved.

During the last 20 years, research has shown that materials which are smaller than the operationally defined particles occupy the entire size range from truly dissolved species, through colloids, and up to small lithospheric materials of defined structure (e.g., clays, feldspars, and quartz). For this discussion, the operational definition of particulate material will be 0.45 μm. Material smaller than 0.45 μm is considered soluble. The term "dissolved" will be used for elements, compounds, and their complexes in true solution-phase. "Nondissolved" includes those aggregates that are in the solution phase but are smaller than operationally defined particles. These aggregates are poorly characterized structurally, compositionally, and thermodynamically (e.g., humic substances, metal oxides). A simplified diagram of the types of material affecting the distribution of solution phase inorganic species is shown in Figure 1.

There is a continuum of interactions among the nondissolved phases and among the truly dissolved phases, as well as between the dissolved and nondissolved phases. This set of interactions and transformations must be understood if we are to adequately describe the availability of inorganic elements over time and space. To facilitate interpretation, we have divided our discussion into the dissolved phase, nondissolved phase, and particulate phase, and finally, the diffusion layer.

Dissolved Phase

The distribution of metals within the dissolved phases can be described from thermodynamic and kinetics considerations that account for the interactions between the free metal ion and various ligands and complexing agents, e.g., Cl^-, OH^-, carbonate, amino acids, and humic substances, that result in the formation of various species. The interactions between the metals and inorganic ligands and structurally well-defined organic ligands, e.g., amino acids and oxylate, are well established and can be described through a series of thermodynamic equations. Interactions of metals with all ligands of this type to form monomolecular complexes may be summarized in the equation:

$$C_M = [M^{2+}] (1 + \Sigma \beta_{ML_n} [L]^n)$$

where C_M is the total metal, M^{2+} is the free metal activity, β_{MLn} is the equilibrium constant for that complexed species, L is the ligand concentration, and n is the stoichiometric coefficient. For polynuclear reactions involving more than one metal ion similar reactions can be formulated, and in fact must be considered when estimating the free metal ion concentration of a given element.

Despite having well-established thermodynamic models for estimating the chemical speciation of truly dissolved species, there are significant problems for which further research is required. For example, the accuracy of model predictions for the distribution of chemical species is controlled by the thermodynamic formation constants of the elements with their inorganic and organic ligands, yet many of these constants are poorly known—in particular, complexes with humic substances and the complexes of organometallics.

While most solution-phase reactions are sufficiently rapid for equilibrium calculations to be valid, there are important exceptions. Metal complexation reactions with humic substances are slow, as are those with several other metals including manganese. Even if adequate formation constants were known, the inability to easily measure directly the free metal species or other species makes validation of the predictions extremely difficult. Without validation and kinetic data relating the interconversions of the free and complexed species of the element, the hypothesis that "free" metal ion concentrations can be directly related to bioavailability is untestable in most cases.

Nondissolved Phase

The composition of the nondissolved phase is also complex. For the most part, metabolically induced changes in elemental speciation are confined to transformations caused by passive alterations of the bulk solution (i.e., pH and redox potential) and extracellular complexing ligands. While the description of chemical speciation in this class of materials is made easier by removing major kinetic processes, it is simultaneously complicated by the fact that this phase includes colloidal and macromolecular sized phases which are poorly characterized and highly variable admixtures.

As with the solution phase, slow reactions must be considered when deciding the appropriateness of the assumption of thermodynamic equilibrium. Dissolution and desorption, in general, are much slower than the respective precipitation and sorption reactions. Hence, metal discharged to a receiving environment may only become available long after its introduction. In addition, binding of metals onto different particles occurs by a number of different mechanisms. Therefore, different chemicals bind to different substrates by different mechanisms and exhibit different desorption behaviors and rates. For example, cadmium is released from detritus more rapidly than lead or copper, while desorption from clay minerals exhibits both fast- and slow-release kinetics. Consequently, adequate quantitative dissolution and desorption information is not available to allow predictive estimates of availability to be made.

The nondissolved phase contains both organic and inorganic materials, but it is felt that the poorly characterized organic mixtures (e.g., humic substances) largely control the chemical transformations and reactions both in bulk solution and at particle surfaces. Because these humic substances are admixtures of numerous naturally occurring organic compounds rather than a single, specific organic compound, chemical structure, nature, and number of binding sites and thermodynamic formation constants may always remain poorly described. Thus, our ability to adequately predict the speciation of metals in the dissolved phase, and the resulting availability to organisms as a result of interactions with this phase, remains poor. Furthermore, humic substances, in their role of complexation of elements, are not fully saturated in the environment and thus provide capacity to bind inorganic species. As a result, aquatic systems have an inherent but variable capacity to buffer against increases in the free metal ion and, consequently, the "availability" of the ion to organisms. Therefore, it is imperative that bioassays which are used to determine the toxicological response to a toxicant load be designed in such a manner that the

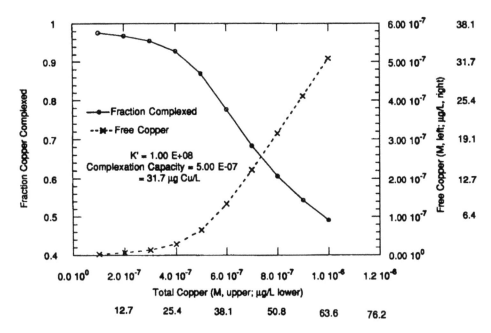

Figure 2. Titration curves of free copper and as a complexed fraction (From Allen, H.E., personal communication.)

ability of the system to protect against adverse impact is not excluded as a result of the experimental design. In addition, the design of toxicity bioassays must consider known interactions that may affect the "availability" of inorganic contaminants.

Special significance must be given to ligands that are present in small concentrations, like humic substances which have a strong affinity for copper and other metals. If a sample is titrated with metal, the speciation changes during the course of the titration; a larger percentage of the metal is noncomplexed at higher concentrations of total metal, as is shown in Figure 2. Metal speciation is a nonlinear function of the total metal concentration which has serious implications for the interpretation and regulatory utilization of bioassay results. Effects determined at high concentrations, for which a high fraction of metal may be bioavailable, are routinely used to establish criteria that are applied at much lower metal concentrations, but because a higher fraction of the metal will be bound at lower total metal concentrations, the criteria will likely be overly protective. It should be noted that the recommended use of water effects ratios suffers from exactly the same problem.[13] It, too, will overestimate bioavailability and may lead to unnecessarily stringent criteria.

Particulate Phase

The particulate phase is a diverse admixture of separate phases including lithospheric minerals (e.g., clays, carbonates), organic and inorganic constituents (e.g., fecal pellets, carbonates, metal oxides, sulfides), and biota. In most cases, these admixtures are poorly defined relative to the distribution of biotic and abiotic material and to the types of phases present. However, in the particulate phase there are processes which determine the solution phase speciation of the inorganic elements, including metabolically mediated uptake or complexation and surface adsorption and precipitation.

Metabolically mediated transformations are poorly described because they involve either transport across membrane boundaries and/or extracellular changes in speciation induced by excreted materials or changes in the physicochemical quality of the solution as a result of metabolism. Abiotic surface transformations are characterized by kinetically controlled "adsorption isotherms", which are usually empirically determined. The commonality of these two phases is that the transformations in dissolved elemental speciation caused by these particles has not been adequately defined thermodynamically or kinetically and, hence, cannot be rigorously described. Thus, it is apparent that our current understanding of the chemical speciation in bulk solution, and the relationship of chemical speciation to bioavailability, are inadequate. Only by understanding the fundamental interactions of chemical speciation and their relation to biota will we be able to predict the bioavailability of inorganic contaminants from first principles. However, in addition to the requirement that we understand thermodynamic equilibrium in bulk solution and relevant reaction kinetics, understanding reactions in the diffusive boundary layers at surfaces is critical to our success in predicting availability.

Diffusion Layer

Our limited understanding of diffusion layer dynamics hinders effective prediction of bioavailability.[14] In fact our ignorance is so acute that, in some situations, we do not know whether reaction or diffusion rates control exchange. We have only started to develop effective models of the physicochemical processes occurring in the diffusion layer of biological surfaces. Further, the critical dissolved-ligand activities within the diffusion layer remain poorly understood. Therefore, kinetics of metal reactions during movement from the bulk water through the microlayer remain difficult to predict except in general terms. For example, the kinetics of metal-ligand exchanges facilitating metal movement into a microlayer containing increased concentrations of metabolic and excretory products such as NH_3, HCO_3^-, and CO_2 are not well understood, even for aluminum (e.g., Playle and Wood[15,16]).

The situation becomes even more difficult to model when transport from a solid phase to the membrane is involved, because the metal must often move into the dissolved phase prior to interaction with the membrane. Consequently, all of the dissolution and desorption reactions and kinetics previously addressed, must be added to the already poorly understood processes just described. Also, in the case of ingested materials, processes associated with intracellular digestion of such particles remain poorly understood. Thus, more critical studies such as those described recently by Decho and Luoma[17] and Luoma et al.[18] on the bioavailability of ingested solid phase-associated metals are essential. Such studies also need to consider cations with a variety of coordination chemistry in organisms with widely varying feeding and digestive processes. Hence, much more basic research on chemical, biochemical, and cytological processes affecting bioavailability of ingested solid-phase metals is needed. Finally, the extensive literature and associated experimental approaches used in similar studies of mammals should be more extensively examined for potential use with aquatic biota.

Microlayer processes at sediment surfaces are also poorly understood. This is unfortunate, as hypotheses regarding the processes thought to control bioavailability of metals in oxic[19] and anoxic[20,21] sediments are presently being formulated. Present methods even fail to accurately determine end members participating in such processes. Luoma[22] suggests that inadequate methods for estimating solid phases of metals, and our poor understanding of sediment-water exchanges, preclude accurate estimation of bioavailability of sediment-associated metals.

CONCLUSIONS AND NEEDS

In order to understand the bioavailability of inorganic chemicals, we must better understand the speciation of chemicals in the aquatic environment. This understanding of speciation must extend from the bulk environment, including both the water column and the sediment, to the site of uptake by or incorporation on or into the organism. We must ascertain the biologically available forms of the chemicals, and develop methods for the analysis of these forms and prediction of their concentrations. This requires that we extend our knowledge in the following areas:

- The biologically active forms of the chemicals to all classes of organisms of concern and through all important routes of exposure, including deposition on or reaction with gill surfaces, uptake by gills or gut, and uptake through biological membranes must be ascertained. It is implicit that the physicochemical mechanisms of uptake be established. If multiple chemical species are available, their relative bioavailabilities also must be established.
- Prediction of speciation requires input parameters of equilibrium constants and binding site concentrations. Present data bases are extensive relative to inorganic ligands and many specific organic ligands, but major deficiencies exist with respect to appropriate values for metal binding by humic substances and heterogeneous solid phases. Furthermore, most work has focused on speciation in the bulk aqueous system, and speciation in microenvironments such as that at the surface of a gill needs to be defined.
- Means to validate speciation predictions are required. The adequacy of any prediction depends on input of appropriate concentrations of all reacting chemical species, in having valid constants for these reactions, and in the reactions attaining equilibrium as predicted by the calculation. Present speciation analytical procedures for solution and solid phases do not provide adequate evaluations of specific chemical species or defined groups of species that can be compared to the results of speciation calculations.
- The kinetics of chemical reactions that are important to the speciation of the chemicals, in most instances, are poorly defined. While it is generally considered that the systems are at equilibrium, many reactions including formation of some inner sphere complexes, dissolution of precipitates, and desorption may be slow relative to mixing and diffusion times. Rates for these chemical reactions must be determined to evaluate the changes in speciation that will occur during transport processes of greatly differing time scales, including advection through mixing zones and diffusion to biological membranes. Finally, the uptake of a metal by an organism forces the system away from equilibrium, but it is not known if this departure from equilibrium is significant and the rates of reequilibration are unknown.

To achieve practical utility, appropriate computational and laboratory or field analytical techniques need to be developed. These techniques must be capable of predicting or determining the speciation of a chemical and of predicting changes in the speciation that occur with time as a result of changes in the environment through advection of the chemical in the bulk water or diffusion of the chemical into an organism.

SUMMARY

In many scientific disciplines, the pursuit of basic knowledge is considered a valuable goal in and of itself. This pursuit also is highly desirable in environmental toxicology and chemistry, but because we are often faced with the conundrum of solving myriads of problems with limited time and budgetary support, we must attempt to make the most effective use of the time and monies allocated for reducing environmental impacts. Yet, criteria and assessment are based on empirical measures, which change as technologies and methodologies advance. However, criteria based on relationships derived from first principles would not vary with technological trends. Such criteria become more defensible and allow environmental stewardship to become more cost effective. Furthermore, conceptual models built upon first principles should enhance the effectiveness of experimental design by pointing to critical measurements necessary for varying environmental parameters. Consequently, more fundamental knowledge of a subject is required to answer basic questions not entirely addressed with empirical observations.

The periodic table of elements represents the basic relationships and trends for all elements known to humans. Without collecting the basic thermodynamic knowledge for elements in environmental settings, we ignore the tremendous predictive capabilities that lie therein, and continue to conduct time and cost-ineffective empirical studies on only one element at a time. Perhaps more disturbing is the fact that analytical constraints require us to study only a very small group of elements, whose chemistries may in no way reflect those of other equally important but methodologically limited chemicals.

Predictability of elemental speciation in aquatic environments based on thermodynamically and kinetically sound theory would facilitate the study of membrane adsorption and transfer phenomena. Elucidation of membrane processes would greatly reduce the requirement for empirical observations of bioavailability by establishing the basic relationships for similar membranes among species and between different membranes within species. Thus, appropriate classes of test organisms could be determined and the number of tests could be optimized.

The quest for fundamental knowledge in environmental toxicology and chemistry will require more and better basic studies. However, more studies are not the only avenue open to environmental toxicologists and chemists. Broadening our sometimes parochial attitude toward scientific research by cross-fertilization from other disciplines may be a valuable means in pointing the way to new approaches to defining and understanding first principles in our research area. Empirical observations and tests have contributed greatly to our knowledge of environmental phenomena and to our ability to assess and regulate our impacts upon the aquatic systems. These contributions should not be minimized. Complete understanding of fundamental principles for elements in the aquatic environment is an attainable goal, but it is a long-term objective. In the interim, empirical tests must continue to be used to test hypotheses and make decisions about our environmental fate. However, increased understanding of these principles will enable us to maximize our predictive capability from both theoretical and empirical knowledge.

REFERENCES

1. Bruland, K.W., Trace elements in sea-water, in Riley, J.P. and Chester, R., Eds., *Chemical Oceanography*, Vol. 8, Academic Press, New York, 1983, 157–220.
2. Windom, H.L., Byrd, J.T., Smith, R.G., and Huan, F., Inadequacy of NASQAN data for assessing metals trends in the nation's rivers, *Environ. Sci. Technol.*, 25, 1137–1142, 1991.
3. Sunda, W. and Guillard, R.R.L., The relationship between cupric ion activity and the toxicity of copper to phytoplankton, *J. Mar. Res.*, 34, 511–529, 1976.
4. Sunda, W. and Gillespie, P.A., The responses of a marine bacterium to cupric ion and its use to estimate cupric ion activity, *J. Mar. Res.*, 37, 761–777, 1979.
5. Allen, H.E., Hall, R.H., and Brisbin, T.D., Metal speciation. Effects on aquatic toxicity, *Environ. Sci. Technol.*, 14, 441–443, 1980.
6. Anderson, D.M., Lively, J.S., and Vaccaro, R.F., Copper complexation during spring phytoplankton blooms in coastal waters, *J. Mar. Res.*, 42, 677–695, 1984.
7. Verweij, W., Glazewski, R., and De Haan, H., Speciation of copper in relation to its bioavailability, *Chem. Spec. Bioavail.*, 4, 43–51, 1992.
8. Cowan, C.E., Jenne, E.A., and Kinnison, R.R., Methodology for determining the relationship between toxicity and the aqueous speciation of a metal, in Poston, T.M. and Purdy, R., Eds., *Aquatic Toxicity and Environmental Fate*, Vol. 9, ASTM STP 921, American Society for Testing and Materials, Philadelphia, 1986, 463–478.
9. Fisher, N.S., On the reactivity of metals for marine phytoplankton, *Limnol. Oceanogr.*, 31, 443–449, 1986.
10. Stackhouse, R.A. and Benson, W.H., The influence of humic acid on the toxicity and bioavailability of selected trace metals, *Aquat. Toxicol.*, 13, 99–108, 1988.
11. Giesy, J.P., Jr., Leversee, G.J., and William, D.R., Effects of naturally occurring aquatic organic fractions on cadmium toxicity to *Simocephalus serrulatus* (Daphnidae) and *Gambusia affinis* (Poeciliidae), *Water Res.*, 11, 1013–1020, 1977.
12. Stackhouse, R.A. and Benson, W.H., Interactions of humic acid with selected trace metals. Influence on bioaccumulation in daphids, *Environ. Toxicol. Chem.*, 8, 639–644, 1989.

13. EPA, Interim Guidance on Interpretation and Implementation of Aquatic Life Criteria for Metals, Office of Science and Technology, U.S. Environmental Protection Agency, Washington, D.C., 1992.

14. Newman, M.C. and Jagoe, C.H., Inorganic ligands and the bioavailability of metals in aquatic environments, this volume, Session 3, Chapter 1, 1993.

15. Playle, R.C. and Wood, C.M., Is precipitation of aluminum fast enough to explain aluminum deposition on fish gills?, *Can. J. Fish. Aquat. Sci.*, 47, 1558–1561, 1990.

16. Playle, R.C. and Wood, C.M., Mechanisms of aluminum extraction and accumulation at the gills of rainbow trout, *Oncorhynchus mykiss* (Walbaum), in acidic soft water, *J. Fish Biol.*, 38, 791–805, 1991.

17. Decho, A.W. and Luoma, S.N., Time-courses in the retention of food material in the bivalves *Potamocorbula amurensis* and *Macoma balthica*. Significance to the adsorption of carbon and chromium, *Mar. Ecol. Prog. Ser.*, 78, 303–314, 1991.

18. Luoma, S.N., Johns, C., Fisher, N.S., Steinberg, N.A., Oremland, R.S., and Reinfelder, J.R., Determination of selenium bioavailability to a benthic bivalve from particulate and solute pathways, *Environ. Sci. Technol.*, 26, 485–491, 1992.

19. Campbell, P.G.C., Lewis, A.G., Chapman, P.M., Crowder, A.A., Fletcher, W.K., Imber, B., Luoma, S.M., Stokes, P.M., and Winfrey, M., Biologically Available Metals in Sediments, National Research Council Canada, NRCC No. 27694, Ottawa, 1988.

20. Di Toro, D.M., Mahony, J.D., Hansen, D.J., Scott, K.J., Hicks, M.B., Mayr, S.M., and Redmond, M.S., Toxicity of cadmium in sediments. The role of acid volatile sulfide, *Environ. Sci. Technol.*, 9, 1487–1502, 1990.

21. Ankley, G.T., Phipps, G.L., Leonard, E.N., Benoit, D.A., Mattson, V.R., Kosian, P.A., Cotter, A.M., Dierkes, J.R., Hansen, D.J., and Mahony, J.D., Acid-volatile sulfide as a factor mediating cadmium and nickel bioavailability in contaminated sediments, *Environ. Toxicol. Chem.*, 10, 1299–1307, 1991.

22. Luoma, S.N., Can we determine the biological availability of sediment-bound trace elements?, *Hydrobiologia*, 176/177, 379–396, 1989.

SESSION 4

ORGANIC TOXICANTS

Chapter 1

Interactions of Organic Pollutants with Inorganic Solid Phases: Are They Important to Bioavailability?

Anne Spacie

INTRODUCTION

The transport and fate of organic pollutants is largely governed by their associations with environmental surfaces. Suspended solids and fine colloids transport sorbed pollutants of all types. Sediments become the final storage site of persistent contaminants such as organochlorine insecticides, PCBs, dioxins, and polycyclic aromatic hydrocarbons. Aquatic organisms "compete" with sediments and suspended phases for accumulation of lipophilic pollutants, making sorption and desorption processes especially important to tissue residue dynamics and toxicity. Many environmental management concerns such as dredging, disposal, groundwater infiltration, and *in situ* remediation hinge on an understanding of sorption and surface interactions.

While organic pollutant behavior has been studied mainly in relation to organic matter, relatively little attention has been given to the influence of natural inorganic phases such as clays, silicates, and metal oxides. Their capacity to sorb organic pollutants is relatively weak compared to organic phases. Nevertheless, the shear bulk of mineral substrates in the environment makes them worthy of closer consideration. This paper briefly reviews the basic assumptions of surface sorption and organic-phase equilibrium partitioning and then examines the roles of surface interactions and particle distribution. The emphasis in this discussion is on specific situations where the inorganic matrix needs to be considered and where recent models of sorptive interaction can be applied.

Sorption or Partitioning?

Surface sorption of a pollutant to a solid at equilibrium may be characterized by a sorption coefficient (m) that expresses the amount of chemical sorbed per mass of solid sorbent (X) compared to its concentration in aqueous solution (C_{aq}). Various isotherms have been proposed, including linear, Langmuir, Freundlich, Frumkin, and nonlinear types, as reviewed by Stumm[1] and others.[2-5] Much of the literature on organic sorption makes use of the Freundlich isotherm:

$$X = m \, C_{aq}^{1/n} \qquad (1)$$
$$\log X = \log m + (1/n) \, C_{aq}$$

because of its adaptability to heterogeneous substrates and ease of equation fitting. The slope of the Freundlich isotherm (m) is interpreted as a measure of the affinity of the pollutant for the solid phase. Three basic features of the various sorption isotherms should be noted for the purposes

1-56670-086-8/94/$0.00+$.50

73

of this discussion. First, all types converge on a simple linear form as $C_{aq} \rightarrow 0$, which is often the case for trace organic pollutants. Second, at higher concentrations, the linear form can seriously overestimate X if the sorption process actually occurs by saturation of active surface sites on the solid. Finally, an empirical fit of sorption data to a particular model does not constitute evidence of the particular mechanism assumed for that model. Empirical fits to sorption data are currently recommended to avoid the limitations of a particular model.[5]

Although many examples of nonlinear sorption patterns are available, the great majority of work on organic transport and fate in aquatic systems has assumed a linear isotherm ($1/n = 1$ in Equation 1).[6,7] A linear sorption model is indistinguishable from simple proportionality—the relationship that would be observed if hydrophobic partitioning were to apply. Such partitioning describes the distribution of a solute between aqueous and organic phases, according to its solubilities in the two solvents:

$$K_p = C_s/C_{aq} = f_{oc} K_{oc} \tag{2}$$

where K_p is the distribution coefficient between sediment or other particulate phase and water, C_s, and C_{aq} are the respective concentrations in solid and water, K_{oc} is the partition coefficient for the organic phase, and f_{oc} is the mass fraction of organic carbon in the solid material. Various attempts to distinguish between monolayer adsorption and organic-carbon partitioning in soils and sediments have been made by interpreting the shapes of isotherms or by measuring entropy changes associated with the sorption process. Recent sorption studies with clays and humic material by Rebhun et al.[8] have demonstrated the linear form of both surface sorption to minerals and partitioning to the organic fraction. Mingelgrin and Gerstl[4] reviewed these approaches and argued for a broader interpretation of sorption that includes two-dimensional surface binding, a three-dimensional boundary layer encompassing various surface interactions, as well as partitioning to bulk organic phases. Karickhoff[9] developed an expression of K_p that incorporates both organic and inorganic sorption components. The organic carbon content of sediment has proven to be a good predictor of bioavailability for hydrophobic organic compounds, both because of their strong affinity for humic material and because natural particles almost always carry a coating of organic matter. Soil/water distribution coefficients based on carbon content (K_{oc}) can be predicted for a wide range of natural particles from octanol-water partition coefficients (K_{ow}) by relationships of the form

$$\log K_{oc} = a \log K_{ow} + b \tag{3}$$

or from calculations based on chemical structure.[10] To a close approximation, K_{oc} is essentially equal to K_{ow}. The equilibrium partitioning approach for predicting steady-state accumulation by infaunal species, recently reviewed by Di Toro et al.,[11] assumes that the amount, but not the type of organic carbon determines bioavailability. It has proven especially useful for reducing the observed variability of field-collected residue data.[12,13] However, a variation of 2X to 3X remains that is not corrected by carbon normalization.[11] Several studies have reported a trend toward higher bioavailability in sediments of low organic carbon content.[13-15] Whether this is evidence of inorganic matrix effects on sorption, kinetic limitations, analytical artifacts, or biotic factors is unclear.

While carbon-based partitioning is likely to be the dominant factor in most situations, inorganic interactions should be considered in cases where f_{oc} is low (<0.002) and better predictability is needed.[11] Clays and other inorganic phases that make up the bulk of soil and sediment can be thought of as "solid supports" for organic matter, much the way solid phases are used in chromatography to hold the active bonded phase. This role as substrate for hydrophobic interactions has been shown experimentally, for example, by sorbing sodium dodecyl sulfate to ferric oxide.[16] The resulting hemicelles of surfactant sorbed to the particles effectively remove hydrophobic organic pollutants from solution according to their solubilities.

Table 1. Comparison of the Percentage of Hydrophobic Organic Chemicals Sorbed to Either Hematite or Kaolinite Coated With Peat Humic Acid

	Hematite	Kaolinite
Anthracene	84%	60%
ibenzothiophene	77%	45%
Carbazole	66%	42%

Source: Interpolated from data of Murphy, E.M., Zachara, J.M., and Smith, S.C., Environ. Sci. Technol., 24, 1507, 1990. With permission.

Recent work by Murphy et al.[17] demonstrates the importance of the mineral surface to the sorption process. They compared the sorption of anthracene, carbazole, and dibenzothiophene on humic material bound to either hematite or kaolinite (f_{oc} = 0.0001–0.005). Experimental K_{oc} values varied from those predicted by K_{ow}, depending on the type of mineral substrate and humic material used (Table 1). They concluded that the orientation and structure of the sorbed humic material was affected by the type of inorganic matrix, and that the humic-mineral interactions were important factors controlling hydrophobic sorption. They also reported nonlinear isotherms for the process, suggesting that binding is by surface adsorption rather than by partitioning to bulk organic phases. Rebhun et al.[8] found significant surface interactions for neutral organic chemicals sorbed to pure clays and clay-humic particles with low to moderate organic content ($f_{oc} < 0.05$).

Two recent models for pollutant interactions that incorporate the variety of sorption processes are the distributed reactivity model (DRM) of Weber et al.[18] and the adsorption-partition model (A-P) of Wang et al.[19] In the DRM approach, the soil or sediment is viewed as a heterogeneous material having different regions where sorption may be linear or nonlinear. The linear portions correspond to a partitioning type of K_p, while the nonlinear portions make use of a Freundlich model. The A-P model treats surface adsorption and partitioning components as additive. This approach predicts that noncompetitive partitioning governs the behavior of chemicals having log $K_{ow} > 3$, but that competitive surface interactions become more important for chemicals of lower hydrophobicity.

Physical Interactions

Inorganic solid phases can influence the behavior of neutral organic pollutants in two significant ways: by determining particle size and by determining the overall proportion of bound material in the system. Particle size, along with density, porosity, and surface area, are critical factors for bioavailability as well as for transport (settling and resuspension) and microbial degradation processes. Smaller particles typically carry greater pollutant concentrations and organic carbon fractions because of the greater surface/volume ratio.[20,21] Deposit-feeding organisms preferentially feed on the finer fractions in a heterogeneous sediment. Both ingestion and direct contact of benthic organisms can thus be affected by the size and structure of particles. Furthermore, their potential exposure to sorbed pollutants depends on a range of sorption/desorption processes.

Initial sorption occurs rapidly (minutes) when a chemical reaches a surface binding site.[1] However, the overall equilibration process for a pollutant with a heterogeneous particle in sediment is typically slow (hours to months). For example, Karickhoff and Morris[22] observed a two-phase desorption process from particles (fitted to a "two-box" model), where the slower phase could require up to several weeks for completion (Table 2). Slow diffusion through the interior regions of the particle was thought to be the delaying factor. Van Hoof and Andren[23] found that aqueous phase PCB equilibration to polystyrene beads could take as much as 96 days. Because

Table 2.   Equilibrium Constants, Times, and Labile Fractions of Organic Pollutants Desorbed From Several Suspended River Sediments

Chemical	Equilibrium constant, log K_p	Equilibration time, days	Labile fraction
Hexachlorobenzene	3.04–4.51	18–28	0.11–0.48
Pentachlorobenzene	3.91–3.34	15–28	0.22–0.29
Pyrene	3.11–3.46	15–21	0.36–0.59
Trifluralin	2.98–1.36	10–21	0.22–0.50

Source: From Karickhoff, S.W. and Morris, K.R., *Environ. Toxicol. Chem.,* 4, 469, 1985. With permission.

of the biphasic characteristic of the desorption process, Eadie et al.[24] applied a two-box model to describe the bioavailable portion of the pollutant. The particular kinetics of the sorption process are most likely related to the size and porosity of the sediment particle.

Equilibration requires diffusion through various aqueous and organic boundaries. Such phases may be interspersed throughout the porous structure of the particle, causing diffusion-limited kinetics to be quite slow. Biological uptake of pollutants can be retarded by slow desorption from particles, as Landrum[25] and Wood et al.[26] have shown experimentally for benthic infauna. Bioavailability to fish can also be limited by slow desorption from suspended solids relative to the brief contact time of respired water at the gill.[27,28]

Two conceptual models are available for describing the effects of particle size and porosity on the movement of sorbates in particles that are viewed as porous spheres:

1. An intraparticle diffusion or radial diffusive penetration model[29,30] where particle radius, porosity, and pollutant diffusivity are used.
2. An intraorganic matter diffusion model[31-33] where nonequilibrium sorption is delayed by diffusion through the three-dimensional structure of the sorbed humics.

These models are attractive because they account for particle structure as well as organic content. The parameters required for the models can be estimated or empirically fitted, as described by Medine and McCutcheon.[34] An analysis of the radial diffusion model (a type of infinite-box model) shows that equilibration time increases with increasing K_{ow} and particle size. For example, a 40-µm-diameter particle requires approximately 100 min to equilibrate for a chemical of log $K_{ow} = 5$, whereas a particle of 800 µm requires over 2 months.[34] Thus, rate constants measured in a particular experimental system should probably be normalized to particle size and structure as well as to organic matter.

Aside from particle size and structure, the dominant effect of inorganic solid phases in sediment/water and soil/water systems is the influence of the overall quantity of solids. Changes in concentration of suspended solids and colloids in the system have an overriding effect on the amount of chemical available for transport, bioaccumulation, and degradation. Such effects become increasingly important for hydrophobic chemicals with high K_p values. Many of the seemingly conflicting results from experimental systems using different ratios of solids can be reconciled by calculating the distributions of chemical in the aqueous, colloidal, and solid phases (see Morel and Gschwend[35] and McKinley and Jenne[36] for recent discussions of the solids-concentration effect). An especially clear presentation of the three-phase distribution of chemical in soil/water and sediment/water systems is given by Pankow and McKenzie.[37] They provide a graphical scheme based on two master variables (ratio of chemical in solids/water vs. ratio in colloid/water) that illustrates changes caused by settling, resuspension, and other mass transfer processes (Figure 1). It can illustrate, for example, the shift that occurs between sorbed and dissolved forms of atrazine during soil erosion from farm fields.[38] Initially in the overland flow, most of the atrazine ($K_{oc} = 2.17$) is sorbed because of the high solids/water mass ratio. But as the runoff continues downstream, dilution leads to a shift of atrazine into the aqueous phase (about 90% in the lower Mississippi).[39] Careful calculation of the relative proportions of solid, colloidal,

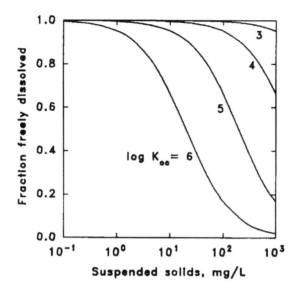

Figure 1. Fraction of freely dissolved chemical as a function of colloidal and suspended solids concentrations and K_{ow}.

and dissolved forms of chemical under study would help to clarify many other cases of environmental fate and transport.

Chemical Interactions

Most common mineral matrices, including silicates, clays, and metal oxides, have surface layers with a reduced coordination number and negative charge. A permanent component of surface charge occurs because of isomorphic substitution of, for example, Al(III) for Si(IV):

$$-Si-Si-Si-Al-Si-Si-$$

In addition, a pH-dependent charge can arise from ionization of functional groups in the lattice, including $Al(OH)_2^{+1}$ (pK$_a$ about 5.0), $(Al-OH-Si)^{+0.5}$ (pK$_a$ about 7.0) and silicate (dissociated only at very high pH).[40] Typically, the zero point of charge, at which negative and positive charges are equal, occurs much below pH 7. Associated organic matter also develops a pH-dependent charge through the dissociation of carboxyl and phenolic groups in humic acids (pK$_a$ approximately 3–5). Because of the overall negative charge on most natural particles, cationic organic pollutants can form strong covalent surface associations. An extreme example is diquat, a +2 quaternary amine that binds tightly to sediment. Both biological uptake and microbial degradation are inhibited by the strong sorption.[41] The s-triazine herbicides may also sorb coulombically through their basic amino acid moiety.[42]

A second chemical characteristic of inorganic solids is their tendency to form surface complexes through ligand exchange. In this process, surface hydroxyl groups are exchanged for another ligand:

$$= AlOH + F^- \rightarrow\ = AlF + OH^- \tag{4}$$

The exchange is competitive and strongly pH dependent because of the loss of hydroxyl in the process. Not only inorganic solutes, such as phosphate, but also organic groups, such as oxalate,

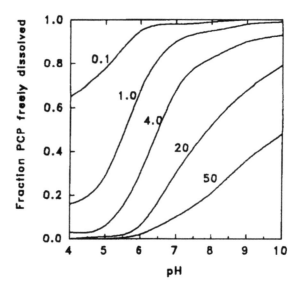

Figure 2. Fraction of freely dissolved pentachlorophenol as a function of suspended solids concentration and pH.[49]

can form surface complexes. Surface-catalyzed reactions, such as the hydrolysis of carboxylic acid esters, can occur on mineral surfaces.[1] For example, parathion undergoes hydrolysis on clay surfaces.[43] Lydy et al.[44,45] illustrated the importance of surface binding through bioavailability tests with parathion-amended sediment and water. When parathion was added directly to dry sediment it bound much more tightly, reducing bioavailability and increasing persistence.

Weak organic acids are an environmentally important group including phenols, cresols, acid dyes, many herbicides, and surfactants. The nonionized forms of such pollutants have generally been found to behave as nonionic organic chemicals by partitioning to organic matter. In most early work, sorption of the anion forms was usually disregarded. More recently, however, significant anionic interactions have been measured. Phenoxyacetic acids have pK_a values of 2 to 3 and sorb well to neutral sediment.[1] Schellenberg et al.[46] found significant anionic sorption at pH levels one unit above the pK_a values of chlorophenols. In particular, pentachlorophenol (PCP) ($pK_a = 4.75$) exists in both neutral and ionic forms at pH 7; sorption of the anion, though weaker, is significant (Figure 2). Increasing the ionic strength of the aqueous phase by the addition of Ca^{++}, K^+, or Li^+ also enhanced sorption. One of several possible mechanisms for anionic sorption is by partitioning into the organic phase along with a counterion, such as K^+.[47]

Jafvert[48,49] measured the sorption of weak organic acids including DNOC, silvex, PCP, DDA, and 2,4-DB in both neutral and anionic forms. The anionic forms bound less strongly (possible competition between hydrophobic attraction and electrostatic repulsion). Their sorption increased as an inverse linear function of pH. Addition of a counterion, Ca^{++}, enhanced anionic sorption, probably by increasing the net surface charge of the sediment.

Weber[50] investigated the behavior of benzidine-based dyes on sediment. Such dyes contain both weakly basic amino groups and strongly acidic sulfonate groups as well as lipophilic portions. Degradation was inhibited by strong sorption to sediment, which is pH dependent. Greater sorption occurred with decreasing pH and with the addition of salts, suggesting an anion-adsorption mechanism rather than simple cation exchange (in which case Ca^{++} or Na^+ would lessen sorption by competition for binding sites).

Surfactants such as linear alkyl sulfonate (LAS) contain both lipophilic and anionic areas, allowing several possible associations with sediment. Depending on carbon chain length, they may form hemicelle structures on mineral surfaces or associate by ion exchange. An environmentally important role of surfactants may be the displacement of neutral organic pollutants from sorption sites. Aronstein et al.[51] showed that low concentrations of two nonionic surfactants promoted the desorption of phenanthrene from soil and led to its biodegradation. In contrast, cationic oligomers that bind to sediments by ion exchange apparently enhance the sorption of naphthalene from solution.[52] They may simply provide additional organic carbon for the partitioning of lipophilic organic chemicals.

Organic cation interactions are more obvious than those of anions because of the strong coulombic attraction for negatively charged soils and sediments. The sorption of benzidine (pK_a = 4.7)[53] and quinoline (pK_a = 4.94)[54] to soils of low organic content is enhanced by protonation at pH values below the pK_a. For such molecules, the cationic exchange capacity of the soil is a better predictor of sorption than organic carbon content. While competitive interactions among cationic sorbates can occur, there is no such competition among pyridine, quinoline, and acridine in their neutral forms.[55] Amine-substituted polycyclic aromatic hydrocarbons, which have some lipophilicity but also cationic substituents, were found to follow carbon-based partitioning, but with somewhat stronger sorption than predicted by K_{oc} alone.[56] Because of their strong attraction for cations, suspended clays can affect the bioavailability and fate of cationic organic molecules, even without the presence of organic matter. Cationic polyelectrolytes used as clarifying aids were significantly less toxic to fish when exposed in the presence of pure suspended clays.[57]

As this review has illustrated, inorganic solid-phase interactions with organic pollutants are considerably more complex than expected from equilibrium partitioning alone. An example of an area where this complexity needs to be addressed is in the testing of sediment toxicity. Methods that make use of either sediment elutriates or pore water are methodologically convenient, and perhaps appropriate for pelagic organisms, but may also produce a variety of artifacts. Changes in ionic strength and pH during sediment manipulations are among the possible pitfalls. Desorption mechanisms that influence both short-term bioavailability and long-term bioaccumulation can be altered or simply overlooked during routine sediment testing.

CONCLUSIONS AND RECOMMENDATIONS

A comprehensive understanding of sediments and sediment-bound pollutants requires a better understanding of inorganic-organic interactions. While K_{oc} serves as a good general predictor for the distribution of neutral hydrocarbons and organochlorines, there is a large group of other organic environmental pollutants that undoubtedly has more complex behavior. Continued research on the soil and sediment interactions of such materials is recommended. In all cases, including neutral and polar contaminants, there is a great need for better kinetic information, which will come first through better definitions of solid, colloidal, and dissolved phases. Once these are carefully quantified, rates of transfer among compartments can be examined more closely. These rates of transfer are especially important for predictions of bioavailability and degradation.

REFERENCES

1. Stumm, W., *Chemistry of the Solid-Water Interface,* John Wiley & Sons, New York, 1992.
2. Elzerman, A.W. and Coates, J.T., Hydrophobic organic compounds on sediments: equilibria and kinetics of sorption, in Hites, R.A. and Eisenreich, S.J., Eds., *Sources and Fates of Aquatic Pollutants,* Adv. Chem. Ser. 216, American Chemical Society, Washington, D.C., 1987, 263–318.
3. Mills, W.B., Porcella, D.B., Ungs, M.J., Gherini, S.A., Summers, K.V., Mok, L., Rupp, G.L., and Bowie, G.L., Water Quality Assessment: a Screening Procedure for Toxic and Conventional Pollutants in Surface and Ground Water, EPA/600/6–85/002a, U.S. Environmental Protection Agency, Athens, GA, 1985.
4. Mingelgrin, U. and Gerstl, Z., Reevaluation of partitioning as a mechanism of nonionic chemicals adsorption in soils, *J. Environ. Qual.,* 12, 1–11, 1983.

5. Kinniburgh, D.G., General purpose adsorption isotherms, *Environ. Sci. Technol.*, 20, 895–904, 1986.
6. Karickhoff, S.W., Brown, D.S., and Scott, T.A., Sorption of hydrophobic pollutants on natural sediments, *Water Res.*, 13, 241–248, 1979.
7. Chiou, C.T., Porter, P.E., and Schmedding, D.W., Partition equilibria of nonionic organic compounds between soil organic matter and water, *Environ. Sci. Technol.*, 17, 227–231, 1983.
8. Rebhun, M., Kalabo, R., Grossman, L., Manka, J., and Rav-Acha, C, Sorption of organics on clay and synthetic humic-clay complexes simulating aquifer processes, *Water Res.*, 26, 79–84, 1992.
9. Karickhoff, S.W., Organic pollutant sorption in aquatic systems, *J. Hydraul. Eng.*, 110, 707–735, 1984.
10. Sabljic, A., On the prediction of soil sorption coefficients of organic pollutants from molecular structure. Application of molecular topology model, *Environ. Sci. Technol.*, 21, 358–366, 1987.
11. Di Toro, D.M., Zarba, C.S., Hansen, D.J., Berry, W.J., Swartz, R.C., Cowan, C.E., Pavlou, S.P., Allen, H.E., Thomas, N.A., and Paquin, P.R., Technical basis for establishing sediment quality criteria for nonionic organic chemicals using equilibrium partitioning, *Environ. Toxicol. Chem.*, 10, 1541–1583, 1991.
12. Bierman, V.J., Jr., Equilibrium partitioning and biomagnification of organic chemicals in benthic animals, *Environ. Sci. Technol.*, 24, 1407–1412, 1990.
13. Lake, J.A., Rubinstein, N.I., Lee, H., II, Lake, C.A., Heltshe, J., and Pavignano, S., Equilibrium partitioning and bioaccumulation of sediment-associated contaminants by infaunal organisms, *Environ. Toxicol. Chem.*, 9, 1095–1106, 1990.
14. McElroy, A.E. and Means, J.C., Factors affecting the bioavailability of hexachlorobiphenyls to benthic organisms, in Adams, W.J., Chapman, G.A., and Landis, W.G., Eds., *Aquatic Toxicology and Hazard Assessment*, STP 971, American Society for Testing and Materials, Philadelphia, 1988, 149–158.
15. Ferraro, S.P., Lee, H., II, Smith, L.M., Ozretich, R.J., and Specht, D.T., Accumulation factors of eleven polychlorinated biphenyl congeners, *Bull. Environ. Contam. Toxicol.*, 46, 276–283, 1991.
16. Holsen, T.M., Taylor, E.R., Seo, Y.-C., and Anderson, P.R., Removal of sparingly soluble organic chemicals from aqueous solutions with surfactant-coated ferrihydrite, *Environ. Sci. Technol.*, 25, 1585–1589, 1991.
17. Murphy, E.M., Zachara, J.M., and Smith, S.C., Influence of mineral-bound humic substances on the sorption of hydrophobic organic compounds, *Environ. Sci. Technol.*, 24, 1507–1516, 1990.
18. Weber, W.J.J., McGinley, P.M., and Katz, L.E., A distributed reactivity model for sorption by soils and sediments. I. Conceptual basis and equilibrium assessments, *Environ. Sci. Technol.*, 26, 1955–1962, 1992.
19. Wang, L., Govind, R., and Dobbs, R.A., Sorption of toxic organic compounds on wastewater solids: mechanism and modeling, *Environ. Sci. Technol.*, 27, 152–158, 1993.
20. Neff, J.M., Bioaccumulation of organic micropollutants from sediments and suspended particulates by aquatic animals, *Fresenius Z. Anal. Chem.*, 319, 132–136, 1984.
21. Evans, K.M., Gill, R.A., and Robotham, P.W.J., PAH and organic content of sediment particle size fractions, *Water Air Soil Pollut.*, 51, 13–31, 1990.
22. Karickhoff, S.W. and Morris, K.R., Sorption dynamics of hydrophobic pollutants in sediment suspensions, *Environ. Toxicol. Chem.*, 4, 469–479, 1985.
23. Van Hoof, P.L. and Andren, A.W., Partitioning and sorption kinetics of a PCB in aqueous suspensions of model particles: solids concentration effect, in Baker, R.A., Ed., *Organic Substances and Sediments in Water*, Vol. 2, Lewis Publishers, Chelsea, MI, 1991, 149–167.
24. Eadie, B.J., Morehead, N.R., and Landrum, P.F., Three-phase partitioning of hydrophobic compounds in Great Lakes waters, *Chemosphere*, 20, 161–178, 1990.
25. Landrum, P.F., Bioavailability and toxicokinetics of polycyclic aromatic hydrocarbons sorbed to sediments for the amphipod *Pontoporeia hoyi*, *Environ. Sci. Technol.*, 23, 588–595, 1989.
26. Wood, L.W., Rhee, G.-Y., Bush, B., and Barnard, E., Sediment desorption of PCB congeners and their bio-uptake by dipteran larvae, *Water Res.*, 21, 875–884, 1987.
27. Eaton, J.G., Mattson, V.R., Mueller, L.H., and Tanner, D.K., Effects of suspended clay on bioconcentration of Kelthane in fathead minnows, *Arch. Environ. Contam. Toxicol.*, 12, 439–445, 1983.
28. Schrap, S.M. and Opperhuizen, A., Relationship between bioavailability and hydrophobicity: reduction of uptake of organic chemicals by fish due to the sorption on particles, *Environ. Toxicol. Chem.*, 9, 715–724, 1990.

29. Gschwend, P.M., Wu, S.-C., Madsen, O.S., Wilkin, J.L., Ambrose, R.B.J., and McCutcheon, S.C., Modeling the Benthos-Water Column Exchange of Hydrophobic Chemicals, EPA/600/3-86/044, U.S. Environmental Protection Agency, Athens, GA, 1986.

30. Wu, S.-C. and Gschwend, P.M., Sorption kinetics of hydrophobic organic compounds to natural sediments, *Environ. Sci. Technol.*, 20, 717–725, 1986.

31. Brusseau, M.L., Jessup, R.E., and Rao, P.S.C., Nonequilibrium sorption of organic chemicals: elucidation of rate-limiting processes, *Environ. Sci. Technol.*, 25, 134–142, 1991.

32. Brusseau, M.L. and Rao, P.S.C., Influence of sorbate structure on nonequilibrium sorption of organic compounds, *Environ. Sci. Technol.*, 25, 1501–1506, 1991.

33. Lee, L.S., Suresh, P., Rao, P.S.C., and Brusseau, M.L., Nonequilibrium sorption and transport of neutral and ionized chlorophenols, *Environ. Sci. Technol.*, 25, 722–729, 1991.

34. Medine, A.J. and McCutcheon, S.C., Fate and Transport of Sediment-Associated Contaminants, EPA/600/D-86/356, U.S. Environmental Protection Agency, Athens, GA, 1987.

35. Morel, F.M.M. and Gschwend, P.M., The role of colloids in the partitioning of solutes in natural waters, in Stumm, W., Ed., *Aquatic Surface Chemistry*, John Wiley & Sons, New York, 1987, 405–422.

36. McKinley, J.P. and Jenne, E.A., Experimental investigation and review of the "solids concentration" effect in adsorption studies, *Environ. Sci. Technol.*, 25, 2082–2087, 1991.

37. Pankow, J.F. and McKenzie, S.W., Parameterizing the equilibrium of chemicals between the dissolved, solid particulate matter, and colloidal matter compartments in aqueous systems *Environ. Sci. Technol.*, 25, 2046–2053, 1991.

38. Squillace, P.J. and Thurman, E.M., Herbicide transport in rivers: importance of hydrology and geochemistry in nonpoint-source contamination, *Environ. Sci. Technol.*, 26, 538–545, 1992.

39. Pereira, W.E. and Rostad, C.E., Occurrence, distributions,and transport of herbicides and their degradation products in the lower Mississippi River and its tributaries, *Environ. Sci. Technol.*, 24, 1400–1406, 1990.

40. Bohn, H., McNeal, B., and O'Connor, G., *Soil Chemistry*, 2nd ed., Wiley-Interscience, New York, 1985.

41. Simsiman, G.V. and Chester, G., Persistence of diquat in the aquatic environment, *Water Res.*, 10, 105–112, 1976.

42. Brown, D.S. and Flagg, E.W., Empirical prediction of organic pollutant sorption in natural sediments, *J. Environ. Qual.*, 10, 382–386, 1981.

43. Saltzman, S., Yaron, B., and Mingelgrin, U., The surface catalyzed hydrolysis of parathion on kaolinite, *Soil Sci. Am. Proc.*, 38, 231–234, 1972.

44. Lydy, M.J., Bruner, K.A., Fry, D.M., and Fisher, S.W., Effects of sediment and the route of exposure on the toxicity and accumulation of neutral lipophilic and moderately water soluble metabolizable compounds in the midge, *Chironomus riparius*, in Landis, W.G. and Van der Schalie, W.H., Eds., American Society for Testing and Materials, Vol. STP 1096, Philadelphia, 1990, 140–164.

45. Lydy, M.J., Lohner, T.W., and Fisher, S.W., Influence of pH, temperature and sediment type on the toxicity, accumulation and degradation of parathion in aquatic systems, *Aquat. Toxicol.*, 17, 27–44, 1990.

46. Schellenberg, K., Leuenberger, C., and Schwarzenbach, R.P., Sorption of chlorinated phenols by natural sediments and aquifer materials, *Environ. Sci. Technol.*, 18, 652–657, 1984.

47. Westall, J.C., Leuenberger, C., and Schwarzenbach, R.P., Influence of pH and ionic strength on the aqueous-nonaqueous distribution of chlorinated phenols, *Environ. Sci. Technol.*, 19, 193–198, 1985.

48. Jafvert, C.T., Sorption of organic acid compounds to sediments: initial model development, *Environ. Toxicol. Chem.*, 9, 1259–1268, 1990.

49. Jafvert, C.T., Assessing the Environmental Partitioning of Organic Acid Compounds, EPA/600/M-89/016, U.S. Environmental Protection Agency, Athens, GA, 1990, 42.

50. Weber, E.J., Studies of benzidine-based dyes in sediment-water systems, *Environ. Toxicol. Chem.*, 10, 609–618, 1991.

51. Aronstein, B.N., Calvillo, Y.M., and Alexander, M., Effect of surfactants at low concentrations on the desorption and biodegradation of sorbed aromatic compounds in soil, *Environ. Sci. Technol.*, 25, 1728–1731, 1991.

52. Podoll, R.T. and Irwin, K.C., Sorption of cationic oligomers on sediment, *Environ. Toxicol. Chem.*, 7, 405–415, 1988.

53. Zierath, D.L., Hassett, J.J., and Banwart, W.L., Sorption of benzidine by sediments and soils, *Soil Sci.*, 129, 277–281, 1980.

54. Zachara, J.M., Ainsworth, C.C., Felice, L.J., and Resch, C.T., Quinoline sorption to subsurface materials: role of pH and retention of the organic cation, *Environ. Sci. Technol.*, 20, 620–627, 1986.

55. Zachara, J.M., Ainsworth, C.C., Cowan, C.E., and Thomas, B.L., Sorption of binary mixtures of aromatic nitrogen heterocyclic compounds on subsurface materials, *Environ. Sci. Technol.*, 21 397–402, 1987.

56. Means, J.C., Wood, S.G., Hassett, J.J., and Banwart, W.L., Sorption of amino substituted and carboxy substituted polynuclear aromatic hydrocarbons by sediments and soils, *Environ. Sci. Technol.*, 16, 93–98, 1982.

57. Cary, G.A., McMahon, J.A., and Kuc, W.J., The effect of suspended soils and naturally occurring dissolved organics in reducing the acute toxicities of cationic polyelectrolytes to aquatic organisms, *Environ. Toxicol. Chem.*, 6, 469–474, 1987.

Chapter 2

Interactions of Organic Chemicals with Particulate and Dissolved Organic Matter in the Aquatic Environment

Frank A. P. C. Gobas and Xin Zhang

INTRODUCTION

It is generally recognized that the fate of organic chemicals in aqueous environments, to a large degree, is controlled by the association or interaction of the chemical with organic matter in the water. Organic chemicals that are freely dissolved in the water, and not bound to or associated with organic matter, are subject to diffusion-controlled processes such as volatilization, diffusion between sediment and water, and bioconcentration in fish and other organisms. Organic chemicals that are associated with organic matter in the water are subject to advective processes such as settling, resuspension, deposition, filter feeding, and dietary accumulation. Also, the rate of transformation of organic chemicals through photolysis or hydrolysis can be affected by the association of the chemical with certain types of organic matter. There are several examples that show that certain organic molecules can act as sensitizers of photochemical reactions, and organic matter in clay can affect the rate of hydrolysis of some organic chemicals. Thus, to predict the fate of organic chemicals in aqueous systems, it is important to know to what extent organic chemicals are associated with organic matter in the water. Furthermore, in order to characterize or estimate the extent a chemical will be in association with organic matter, it is important to understand the nature of the interactions between organic chemicals and organic matter. The objective of this paper is to summarize and discuss the molecular mechanisms of chemical interaction with organic matter in aqueous systems.

We will first discuss existing mechanistic models for the interactions of organic chemicals with dissolved and particulate organic matter in aqueous systems. Then, we will briefly review some laboratory and field observations regarding organic chemical interactions with particulate and dissolved organic matter in water. Finally, we will compare the experimental findings to the model predictions in order to explore the nature of organic chemical interactions with organic matter and to formulate new models and hypotheses that can be investigated in future research.

MODELS OF ORGANIC CHEMICAL INTERACTIONS WITH PARTICULATE AND DISSOLVED ORGANIC MATTER IN AQUEOUS SYSTEMS

Organic matter in the water column can take several forms. Often, a distinction is made between dissolved organic matter, which includes a variety of substances such as other pollutants, alcohols, ketones, carboxylic acids, phenols, proteins, and fulvic and humic acids, and particulate or suspended organic matter, including some humic acids, microorganisms, ligneous matter, litter from plants, and small planktonic organisms. However, organic matter may also be present as emulsified oils, or as microlayers or "slick" on the water surface. The chemical nature of these materials is often in considerable doubt, and discrimination by analytical methods is difficult. Certain parameters, such as the total quantity of organic carbon (TOC), are relatively easy to measure. However, discrimination between dissolved and suspended matter, which is often done by filtration using submicron filters, centrifuging, dialysis, selective absorption, or headspace

1-56670-086-8/94/$0.00+$.50

analysis, is more difficult. As a result, there is only an operational definition between "dissolved" and "particulate" organic matter. Usually, organic matter with an internal diameter less than 1 to 0.1 μm is considered to be dissolved, and organic matter of larger diameters are considered to be particles. There is little doubt but that a significant proportion of "dissolved" organic matter is made up of suspended particles with diameters of less than 0.1 μm. These very small particles are operationally defined as dissolved. Although such particles are small, they are still very large when compared to molecular dimensions of chemicals, which are typically 0.2 nm, or roughly 1/1000th of the size of the smallest particles which can be filtered out of solution. The nonsuspended or nonparticulate fraction of the dissolved organic matter are organic molecules that are in solution. These truly dissolved organic molecules can be viewed as cosolvents, and may in fact behave towards organic chemicals in the environment in a similar fashion as cosolvents in pharmaceutical drug formulations and chemical engineering applications.

Sorption to Particulate Organic Matter

A comprehensive review of organic chemical sorption in aquatic systems has been presented by Karickhoff.[1] In essence, Karickhoff's sorption theory assumes that the chemical is dissolved in a solid solution form in and on the organic matter of the particle. Sorption is therefore often viewed as a partitioning process of the chemical between the water and the organic content of particulate matter, which can be represented by an organic carbon-based sorption coefficient K_{OC} (liters per kilogram):

$$K_{OC} = C_{OC}/C_{WD} = C_P/(OC \cdot C_{WD}) = K_P/OC \qquad (1)$$

where C_{OC} is the chemical concentration in the organic matter of the particles (nanograms per kilogram of organic matter), C_{WD} is the freely dissolved chemical concentration in the water (nanograms per liter), C_P is the chemical concentration in the particulate matter (nanograms per kilogram of particle), OC is the fraction of organic matter in the particle (kilograms of organic matter per kilogram of particle) and K_P is the sorption coefficient of the chemical to the particulate matter (liters of water per kilogram of particle).

Karickhoff's theory is often applied to express the fraction of the total chemical concentration in the water that is sorbed to particulate matter and that is freely dissolved in aqueous systems containing particulate organic matter. For example, a simple mass balance of the chemical in the water shows that:

$$V_{WT} \cdot C_{WT} = V_W \cdot C_{WD} + M_P \cdot OC \cdot C_{OC} \qquad (2)$$

where V_{WT} is the total volume (liters) of the suspension (i.e., water and particulate matter), C_{WT} is the chemical concentration (nanograms per liter) in the suspension, V_W is the volume (liters) of the water in the suspension, C_{WD} is the freely dissolved chemical concentration in the water of the suspension or the freely dissolved chemical concentration in the suspension, M_P is the mass (kilograms) of particulate matter in the suspension, and C_{OC} is the chemical concentration (nanograms per kilogram) in the organic matter of the particles in suspension, i.e., C_P/OC. Since the fraction of particulate organic matter in natural waters is usually small, i.e., approximately 10^{-5} to 10^{-7} (kg/kg), V_{WT} and V_W are approximately equal. Substitution of Equation 1 in Equation 2 then results in an expression for the fraction of freely dissolved chemical F_{DW}, i.e., C_{WD}/C_{WT}:

$$F_{DW} = 1/(1 + \phi_P \cdot OC \cdot K_{OC}) \qquad (3)$$

where ϕ_P is the concentration of particles in the water (in kilograms of particles per liter of water). The fraction of chemical sorbed to the particulate matter, F_{SW}, is thus $1 - F_{DW}$. Various authors

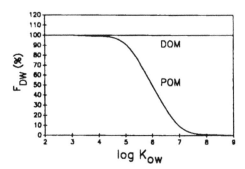

Figure 1. Estimates of the fraction (in %) of the total chemical concentration in the water that is freely dissolved F_{DW} for organic substances of different K_{OW} in water containing a concentration of, respectively, particulate (POM) and dissolved organic matter (DOM) of 10^{-6} g/ml.

have investigated the magnitude of K_{OC} and its relationship with the octanol-water partition coefficients. We will review this work later. However, there are two assumptions that are often used to approximate the value of K_{OC}. Karickhoff's empirical work[2] suggests that K_{OC} is proportional to 0.4 to 0.6 of K_{OW}. DiToro[3] suggested that, within experimental error, K_{OC} equals K_{OW}, such that Equation 3 can be rewritten as

$$F_{DW} = 1/(1 + \phi_P \cdot OC \cdot K_{OW}/d_{OC}) \tag{4}$$

where d_{OC} is the density of the organic matter (in kilograms per liter).

Application of Equation 4 in a typical environmental situation where the concentration of particulate organic matter, i.e., $\phi_P \cdot OC$, for example, is 10^{-6} kg/l, shows that for a chemical with a log K_{OW} of 6, approximately 50% of the chemical is freely dissolved and 50% is sorbed. Figure 1 illustrates that if K_{OW} increases, the fraction of dissolved chemical in the suspension drops sharply, e.g., to 0.99% for a chemical with a log K_{OW} of 8. This illustrates that particulate organic matter, which in natural water typically ranges between 10^{-5} to 10^{-7} kg/l, can substantially reduce the fraction of freely dissolved and bioavailable chemical in the water, particularly if K_{OW} is large.

Interaction with Organic Cosolvents

It has been suggested that freely dissolved organic molecules or organic cosolvents in the water column can affect the extent to which organic chemicals are freely dissolved and "sorbed or bound" in water.[4] Organic cosolvents are known to increase the solubility of the chemical in the water, possibly by forming aggregates or molecular complexes with hydrophobic substances. Various polar organic solvents (e.g., alcohols and glycols) in aqueous systems are known to increase the solubility of hydrophobic solutes above their solubilities in pure water.[4-6] The solubility enhancement increases exponentially with the volume fraction of organic cosolvent. Studies of the effect of cosolvents on chemical solubility have shown that the logarithm of the solubility of a chemical, S (mole per mole), is often a linear function of the volume fraction of cosolvent, θ, extending from the solubility of the chemical in pure water S_W (mole per mole) to the solubility in the pure cosolvent S_C (mole per mole):

$$\ln S = \ln S_W + \theta \cdot (\ln S_C - \ln S_W) \tag{5}$$

This relationship appears to be maintained for several solutes and various organic cosolvents.[6,7] Small deviations from this linear behavior have been observed by Dickhut et al.[4] due to non-ideal solution behavior.

To explore the possible role of organic solvents on the bioavailability of hydrophobic chemical substances, we can invoke the cosolvent theory to estimate the effect of freely dissolved organic molecules on the fraction of freely dissolved chemical in the water column, F_{DW} or S_W/S. We will assume that only the fraction of freely dissolved chemical in the water, i.e., S_W, is available for uptake across respiratory surfaces (e.g., gill membrane). Organic solvents may thus raise the apparent solubility(s) of organic chemicals in water, but reduce the fraction of freely dissolved and bioavailable chemical in the water. Considering the absence of information regarding the nature of freely dissolved organic matter, it is reasonable to assume for illustrative purposes that the freely dissolved organic matter has phase characteristics similar to octanol. In that case, S_C equals the solubility in octanol S_O (mol/mol) and Equation 5 can be rewritten to

$$\ln S = \ln S_W + \theta \cdot (\ln S_O - \ln S_W). \qquad (6)$$

From Equation 6 it follows that the fraction of freely dissolved chemical in the water F_{DW} can be estimated from θ and $\ln K_{OW}$ (i.e., $\ln S_O - \ln S_W$):

$$\ln F_{DW} = \ln (S_W/S) = \theta \cdot (\ln S_W - \ln S_O) = -\theta \cdot \ln K_{OW} \qquad (7)$$

In natural waters, the volume fraction of dissolved organic carbon, θ, is usually in the range of 10^{-5} to 10^{-7}. For example, if θ is 10^{-6}, i.e., approximately 1 mg of organic carbon per liter of water, and K_{OW} is 10^{-6}, then $\ln F_{DW}$ is approximately -1.38×10^{-5}, and F_{DW} is 0.99999.

Equations 4 and 7 suggest that particulate organic matter and organic cosolvents may have a distinctively different tendency to "bind" organic chemicals. This is further illustrated in Figure 1. Whereas low concentrations of particulate organic matter can significantly reduce the fraction of dissolved and bioavailable chemical, whereas a similar concentration of dissolved organic molecules (or cosolvent) would have virtually no effect on the fraction of freely dissolved chemical in the water phase. We will now test the theory against experimental data.

EXPERIMENTAL OBSERVATIONS

A large number of studies in the laboratory and in the field have been carried out to measure the effect of particulate and dissolved organic matter on the fraction of freely dissolved chemical in the water. Many of these studies involve biological organisms, such as fish and *Daphnia* species. These studies are particularly useful because biological organisms are probably the best "tools" to measure the bioavailable chemical concentration in the water. The bioavailable chemical concentration is often considered to be the "freely" dissolved chemical concentration in the water, as only freely dissolved organic molecules are believed to be able to permeate through the gill membrane. The fraction of dissolved chemical in the water is therefore often referred to as the chemical's bioavailability. If the results of experimental work regarding the effect of particulate and dissolved organic matter on the fraction of dissolved and bioavailable chemical in the water are compiled, the following trends emerge.

Particulate Organic Matter

1. Laboratory experiments of chemical sorption to particulate matter have shown that if experiments are conducted in which a chemical is first adsorbed and then desorbed, some of the chemical becomes "resistant" to desorption.[8] The remaining chemical is labile or "reversibly" sorbed. The reasons for this difference are not entirely clear, but the current view is that the labile chemical is associated with the outer and more accessible surfaces of the sorbent. The resistant material may have migrated into the depths of the sorbent phase and

become trapped.[3] This is occasionally termed as "irreversible" sorption. However, it is probable that given enough time, the material will desorb, thus the process is not inherently irreversible. Most experiments have been done with soils, thus there is an element of doubt as to the applicability of these findings to suspended matter. As the effect relates to the kinetics of sorption to particulate matter, it is not expected to have a significant effect on the values of equilibrium sorption coefficients. However, it may become important in situations in which the chemical in solution is exposed to rapidly changing sorbent concentrations. Such situations may be encountered in mixing zones where a waste water containing a high suspended sediment concentration is rapidly diluted. In that case, it is possible that the sorbed chemical will be slow to dissolve and adjust to the new suspended solids concentration. It may tend to remain associated with the solids and display an abnormally high apparent partition coefficient. Conversely, if in a water quality analysis, reliance is placed on the suspended matter in the receiving water to reduce the amount of dissolved chemical, i.e., reduce its bioavailability, then it should be recognized that this sorption process may take some time. There may be a fairly rapid initial sorption, followed by a more prolonged final sorption.

2. In laboratory experiments of chemical sorption to particulate matter, several authors have observed a "solids concentration" effect, which was first reported by O'Connor and Connolly[9] and refers to the drop of the sorption coefficient, K_P, with increasing sorbent concentration. It has been the subject of considerable discussion and controversy because it apparently contravenes the laws of thermodynamics. Several authors have provided data that suggest that the "solids concentration effect" is the result of analytical problems associated with the difficult analysis of the chemical concentration in the water.[10,11] These authors argue that as a result of adding particulate matter to water, a so-called "third phase" is introduced in the water. This "third phase" is poorly defined, but may consist of nonsettling particles or dissolved molecules which cannot be removed by filtration or centrifugation. Chemicals sorbing to "third phase" materials enhance the apparent solubility of the chemicals in the water, thus causing an apparent drop of the sorption coefficient. One of the most important implications of the "solids concentration" effect is that in nepheloid layers at the sediment-water interface, sorption coefficients could be lower than in the overlying water due to the high particle concentration. Lower sorption coefficients would result in a higher bioavailability of hydrophobic organic chemicals to benthic organisms. There are convincing data that both support and reject the existence of the "solids concentration" effect. On one hand, experiments with glass beads,[12] which may eliminate the introduction of a "third phase" into the water, tend to support the existence of the "solids concentration" effect. On the other hand, recent experiments using headspace analysis suggest that there is no "solids concentration" effect.[13]

Dissolved Organic Matter

1. In laboratory experiments with fish[14,15] and *Daphnia* species[16] the addition of artificial dissolved organic matter, typically Aldrich humic acids, is known to reduce the bioavailability of hydrophobic organic chemicals considerably. For commercially produced humic acids, K_{OC} values tend to range between 0.1 of K_{OW} for a given compound[17-21] to 1.0 of K_{OW}.[22] For example, McCarthy and Jiminez[16] reported the following correlation for the interaction of 5 PAHs to Aldrich humic acids:

$$\log K_{OC} = 1.03.\log K_{OW} - 0.49 \qquad (8)$$

which is similar to the correlation reported by Karickhoff et al.,[23] i.e.,

$$\log K_{OC} = 1.0.\log K_{OW} - 0.21 \qquad (9)$$

and suggests that the commercially extracted humic acids behave like particulate organic matter with an "octanol-like" composition. These results are further supported by recent experiments of Resendes et al.[13] using headspace analysis, who also show that commercial humic acids act as particles with an octanol-like composition with regards to organic chemical sorption.

2. Laboratory and field experiments with naturally occurring dissolved organic matter shows that chemical sorption to natural dissolved organic matter, as expressed by the K_{OC}, can be considerably smaller than that to commercially available dissolved organic matter. For example, Landrum et al.[18] found that K_{OC} values for benzo-a-pyrene and p,p'-DDT are approximately one order of magnitude lower than the K_{OC} for Aldrich humic acids, and Evans[24] showed that K_{OC} for natural dissolved organic matter from different lakes is approximately 2% of K_{OW}. Eadie et al.[25] found similar results in the Great Lakes.

3. Field studies[18,24,26,27] show that there are considerable variations in the partition coefficients for several organic chemicals to dissolved organic matter from different sources, although the concentration of dissolved organic matter in the different waters was approximately similar. These findings agree with the observations of Gauthier et al.,[28] who measured K_{OC} values in humic materials that varied for the same chemical substance by as much as a factor of ten, depending upon the humic material. These results indicate that the structure and composition of the humic materials are important factors controlling the binding of hydrophobic organic chemicals to dissolved organic matter in the water phase.

4. In a study in the Great Lakes, Eadie et al.[25] observed that particulate organic matter in Great Lakes water has a similar sorption tendency as octanol, whereas dissolved organic matter tends to have a smaller sorption effect. Also, the authors observed that different organic chemicals varied widely in their ability to associate with the dissolved organic matter. These findings further illustrate that dissolved organic matter has a chemical sorption affinity that is considerably less than that of particulate organic matter.

DISCUSSION

The experimental observations suggest that in absence of experimental artifacts due to slow desorption kinetics and difficulties in the measurements of "freely dissolved" chemicals concentrations in water containing "third phase" materials, chemical partitioning to particulate organic matter is well described by Karickhoff's chemical partitioning theory. Consequently, Equation 4 is an adequate method for estimating the bioavailability of hydrophobic organic chemicals in aqueous systems containing particulate organic matter. However, since sorption equilibria are slow to establish, abnormally low or high (apparent) sorption coefficients may be encountered in situations where concentrations or particle densities are rapidly changing and the sorption equilibrium has not been reached.

The results of the experiments with dissolved organic matter demonstrate that "dissolved" organic matter from commercial sources (e.g., Aldrich humic acid) tends to behave like "particulate" organic matter. Sorption coefficients of hydrophobic organic chemicals to dissolved organic matter in the field are often considerably less than those to Aldrich humic acids. Aldrich humic acids may therefore be a poor model for studying the interaction of organic chemicals with natural dissolved organic matter. A prevalent result in many sorption studies is that dissolved organic matter in natural waters shows a wide range of sorption coefficients, and that K_{OW} is a poor predictor of organic chemical sorption to naturally occurring dissolved organic matter. Currently, it is unknown what causes the differences in sorption coefficients among different types of dissolved organic matter. However, based on current experimental evidence, the following explanations can be hypothesized.

Particle Size

The cosolvent theory outlined above suggests that at typical environmental concentrations of 10^{-5} to 10^{-7} kg/l, organic matter in "freely dissolved" form can not significantly reduce the fraction of dissolved or bioavailable chemical below 1.0. On the other hand, there is good experimental evidence that the same concentrations of organic matter "in particulate form" can have a large effect on the fraction of dissolved and bioavailable chemical in the water phase. In the light of these two theories, it is important to be able to distinguish between organic matter in freely dissolved and particulate form. However, current methods of separating particulate from dissolved organic matter (e.g., filtration and centrifugation) can only separate particulate organic carbon from total organic carbon up to a certain particle size (e.g., particles with internal diameters exceeding 0.1 μm). As a result, it is likely that a significant fraction of the particulate organic matter, i.e., the smaller particles with diameters less than 0.1 μm, is considered to be "dissolved," while in reality it consists of very small particles. This fraction of so-called "dissolved" organic matter may be responsible for virtually all of the sorptive capacity of the "dissolved organic matter". The fraction of particulate organic matter in the so-called "dissolved" organic matter may thus determine the sorptive capacity of the "dissolved" organic matter. The fractions of particulate matter in the "dissolved" organic matter content of natural waters are likely to vary between different natural waters, which provides an explanation for the differences in the sorptive capacities of dissolved organic matter between natural waters. In other words, dissolved organic matter from lake A may have a lower sorptive capacity than dissolved organic matter from lake B, because it contains less smaller inseparable (e.g., unfilterable) particles than the dissolved organic matter fraction from lake B.

Chemical-Organic Matter Interaction

An additional explanation for the differences in chemical sorption coefficients to various types of dissolved organic matter is that interactions of the chemical with the dissolved matter vary considerably, depending on the chemical nature of the dissolved organic matter. Gauthier et al.[28] reached a similar conclusion based on experiments with different types of organic matter. The thermodynamic basis for differences in sorption coefficients follows when the sorption coefficient is expressed as an equilibrium partition coefficient. It can then be shown that at dilute solute and sorbent concentrations, which often occur under field conditions, the sorption coefficient to dissolved organic matter, K_{DOM}, can be viewed as the ratio of the chemical's solubilities in the dissolved organic matter, S_{DOM}, and that in water, S_W. That is, K_{DOM} equals S_{DOM}/S_W. As we discussed earlier, the octanol-water partition coefficient K_{OW} can be expressed in a similar manner as S_O/S_W. Differences between sorption and octanol-water partition coefficients are thus largely a reflection of the differences between the chemical solubilities in the dissolved organic matter and in octanol, i.e., K_{DOM}/K_{OW} equals S_{DOM}/S_O. Figure 2 shows partition coefficients of a series of linear alcohols between water and several organic solvents, including octanol, and illustrates that partition coefficients of the same chemical can vary by almost two orders of magnitude, depending on the organic solvent. Differences in phase compositions between dissolved organic matter from different aquatic systems may have a similar effect on chemical interactions to dissolved organic matter, causing chemical sorption coefficients of dissolved organic matter to differ between locations. If these differences in sorption coefficients between different types of organic matter are indeed large, then there may not be a single solvent, such as octanol, that could be used as a model to characterize or estimate chemical sorption to various types of dissolved organic matter. However, if interaction of hydrophobic organic chemicals with dissolved organic matter has, as the cosolvent theory suggests, little or no effect on the chemical's bioavailability and the fraction of freely dissolved chemical in natural waters, then there may be little merit in more detailed studies of chemical-dissolved organic matter interactions in terms of improving chemical fate models of hydrophobic organic chemicals in aquatic systems.

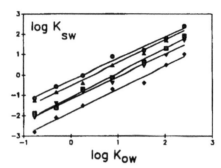

Figure 2. Logarithms of the partition coefficients (log K_{sw}) of some linear alcohols between ether/water (•), chloroform/water (▲), benzene/water (■), carbon tetrachloride/water (▼), and hexane/water (♦) vs. the logarithm of the octanol-water partition coefficient. (Data are from Dunn, W.J., Grigoras, S., and Johansson, E., *Partition Coefficient: Determination and Estimation*, Dunn, W.J., III, Block, J.H., and Pearlman, R.S., Eds., Pergamon Press, New York, 1986.)

REFERENCES

1. Karickhoff S.W., Organic pollutant soprtion in aquatic systems, *J. Hydraul. Eng.*, 6, 707–735, 1984.
2. Karickhoff S.W., Semi-empirical estimation of sorption of hydrophobic pollutants on natural sediments and soils, *Chemosphere*, 10, 833–846, 1981.
3. DiToro, D.M., A particle interaction model of reversible organic chemical sorption, *Chemosphere*, 15, 1503–1538, 1985.
4. Dickhut, R.M., Armstrong, D.E., and Andren, A.W., The solubility of hydrophobic aromatic chemicals in organic solvent/water mixtures. Evaluation of four mixed solvent solubility estimation methods, *Environ. Toxicol. Chem.*, 10, 881–889, 1991.
5. Dickhut, R.M., Andren, A.W., and Armstrong, D.E., Naphthalene solubility in selected organic solvent/water mixtures, *J. Chem. Eng. Data*, 34, 438–443, 1989.
6. Yalkowsky, S.H. and Roseman, T.J., Solubilization of drugs by cosolvents, in Yalkowsky, S.H., Ed., *Techniques of Solubilization of Drugs*, Marcel Dekker, New York, 1981, 91–134.
7. Yalkowsky, S.H. and Rubino, J.T., Solubilization by cosolvents. I. Organic solutes in propylene glycol-water mixtures, *J. Pharm. Sci.*, 74, 416–421, 1985.
8. DiToro, D.M. and Horzempa, L.M., Reversible and resistant components of PCB adsorption-desorption. Isotherms, *Environ. Sci. Technol.*, 16, 594–602, 1982.
9. O'Connor, D.J. and Connolly, J.P., The effect of concentration of desorbing solids on the partition coefficient, *Water Res.*, 14, 1517–1523, 1980.
10. Gschwend, P.W. and Wu, S.C., On the consistancy of sediment-water partition coefficients of hydrophobic organic pollutants, *Environ. Sci. Technol.*, 19, 90–96, 1985.
11. Voice, T.C. and Weber, W.J., Sorbent concentration effects in liquid/solid partitioning, *Environ. Sci. Technol.*, 19, 789–798, 1985.
12. DiToro, D.M., Personal communication, 1991.
13. Resendes, J., Shiu, W.Y., and Mackay, D., Sensing the fugacity of hydrophobic organic chemicals in aqueous systems, *Environ. Sci. Technol.*, 26, 2381–2387, 1992.
14. Black, M.C. and McCarthy, J.F., Dissolved organic macromolecules reduce the uptake of hydrophobic organic contaminants by the gills of rainbow trout (*Salmo gairdneri*), *Environ. Toxicol. Chem.*, 7, 593–600, 1988.
15. Johnsen, S., Kukkonen, J., and Grande, M., Influence of natural aquatic humic substances on the bioavailability of benzo(a)pyrene to Atlantic salmon, *Sci. Total Environ.*, 81/82, 691–702, 1989.
16. McCarthy, J.F. and Jimenez, B.D., Reduction in bioavailability to bluegills of polycyclic aromatic hydrocarbons bound to dissolved humic material, *Environ. Toxicol. Chem.*, 4, 511–521, 1985.
17. Carter, C.W. and I.H. Suffet. Binding of DDT to dissolved humic materials, *Environ. Sci. Technol.*, 16, 735–740, 1982.
18. Landrum, P.F., Nihart, S.R., Eadie, B.J., and Gardner, W.S., Reverse-phase separation method for determining pollutant binding to Aldrich humic acid and dissolved organic carbon of natural waters, *Environ. Sci. Technol.*, 18, 187–192, 1984.

19. Hassett, J.P. and Milicic, E., Determination of equilibrium and rate constants for the binding of a polychlorinated biphenyl congener by dissolved humic substances, *Environ. Sci. Technol.*, 19, 638–643, 1985.
20. Yin, C. and Hassett, J.P., Gas-partitioning approach for laboratory and field studies of mirex fugacity in water, *Environ. Sci. Technol.*, 20, 1213–1217, 1986.
21. Chiou, C.T., Kile, D.E., Brinton, T.I., Malcolme, R.L., Leenheer, J.A., and MacCarthy, P., A comparison of water solubility enhancements of organic solutes by aquatic humic materials and commercial humic acids, *Environ. Sci. Technol.*, 21, 1231–1234, 1987.
22. Landrum P.F., Reinhold, M.D., Nihart, S.R., and Eadie B.J., Predicting the bioavailability of organic xenobiotics to *Pontoporeia hoyi* in the presence of humic and fulvic materials and natural dissolved organic matter, *Environ. Toxicol. Chem.*, 4, 459–467, 1985.
23. Karickhoff, S.W., Brown, D.S., and Scott T.A., Sorption of hydrophobic pollutants on natural sediments, *Water Res.*, 13, 241–248, 1979.
24. Evans, H.E., The binding of three PCB congeners to dissolved organic carbon in fresh waters, *Chemosphere*, 17, 2325–2338, 1988.
25. Eadie, B.J., Moorehead, N.R., and Landrum, P.F., Three-phase partitioning of hydrophobic organic compounds in Great Lakes waters, *Chemosphere*, 20, 161–178, 1990.
26. Moorehead, N.R., Eadie, B.J., Lake, B., Landrum, P.F., and Berner, D., The sorption of PAH onto dissolved organic matter in Lake Michigan, *Chemosphere*, 15, 403–412, 1986.
27. Kukkonen, J., Oikari, A., Johnsen, S., and Gjessing, E., Effects of humus concentrations on benzo-a-pyrene accumulation from water to *Daphnia magna*: comparison of natural waters and standard preparations, *Sci. Total Environ.*, 79, 197–207, 1989.
28. Gauthier, T.D., Seltz, W.R., and Grant, C.L., Effects of structural and compositional variations of dissolved humic materials on pyrene K_{OC} values, *Environ. Sci. Technol.*, 21, 243–248, 1987.
29. Dunn, W.J., Grigoras, S., and Johansson, E., Principal component analysis of partition coefficient data. An approach to a new method of partition coefficient estimation, in *Partition Coefficient: Determination and Estimation*, Dunn, W.J., III, Block, J.H., and Pearlman, R.S., Eds., Pergamon Press, New York, 1986, 21–35.

Chapter 3

Synopsis of Discussion Session: Influences of Particulate and Dissolved Material on the Bioavailability of Organic Compounds

I. H. (Mel) Suffet (Chair), Chad T. Jafvert, Jussi Kukkonen, Mark R. Servos, Anne Spacie, Lisa L. Williams, and James A. Noblet*

INTRODUCTION

The two basic routes of exposure by which organisms accumulate organic pollutants (bioavailability) are (1) transport across biological membranes exposed to the aqueous phase, and (2) direct ingestion of contaminated food particles. As our purview is to address the effects of particulate and dissolved material within the water and sediment environments on the bioavailability of organic chemicals, we will focus almost exclusively on how these materials influence the rate and magnitude of direct chemical transport across membranes from the external aqueous phase. Although contaminants associated with ingested food particles must cross biological membranes within the gut, digestive (catabolic) processes acting therein make this a unique route of exposure, separate from other routes in which membranes are in direct contact with the aqueous phase. In many instances, exposure via ingestion may contribute significantly to the overall uptake of a contaminant from the environment,[1,2] although for most species its relative importance is poorly understood.

The role that sorption to organic or inorganic materials plays in controlling the bioavailability of ingested particle-associated contaminants is poorly understood. Additionally, while a discussion of mechanisms implies a discussion of mass transport from the source of the pollution, as well as the uptake kinetics and pharmacokinetics within the organism, we have emphasized the equilibrium relationships of these interactions and the conditions under which equilibrium assumptions may not be valid. A more thorough discussion of the kinetic aspects of bioavailability is the topic of another chapter.

The potential uptake of a chemical is influenced by the total concentration as well as the bioavailability of the chemical in the matrix in which the organism is exposed.[3] Mass transfer processes such as bioturbation, resuspension, sedimentation, eddy currents, etc., will influence the environmental availability by altering the amount and/or environmental distribution of the chemicals. The environmental bioavailability of the contaminants to which organisms are exposed is affected by many factors, including association with solid and dissolved organic phases.[4-9] Because it is the freely dissolved form of the contaminant which is transported across biological membranes, a reduction in the freely dissolved concentration (i.e., aqueous activity or fugacity) of the contaminant translates directly to reduced bioavailability. The characteristics of the solid or dissolved organic matter, as well as the chemical and physical properties of the contaminant, influence the mechanism and relative strength of the association. Therefore, an understanding of the mechanisms which control the sorption of organic chemicals to the solid and dissolved matrices is essential for understanding the magnitude of exposure as a function of environmental conditions.

*Post-meeting participant, from the Department of Environmental Health Sciences, UCLA-School of Public Health, Los Angeles, CA, 90024.

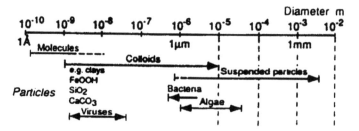

Figure 1. The size spectrum of waterborne particles. (From Stumm, W. and Morgan, J. J., *Aquatic Chemistry,* Wiley-Interscience, New York, 1981. With permission.)

FUNDAMENTAL AND EMPIRICAL RELATIONSHIPS

To predict the accumulation of organic pollutants transported across biological membranes, the freely dissolved concentration in the water column, $C_{w,aq}$, or the freely dissolved concentration in the pore water, $C_{s,aq}$, and the distribution of the organic pollutant among the various environmental phases ("compartments") in the system must be determined or predicted. Sediment-water systems are extremely heterogeneous, resulting in a wide variety of sorptive "sites" or phases within this complex assemblage of matrices. One classification scheme for these sorptive materials is based on size. Accordingly, materials range in size from simple dissolved molecules, to complex dissolved material (macromolecules), to colloids, to large suspended particles, to bedded sediment aggregates of organic and inorganic material (Figure 1). The heterogeneous particles will have different capacities for the same organic pollutant, resulting in different sorptive phases on the same particle. For example, Weber et al.[10] have shown that distributed sorption by sediment does occur.

Equilibrium Partitioning

The aqueous concentration, C_{aq}, at equilibrium is related to the distribution coefficients, K_i of the various phases (i) according to the mass balance equation:

$$\frac{C_{aq}}{C_T} = \frac{1}{1 + \Sigma K_i f_i} \tag{1}$$

where, f_i is the mass of each phase (kilograms) per volume of water (liters) in the system, and C_T is the total concentration in the system, normalized to the aqueous phase volume. Although there is a continuum of possible size fractions of the sorptive phases, three fractions have been described most by researchers: dissolved organic matter (DOM) (measured as dissolved organic carbon, DOC), suspended solids (SS), and bulk sediments (SED). The definitions of these materials as sorptive "fractions" or "phases" are operational, and designations for the different sizes of the fractions are a matter of convention. For example, DOM is often defined as that material which passes through a 0.45-μm filter, which includes both truly dissolved organic species and some colloidal material. Estimating C_{aq} depends on the accurate measurement of K_i and f_i values for each phase. It must be remembered that it is not necessary to base the summation term on only one bulk property of each size fraction (e.g., the organic carbon content); several types of sorptive sites may exist within each size fraction (e.g., clay minerals, organics, oxides). Heterogeneous particles will contain regions of differing affinity within each individual particle for the same organic pollutant.[10,11]

The sediment-water system can be represented by a high solids content sediment layer and a low solids content water layer, separated by a phase boundary (Figure 2); although the boundary is never so distinct in most natural systems. Each layer contains analogous components of water,

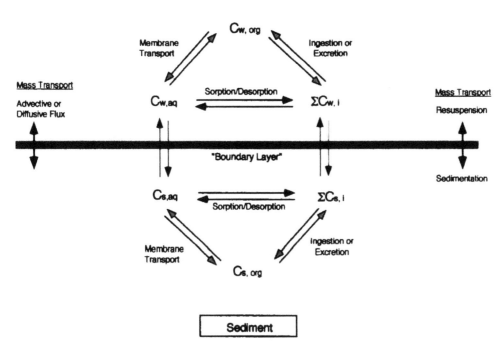

Figure 2. An equilibrium partitioning approach to the distribution of organic pollutants in water-sediment systems. The first subscript indicates the environmental compartment (w = water, s = sediment) and the second subscript refers to the phase within that compartment, e.g., aq = aqueous (freely dissolved), org = organism (biota), and i = any pollutant sorbing phase (e.g., bulk sediment, suspended sediment, colloids, DOM).

biota, and a plethora of nonliving sorptive phases. Exchange is relatively rapid among components in the "water layer", but intrasediment and water-sediment exchange may be much slower. Therefore, equilibrium cannot be assumed across the sediment-water boundary.[12] Furthermore, the quantity and nature of sorbents in the sediment layer will differ from those in the water column above, so that $C_{w,aq}$ is not necessarily equal to $C_{s,aq}$ for the same mass of chemical and the same solids mass/water volume ratio. As a result, Equation 1 should be applied separately to the two "layers" if the equilibrium assumption is made.

Figure 2 shows the partitioning of an organic pollutant within and between two multiphase systems—one on each side of the sediment-water interface. The associated contaminant concentrations are $C_{w,aq}$, $C_{w,org}$, and $\Sigma C_{w,i}$, or $C_{s,aq}$, $C_{s,org}$, and $\Sigma C_{s,i}$, assuming the interactions are at steady-state or equilibrium in the water column and sediment domains, respectively.[4,13–17] $C_{w,org}$ and $C_{s,org}$ are the pollutant concentrations associated with the biota of the aqueous and sediment layers, respectively. $C_{w,i}$ and $C_{s,i}$ are the contaminant concentrations in each of the sorbing phases, i, of the water column (e.g., suspended sediments) and bulk sediment environments, respectively. Figure 3 shows the simplest case approximation of two homogeneous sorbing phases in each layer, $\Sigma C_{w,i} = \{C_{w,ss} + C_{w,doc}\}$ and $\Sigma C_{s,i} = \{C_{s,sed} + C_{s,doc}\}$, where $C_{w,ss}$, $C_{s,sed}$, $C_{w,doc}$, and $C_{s,doc}$ are the contaminant concentration in the suspended sediment, bulk sediment, water column DOM, and pore water DOM phases, respectively. Since natural surface waters usually contain significant DOM (1 to 30 mg/l),[18] it is always necessary to include $C_{w,doc}$ and $C_{s,doc}$ terms in the $\Sigma C_{w,i}$ and $\Sigma C_{s,i}$ sums, respectively. Thus, the physical nature of the water column is best described by, at

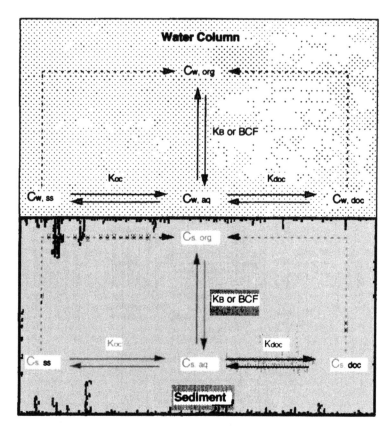

Figure 3. A simple four-phase model describing the bioavailability of organic pollutants in aquatic systems. The freely dissolved fractions of the contaminant ($C_{w,aq}$ and $C_{s,aq}$) are the most bioavailable. However, the contribution from bound fractions ($C_{w,ss}$, $C_{w,doc}$, $C_{s,ss}$, and $C_{s,doc}$), indicated by the dashed lines, may become important for certain organisms via ingestion as part of the food supply.

minimum, a three-phase system consisting of water, suspended solids (>0.45 μm), and dissolved organic matter (soluble and colloidal fraction <0.45 μm).[3,4,6,13–15,17]

Although only the freely dissolved phases are bioavailable via external biological membranes, at thermodynamic equilibrium the final concentration of the chemical in the organism is independent of the route by which the organism is exposed to the chemical.[19] This results from the fact that, at equilibrium, the chemical potentials (or fugacities) in all phases (SED, SS, DOM, water, and biota) are equal. As stated previously, however, catabolic reactions in the gut may render ingested particle or DOM-associated contaminants bioavailable (hence the dashed lines in Figure 3).

The sediment-water partition coefficient (K_{sed}) can be normalized to the organic carbon content of the sediment,

$$K_{oc} = \frac{K_{sed}}{f_{oc}} = \frac{C_{s,sed}}{[(C_{s,aq} + C_{s,doc}) \times (f_{oc})]} \tag{2}$$

where f_{oc} is the mass fraction of organic carbon in the sediment expressed as a decimal. Spacie[20] describes the factors affecting solid-phase equilibrium conditions. Organic carbon-based partitioning in sediments (i.e., partitioning described by K_{oc}), is presently the best general approach to define the interactions between the sediments and organic pollutants with octanol-water partition coefficients (K_{ow}) greater than 2 (Table 1). The use of the organic carbon normalized partition

Table 1.   Compounds of Interest for the Study of Bioavailability in
            Water

Nonpolar Organics	Examples
$\log K_{ow} > 5$	Polychlorinated biphenyls
	Perylene
	p,p'-DDT
$2 < \log K_{ow} < 5$	Lindane
	Anthracene
	Atrazine
$\log K_{ow} < 2$	Toluene
	Methylene chloride
	Acetone

Polar Organics	Examples
Organic acids	2,4-D
	Chlorophenols
	Nitrophenols
Organic bases	Acridine
	Aniline
	Quinoline

coefficient, K_{oc}, has been shown to successfully predict the freely dissolved species concentrations to within better than a factor of 10 when the f_{oc} is greater than 0.003 (i.e., sediment OC $> 0.3\%$).[21]

Experimental evidence elucidating the mechanisms that affect local microenvironmental conditions in the sediment is limited. These conditions, which include mineral matrix effects and the chemical structure of the organic matter, affect sorption capacity (i.e., fugacity capacity). Variability in the nature and properties of organic carbon increase the uncertainty in the prediction of C_{aq} from K_{oc}. It may be possible to improve upon our present predictive capability by increasing our understanding of the chemical characteristics and binding properties of different forms of organic carbon. For example, the K_{oc} term in Equation 2 might be subdivided according to size fractions or functional classes of humic material.

The K_{oc} partitioning approach is also applicable to organic matter-pollutant interactions in the water column or pore water (e.g., colloids and suspended sediment). However, the predictive ability of this approach modified for the DOC phase commonly varies by more than a factor of 10.[5,22,23] Some of this variation can be explained by the molecular characteristics of the DOM such as aromaticity, molecular size distribution, or hydrophobic acid content,[6,22,24] as well as water quality conditions (e.g., oxic vs. anoxic conditions).[25] It is evident that accurate prediction of the water column aqueous activity will require an understanding of the extent to which different classes of organic chemicals partition to the various fractions and forms of the DOM.

As previously stated, equilibrium is achieved when the free energy, chemical potential, or fugacity of a chemical in all environmental compartments (i.e., sediment, DOM, water, and biota) are equal. If a chemical exhibits equivalent partitioning behavior toward lipid and organic carbon, then at equilibrium the organic carbon normalized concentration (C_{oc}) of the chemical will be the same as the lipid normalized organism concentration, $C_{s,org}$ or $C_{w,org}$, for sediment-dwelling or water-column organisms, respectively. Experimentally measured partition coefficients for organism lipid and sediment organic carbon generally agree within an order of magnitude.[19] Some of this variation can be explained by the molecular characteristics of DOM such as aromaticity, molecular size distribution, or hydrophobic acid content.[26,27] For many nonionic organic compounds this could be extended, and the K_{ow} used to predict K_{oc} and the bioconcentration factor (K_B or BCF = C_{biota}/C_{water}). The simplest approach is to assume that $K_{ow} = K_{oc}$ = lipid normalized K_B (BCF), or that organic carbon, lipid, and octanol are all equivalent in terms of partitioning. More accurate predictions of K_B requires recognition that the above assumption is not valid, and that these parameters are generally related by $K_B = F_1(K_{oc}) = F_2(K_{ow})$, where F_1 and F_2 are empirically determined mathematical functions.[28]

Sorption to Bulk Sediment

DiToro et al.[29] have recently reviewed the available information related to bioavailability of nonionic organic contaminants in bulk sediment under near-equilibrium conditions. They conclude that there is essentially no relationship between sediment chemical concentrations (total dry weight basis) and biological effects. However, the chemical concentrations in the pore water ($C_{s,aq}$), or the sediment chemical concentrations on an organic carbon normalized basis at which biological effects occur, vary only by a factor of around 2 for most sediments and chemicals. Furthermore, the observed biological effects concentration determined in water-only exposures is the same as those predicted or measured in sediments.[29] Similarly, DeWitt et al.[21] found that K_{oc} values were able to predict free concentrations of fluoranthene within better than a factor of 10 when the f_{oc} is greater than 0.003 (0.3%). Hence, organic carbon normalized partitioning is presently the best general approach to estimate near-equilibrium interactions between the sediments and nonpolar, nonionic organic pollutants with K_{ow} greater than 2 to 3. This applies to suspended sediments in the water column as well as the pore water environment of superficial sediments.

Sorption to Colloidal and Dissolved Material

Unlike K_{oc}'s of bedded or suspended sediments, the organic carbon normalized partition coefficient, K_{oc}'s, of a given pollutant to natural DOM varies among the waters from different sources.[4,13,14,22,30-34] The affinity of DOM for sorption of organic pollutants has been related to various chemical properties of DOM. In a series of studies,[6,22,35] the aromaticity of DOM in natural water was found to correlate to the observed binding capacities. In other studies, the polarity of the DOM fraction[25,36-38] and molecular size[36,37] were found to correlate to the binding capacities.

Kukkonen et al.[24] have shown differences in binding capacities of benzo(a)pyrene (BaP) and polychlorinated biphenyls (PCB) congeners for three XAD-8 resin DOM fractions of differing polarity from one natural source (Figure 4). Most importantly, it was also shown that the carbon-normalized binding coefficients, K_{oc}'s, of these chemicals in the original water can be calculated as a sum of the relative K_{oc}'s of the three fractions. In addition to the differences in binding among the different fractions, it was shown that BaP is bound to both of the operationally defined hydrophobic fractions (neutral and acidic), while the affinity of PCB was mainly to the hydrophobic neutral fraction.[24] It is also possible to explain earlier observed differences in K_{oc} values for compounds with similar K_{ow} values (PAHs vs. PCBs), through differences in binding to the different DOM fractions, and the resulting apparent solubility enhancement.

DOM can also be fractionated by apparent molecular weight (for example, by ultrafiltration) in order to investigate the binding capacities of the resulting fractions. It has been shown that in oxic waters, the larger molecules and colloids have the highest binding capacities.[25,36-38] This phenomenon is illustrated in Table 2, which shows that when removing the greater than 100,000 Dalton (Da) fraction from water (about 10% of total DOC), about 45% of the binding capacity of DOM is removed.[38] However, Hunchak-Kariouk[25] has recently shown that in anoxic sediments this may not hold true. In a study of the binding of 2,2',4,4'-tetrachlorobiphenyl with DOC under anoxic conditions (i.e., Eh < −350 mV), the smaller molecular weight fraction (<5000 Da) exhibited the highest binding capacity. However, when the anoxic water was aerated to an oxic condition (i.e., Eh > +200 mV), the higher molecular weight fraction (>5000 Da) once again had the highest binding capacity. In a subsequent study, Hunchak-Kariouk and Suffet[39] observed that the $K_{s,doc}$ (sediment water-DOC partition coefficient) is a more sensitive function of DOC

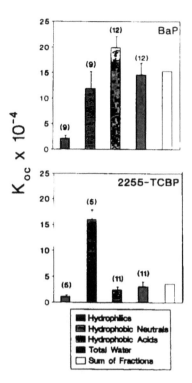

Figure 4. The K_{oc} of benzo(a)pyrene (BaP) and 2,2′,5,5′ tetrachlorobiphenyl (2255-TCBP) to DOM fractions and total unfractionated water. Mean ± SD and (n) are shown on charts. (From Kukkonen, J., Mc-Carthy, J. F., and Oikari, A., *Arch. Environ. Contam. Toxicol.,* 19, 551–557, 1990. With permission.)

Table 2. Partition Coefficients (K_{oc}; Mean ± SD, n = 6) of Benzo(a)pyrene to Different Molecular Size Fractions of DOM Obtained by Ultrafiltration

MW Fraction	DOC (mg/l)	% of total DOC	K_{oc}
Filtered (0.45 µm) unfractionated water	49.9	100.0	$1.34 \pm 1.5 \times 10^5$
<100,000 Dalton	44.6	89.4	$7.2 \pm 2.0 \times 10^4$
< 30,000 Dalton	34.9	69.9	$3.1 \pm 1.0 \times 10^4$
< 10,000 Dalton	24.8	47.5	$0.5 \pm 0.2 \times 10^4$
< 5,000 Dalton	17.2	34.5	$< 10^3$
< 1,000 Dalton	5.0	10.0	$< 10^3$
< 500 Dalton	1.7	3.4	not analyzed

Note: Water source: Hyde County, NC, U.S.
From Kukkonen, J., *Effects of Dissolved Organic Material in Fresh Waters on the Binding and Bioavailability of Organic Pollutants,* University of Joensuu, Finland, 1991. With permission.

concentration in anoxic as compared to oxic conditions (Figure 5). Therefore, to understand the actual environment under study, samples must be maintained oxic or anoxic after sampling and throughout analysis.

If the partitioning of the chemical to lipid is the same as to the solid or dissolved organic carbon, at equilibrium the organic carbon normalized contaminant concentration (micrograms per kilogram of carbon) will be approximately the same as the lipid normalized contaminant concentration (micrograms per kilogram of lipid) in the organism. Indeed, Bierman[19] has shown that for a range of organisms (both benthic and pelagic) and log K_{ow} 4 to 8, partition coefficients for organism lipid and sediment organic carbon often agree within an order of magnitude. Because a direct measure of the chemical activity in the water is not required, this approach would be a

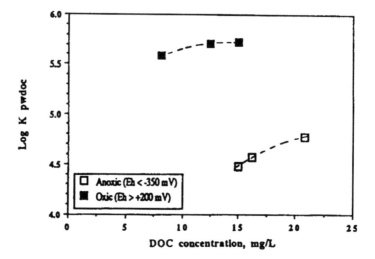

Figure 5. Influence of DOC concentration on the calculated K_{pwdoc} (i.e., $K_{a,doc}$ herein) values for oxic and anoxic pore waters extracted from the same sediment sample. (From Hunchak-Kariouk, K. and Suffet, I. H., *Finn. Humus News*, 3, 157–170, 1991. With permission.)

Table 3.   Analytical Methods Used in Determining the Association of Organic Chemicals to Dissolved Organic Matter

Physical Separation Methods	Solubility Enhancement Measurements
• Liquid–liquid extraction	• Apparent water solubility
• Equilibrium dialysis	• Fluorescence quenching
• Reverse-phase chromatography	• Headspace-free equilibration
• Ultrafiltration	
• Molecular size exclusion chromatography	

very useful and practical tool for predicting the bioavailability of nonionic organic chemicals in sediments. For sediments, the equilibrium assumption appears to apply for a wide variety of chemicals and sediment types.[29] However, this assumption does not seem to apply as well to DOM in sediment pore water[13,17] and in the water column.[4,14,22,30-34]

METHODS OF DETERMINING AQUEOUS CONCENTRATION IN THE PRESENCE OF DISSOLVED ORGANIC MATTER

When an appreciable amount of DOM is present in the water, estimation of C_{aq} for very hydrophobic chemicals (log $K_{ow} > 5$) is problematic. Several approaches have been introduced to determine the freely dissolved concentration or activity of such chemicals in the aqueous phase (Table 3). All of the methods described to date have limitations to their universal application, especially for estimations or calculations in natural environments. The most desirable technique is one that avoids the operational definitions of the phase or fraction and measures the aqueous activity directly. A summary of these methods is provided below, along with their major limitations:

- In liquid-liquid extraction procedures,[40-42] the freely dissolved pollutant is extracted from equilibrated solution with solvent. The extraction procedure, however, may be so efficient that weak pollutant-DOM interactions are broken and the amount of freely dissolved pollutants is overestimated.
- In equilibrium dialysis,[30] the DOM sample is closed within a dialysis bag and immersed in pollutant containing DOM-free water. The technique is limited to pollutants that can pass through the dialysis

to the organic carbon content of the sediment layer (for neutral nonpolar compounds, specifically). Therefore, the denominator of Equation 1 can be approximately expressed as $(1 + K_{oc}f_{oc})$, where K_{oc} is the organic carbon normalized partition coefficient, and f_{oc} is the mass of organic carbon in the sediment per water volume.[60] For the more hydrophobic compounds, similar calculations are valid for the solid matrix, however, in these systems diffusional processes generally limit attainment of equilibrium. This often results in layered contamination, where relatively clean recent and older sediments are separated by a layer of contaminated sediments. In boreal peat bogs, this layering effect is often evident because of the low turnover or scour potential of the system. Essentially, instead of two compartments (water and sediments), slow contaminant fluxes between several sediment compartments may limit transport, and hence, bioavailability. Similar conditions may often exist in the water column, where transport is limited by mixing barriers such as a thermocline or a salt wedge. While lake turnover events disrupt this effect in the water column, this effect is disrupted at the water-sediment interface by natural changes in hydraulic scour (due to sedimentation itself, for example), storm events, dredging, and bioturbation.

Many environmentally significant compounds display weak association to natural materials. These compounds may be small nonhydrogen-bonding compounds such as benzene, methylene chloride, or mono- or dichloroethanes; hydrogen-bonding polar compounds such as nitrobenzene or bis(2-chloroethyl)ether; or ionizable compounds such as phenols and cresols, or even the phenoxyacetic acid herbicides which are essentially completely ionized at neutral pH (pK_a of ~3 to 4). While sorption of these compounds may be very significant in bedded sediments or soils present in agricultural runoff, a quantitative estimation of their environmental distribution is somewhat problematic. This may become even more significant as more polar compounds replace hydrophobic compounds as pesticides and for other uses, since it is their discharge that requires regulation. At present, very little is known regarding their mass transfer rates in the environment.

Physiological and behavioral characteristics of biota may contribute to the nonequilibrium distribution of chemical between the biota and the other phases. Organism growth and metabolic transformations may prevent the chemical concentration in the organism from approaching equilibrium with the surrounding medium (water and/or sediment). Tropic transfer may cause the chemical activity in the organism to exceed that in the water (bioconcentration).[2] Benthic organisms may have widely different levels of exposure because of selectivity in the size and nature of the particles they ingest.[61]

In summary, future studies are needed to define mass transfer from the solid phase to the pore water, and mass transport from the pore water to the overlying water column. Moreover, studies that test hypotheses of processes controlling mass transport at specific sites should be conducted in order to have a better understanding of the relative importance of specific physical, chemical, and biological mechanisms, and to develop better predictive tools (i.e., models).

CONCLUSIONS

1. The most appropriate measure of the bioavailable fraction of an organic chemical is the aqueous activity. This applies to the water column as well as the pore water environment of superficial sediments.

2. Organic carbon normalized partitioning is presently the best general approach to estimate the short-term interactions between the sediments and nonpolar organic pollutants with log K_{ow} greater than 2.

3. Partitioning to dissolved organic matter (DOM) generally correlates poorly with the organic carbon content of the DOM, as variations in K_{oc} for DOM exceed a factor of 10. These constants are conditional and site specific, as local water quality may determine the character, and hence, the binding properties of the DOM. For optimization of the analysis system, maintenance of water quality is suggested (e.g., oxic/anoxic conditions, pH, ionic strength, etc.).

4. Improved methodologies are required for characterizing partition coefficients, particularly among aqueous phase components. At present, it is not clear what to measure in order to predict K_{oc} for the continuum of material that is DOM. Characterization of the operationally defined dissolved organic carbon (DOC) phase—those organic materials that pass through a 0.45-μm filter—should be performed when studying bioavailability, particularly when the log K_{ow} of the compound of interest is greater than 5. Suggestions for characterization include size fractionation, lipophilicity, and aromaticity.

5. Observed variations in partition coefficients with solids concentration, the so-called "solids effect", can be explained by a combination of kinetic effects and artifacts in the measurement of the aqueous activity.

6. A better understanding of mass transfer processes between the soluble phases ($C_{w,aq}$ or $C_{s,aq}$) and the solid phases ($C_{w,l}$ and $C_{s,l}$), is needed for better predictions of bioavailability. Bioconcentration and bioaccumulation of organic contaminants is a dynamic process.

7. A better understanding of the relative importance of various uptake routes for benthic organisms is necessary. The exact role that sorption to organic or inorganic materials plays in controling bioavailability via ingestion is poorly understood and requires future study.

8. A better understanding of factors controling the association of organic acids and bases to soluble and solid organic phases is necessary.

REFERENCES

1. Thomann, R.V. and Connolly, J.P., Model of PCB in the Lake Michigan lake trout food chain, *Environ. Sci. Technol.*, 18, 65–71, 1984.
2. Connolly, J.P. and Pedersen, C.J., A thermodynamic-based evaluation of organic chemical accumulation in aquatic organisms, *Environ. Sci. Technol.*, 22, 99–103, 1988.
3. Landrum, P.F., Hayton, W.L., Lee, H., II, McCarty, L., Mackay, D., and McKim, J., this volume, Session 6, Chapter 3, 1994.
4. Henry, L., Friant, S.L., and Suffet, I.H., Sorption of chlorinated hydrocarbons in the water column by dissolved and particulate material, in Adv. Chem. Ser., Vol. 219, Suffet, I.H. and McCarthy, P., Eds., *Aquatic Humic Substances: Influence on Fate and Treatment of Pollutants*, American Chemical Society, Washington, D.C., 1989, 159–171.
5. Landrum, P.F., Nihart, S.R., Eadie, B.J., and Gardner, W.S., Reverse-phase separation method for determining pollutant binding to Aldrich humic acid and dissolved organic carbon of natural waters, *Environ. Sci. Technol.*, 18, 187–192, 1984.
6. McCarthy, J.F., Roberson, L.E., and Burris, L.W., Association of benzo(a)pyrene with dissolved organic matter: prediction of K_{dom} from structural and chemical properties of the organic matter, *Chemosphere*, 19, 1911–1920, 1989.
7. Landrum, P.F., Reinhold, M.D., Nihart, S.R., and Eadie, B.J., Predicting the bioavailability of organic xenobiotics to *Pontoporeia hoyi* in the presence of humic and fulvic materials and natural dissolved organic matter, *Environ. Toxicol. Chem.*, 4, 459–467, 1985.
8. McCarthy, J.F., Jimenez, B.D., and Barbee, T., Effects of dissolved humic material on accumulation of polycyclic aromatic hydrocarbons: structure-activity relationships, *Aquat. Toxicol.*, 7, 15–24, 1985.
9. Leversee, G.L., Landrum, P.F., Geisy, J.P., and Fannin, T., Humic acid reduced bioaccumulation of some polycyclic hydrocarbons, *Can. J. Fish. Aquat. Soc.*, 40(Suppl.2), 63–69, 1983.
10. Weber, W.J., McGinley, P.M., and Katz, L.E., A distributed reactivity model for sorption by soils and sediments. 1. Conceptual basis and equilibrium assessments, *Environ. Sci. Technol.*, 26, 1955–1962, 1992.
11. Karickhoff, S.W., Organic pollutant sorption in aquatic systems, *J. Hydraul. Eng.*, 110, 707–735, 1984.
12. Batley, G.E. and Giles, M.S., Solvent displacement of sediment interstitial water before trace metal analysis, *Water Res.*, 13, 879–886, 1979.
13. Caron, G. and Suffet, I.H., Binding of nonpolar pollutants to dissolved organic carbon: environmental fate modeling, in Adv. Chem. Ser., Vol. 219, Suffet, I.H. and MacCarthy, P., Eds., *Aquatic Humic Substances: Influence on Fate and Treatment of Pollutants*, American Chemical Society, Washington, D.C., 1989, 117–130.

14. Caron, G., Suffet, I.H., and Belton, T., Effects of dissolved organic carbon on the environmental distribution of nonpolar organic chemicals, *Chemosphere*, 14, 993–1000, 1985.

15. Gschwend, P.M. and Wu, S., On the constancy of sediment-water partition coefficients of hydrophobic organic pollutants, *Environ. Sci. Technol.*, 19, 90–96, 1985.

16. Voice, T.C. and Weber, W.J., Sorbent concentration effects in liquid/solid partitioning, *Environ. Sci. Technol.*, 19, 789–796, 1985.

17. Baker, J.E., Capel, P.D., and Eisenreich, S.J., Influences of colloids on the sediment water partition coefficients of polychlorinated-biphenyl congeners in natural waters, *Environ. Sci. Technol.*, 20, 1136–1143, 1986.

18. Thurman, E.M., *Organic Geochemistry of Natural Waters*, Martinius Nijhoff, Dordecht, 1985.

19. Bierman, V.J., Equilibrium partitioning and biomagnification of organic chemicals in benthic animals, *Environ. Sci. Technol.*, 24, 1407–1412, 1990.

20. Spacie, A., this volume, Session 4, Chapter 1, 1994.

21. DeWitt, T.H., Ozretich, R.J., Schwarts, R.C., Lamberson, J.O., Scholts, D.W., Disworth, G.R., Jones, J.K.P., Hoselton L., and Smith, L.M., The influence of organic matter quality on the toxicity and partitioning of sediment-associated fluoranthene, *Environ. Toxicol. Chem.*, 11, 197–208, 1992.

22. Kukkonen, J. and Oikari, A., Bioavailability of organic pollutants in boreal waters with varying levels of dissolved organic material, *Water Res.*, 25, 455–463, 1991.

23. Servos, M.R. and Muir, D.C.G., Effect of suspended sediment concentration of the sediment to water partition coefficient for 1,3,6,8-tetrachlorodibenzo-*p*-dioxin, *Environ. Sci. Technol.*, 23, 1302–1306, 1989.

24. Kukkonen J., McCarthy, J.F., and Oikari, A., Effects of XAD-8 fractions of dissolved organic carbon on the sorption and bioavailability of organic micropollutants, *Arch. Environ. Contam. Toxicol.*, 19, 551–557, 1990.

25. Hunchak-Kariouk, K., Influence of Anoxic Pore Water Dissolved Organic Matter on the Fate and Transport of Hydrophobic Organic Pollutants, Ph.D. thesis, Drexel University, Philadelphia, 1992.

26. Garbarini, D.R. and Lion, L.W.. Influence of the nature of soil organics on the sorption of toluene and trichloroethylene, *Environ. Sci. Technol.*, 20, 1263–1269, 1986.

27. Grathwohl, P., Influence of organic matter from soils and sediments from various origins on the sorption of some chlorinated aliphatic hydrocarbons: implications on K_{oc} correlations, *Environ. Sci. Technol.*, 24, 1687–1692, 1990.

28. Chessels, M., Hawker, D.W., and Connell, D.W., Influence of solubility in lipid on bioconcentration of hydrophobic compounds, *Ecotoxicol. Environ. Saf.*, 23, 260–273, 1992.

29. DiToro, D.M., Zarba, C.S., Hansen, D.J., Berry, W.J., Swartz, R.C., Cowan, C.E., Pavlou, S.P., Allen, H.E., Thomas N.A., and Paquin, P.R., Technical basis for establishing sediment water quality criteria for nonionic organic chemicals using equilibrium partitioning, *Environ. Toxicol. Chem.*, 10, 1541–1583, 1991.

30. Carter, C.W. and Suffet, I.H., Binding of DDT to dissolved humic material, *Environ. Sci. Technol.*, 16, 735–740, 1982.

31. Carter, C.W. and Suffet, I.H., Interaction between dissolved humic and fulvic acids and pollutants in the aquatic environment, in Symp. Ser., Vol. 225, Swann, R.L. and Eschenroeder, A., Eds., *Fate of Chemicals in the Environment*, American Chemical Society, Washington, D.C., 1983, 217–229.

32. Whitehouse, B., The effects of dissolved organic matter on the aqueous partitioning of polynuclear aromatic hydrocarbons, *Estuarine Coastal Shelf Sci.*, 20, 393–402, 1985.

33. Morehead, N.R., Eadie, B.J., Lake, B., Landrum, P.F., and Berner, D., The sorption of PAH onto dissolved organic matter in Lake Michigan waters, *Chemosphere*, 15, 403–412, 1986.

34. Kukkonen, J., Oikari, A., Johnsen, S., and Gjessing, E., Effects of humus concentration of benzo(a)pyrene accumulation from water to *Daphnia magna*. Comparison of natural waters and standard preparations, *Sci. Total Environ.*, 79, 197–207, 1989.

35. Gauthier, T.D., Shane, E.C., Guerin, W.F., Seitz W.R., and Grant, C.L., Fluorescence quenching method for determining equilibrium constants for polycyclic aromatic hydrocarbons binding to dissolved humic materials, *Environ. Sci. Technol.*, 20, 1162–1166, 1986.

36. Saint-Fort, R. and Visser, S.A., Study of the interactions between atrazine, diazinon, and lindane with humic acids of various molecular weights, *J. Environ. Sci. Health*, A23, 613–624, 1988.

37. Jota, M.A.T. and Hassett, J.P., Effects of environmental variables on binding of a PCB congener by dissolved humic substances, *Environ. Toxicol. Chem.*, 10, 483–491, 1991.

38. Kukkonen, J., *Effects of Dissolved Organic Material in Fresh Waters on the Binding and Bioavailability of Organic Pollutants*, University of Joensuu, Finland, 1991, 39.

39. Hunchak-Kariouk, K. and Suffet, I.H., Binding by hydrophobic organic pollutants to humic substances. *Finn. Humus News*, 3, 157–170, 1991.

40. Gjessing, E.T. and Bergling, L., Adsorption of PAH's to aquatic humus, *Arch. Hydrobiol.*, 92, 24–30, 1981.

41. Johnsen, S., Interactions between polycyclic aromatic hydrocarbons and natural aquatic humic substances, *Sci. Total Environ.*, 67, 269–278, 1987.

42. Hassett, J.P. and Milicic, E., Determination of equilibrium and rate constants for binding of polychlorinated biphenyl congener by dissolved humic substances, *Environ. Sci. Technol.*, 19, 638–643, 1985.

43. Means, J.C. and Wijayaratne, R., Role of natural colloids in the transport of hydrophobic pollutants, *Science*, 215, 968–970, 1982.

44. Chiou, C.T., Kile, D.E., Brinton, T.I., Malcolme, R.L., Leenheer, J.A., and MacCarthy, P., A comparison of water solubility enhancements of organic solutes by aquatic humic materials and commercial humic acids, *Environ. Sci. Technol.*, 21, 1231–1234, 1987.

45. Yin, C. and Hasset, J.P., Gas partitioning approach for laboratory and field studies of mirex fugacity in water, *Environ. Sci. Technol.*, 20, 1213–1217, 1986.

46. Maguire, R.J. and Tkacz, R.J., Potential underestimation of chlorinated hydrocarbon concentrations in freshwater, *Chemosphere*, 19, 1277–1287, 1989.

47. Suffet, I.H. and MacCarthy, P., Introduction, Adv. Chem. Ser., Vol. 219, *Aquatic Humic Substances: Influence on Fate and Treatment of Pollutants*, American Chemical Society, Washington, D.C., 1989.

48. Aiken, G.R., McKnight, D.M., Wershaw, R.L., and MacCarthy, P., in *Humic Substances in Soil, Sediment, and Water: Geochemistry, Isolation and Characterization;* Aiken, G.R., McKnight, D.M., Wershaw, R.L., and MacCarthy, P., Eds., Wiley-Interscience, New York, 1985, 1–9.

49. Malcolm, R.L., in *Humic Substances in Soil, Sediment, and Water: Geochemistry, Isolation and Characteristics,* Aiken, G.R., McKnight, D.M., Wershaw, R.L., and MacCarthy, P., Eds., Wiley-Interscience, New York, 1985, 191–209.

50. O'Connor, D.J. and Connolly, J.P., The effect of concentration of adsorbing solids on the partition coefficient, *Water Res.*, 14, 1517–1523, 1980.

51. Voice, T.C., Rice, C.P., and Weber, W.J., Effect of solids concentration on the sorptive partitioning of hydrophobic pollutants in aquatic systems, *Environ. Sci. Technol.*, 17, 513–518, 1983.

52. DiToro, D.M., A particle interaction model of reversible organic chemical sorption, *Chemosphere*, 14, 1503–1538, 1985.

53. Mackay, D. and Powers, B., Sorption of hydrophobic chemicals from water. A hypothesis for the mechanism of particle concentration effect, *Chemosphere*, 16, 745–757, 1987.

54. Narine, D.R. and Guy, R.D., Binding of diquat and paraquat to humic acid in aquatic environments, *Soil Sci.*, 133, 356–363, 1982.

55. Jafvert, C. T., Sorption of organic acid compounds to sediments. Initial model development, *Environ. Toxicol. Chem.*, 9, 1259–1268, 1990.

56. Jafvert, C.T., Westall, J.C., Grieder, E., and Schwarzenbach, R.P., Distribution of hydrophobic ionogenic organic compounds between octanol and water: organic acids, *Environ. Sci. Technol.*, 24, 1795–1803, 1990.

57. McKim, J., this volume, Session 6, Chapter 2, 1993.

58. Jafvert, C.T. and Weber, E.J., Sorption of Ionizable Organic Compounds to Sediments and Soils, Off. Res. Dev., EPA/600/S3–91/017, U.S. Environmental Protection Agency, Washington, D.C., 1991.

59. Crosby, D., this volume, Session 5, Chapter 1, 1994.

60. Karickhoff, S.W., Semi-empirical estimation of sorption of hydrophobic pollutants in natural sediments and soils, *Chemosphere*, 10, 833–846, 1981.

61. Lee H., II, A clam's eye view of the bioavailability of sediment-associated pollutants, in *Organic Substances and Sediments in Water,* Vol. III, Baker, R., Ed., Lewis Publishers, Chelsea, MI, 1991, 73–93.

62. Stumm, W. and Morgan, J.J., *Aquatic Chemistry,* Wiley-Interscience, New York, 1981.

SESSION 5

DYNAMIC ENVIRONMENTAL FACTORS

Chapter 1

Photochemical Aspects of Bioavailability

Donald G. Crosby

INTRODUCTION

Imagine that we are looking down through shallow lake water or seawater—we can see the sand ripples and shell fragments on the bottom, so light obviously is penetrating. We are looking through the surface organic microlayer, through the water column, and through the thin haze of suspended particles to the sediment surface. UV energy will be available to interact with chemicals associated with any of these environmental phases, so the possibilities for photodegradation are extensive.

The connection between photochemistry and bioavailability may not be obvious at first glance. However, presuming that any force which alters the chemical structure, physical properties, or concentration of a substance may affect its bioavailability, accepting that sunlight is ubiquitous, and perceiving the growing body of literature on environmental photochemistry allows one to recognize the relationship, at least for aquatic environments.

The ultraviolet (UV) component of sunlight (290 to 400 nm) is the source of photochemical energy for most reactions important to bioavailability. Despite a common misconception, radiation in this range readily penetrates clear seawater and fresh water to a depth of many meters (Figure 1),[1] so we are dealing with more than just a surface phenomenon. About half of the energy comes from open sky rather than direct sunlight, although intensity is governed primarily by solar angle (time of day and season), the resulting surface reflection, and the shading and dispersion due to suspended particles.[2] At noon on a cloudless summer day, surface water can be shown to receive as much as 1 kW/m^2 of sunlight energy (2 mol of photons at 300 to 500 nm).

Natural waters are rich in chemical reagents: oxygen, halides, nitrate and nitrite, transition metal ions, dissolved organic carbon (DOC), and others. Hydroxide or hydronium ions may dominate, depending upon the pH. These reagents are free to attack light-energized solutes, and so most photodegradation reactions involve oxidation (by reactive oxygen species), reduction (by metals or DOC), or nucleophilic displacement by substituents.[3] A particular substrate molecule may undergo any or all of these reactions, as well as others, and the resulting photoproducts, themselves, often are further degraded into smaller organic and inorganic fragments. Chemical microenvironments dictate the relative importance of particular reaction types; oxidation predominates in well-aerated surface water, reduction in organic-rich anaerobic water, and hydrolysis under alkaline conditions. Wastewater often contains these and other reagents at elevated concentrations.

UV energy must be absorbed in order for chemical change to occur. When the substrate of interest is the absorbing species, the resulting reaction is termed "direct" photodegradation; when absorption is by some other species, followed by a transfer of energy to the substrate, the term

1-56670-086-8/94/$0.00+$.50
© 1994 by CRC Press, Inc.

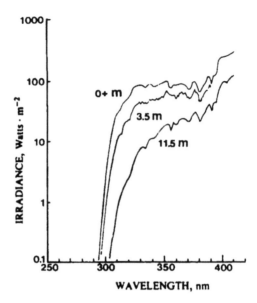

Figure 1. Penetration of ultraviolet radiation into seawater. (Adapted from Baker K. S. and Smith, R. C., *The Role of Solar Ultraviolet Radiation in Marine Ecosystems*, Plenum Press, New York, 1982.)

is "sensitized" photodegradation; and when UV absorption results in generation of a reagent which attacks the substrate, the term is "indirect" photodegradation. Energy absorption (A) is described by the Lambert-Beer expression

$$A = \varepsilon Cl \tag{1}$$

where C is molar concentration, l is the length of the light path (depth), and the proportionality constant ε (molar absorptivity or extinction coefficient) is a measure of the electronic unsaturation of the solute. Although both depth and concentration obviously are important, the *rate* of a resulting photochemical reaction, as measured by substrate half-life ($t^{1/2}$), is strongly dependent upon sunlight radiation intensity (I_λ), at the wavelengths absorbed by the chemical:

$$t_{1/2} = \frac{4.17 \times 10^{20}}{2.303\phi\Sigma\varepsilon_\lambda I_\lambda}$$

where ϕ represents the efficiency of the photochemical process (the quantum yield).

CLEAR WATER

Relatively clean, clear water is the medium in which most aquatic photodegradation reactions are known to take place. The common wood preservative and disinfectant, pentachlorophenol (PCP), provides an example.[4] PCP absorbs UV energy from sunlight (λ_{max}320 nm). As shown in Figure 2 (Reaction 1), its initial photodegradation reaction is with hydroxide ion (from water) to replace each ring Cl by OH. The resulting diols can then be oxidized by atmospheric oxygen to quinones and acids (Reaction 2), or the PCP can be reduced by abstraction of hydrogens from other molecules (Reaction 3) to provide tetra- and trichlorophenols which also undergo reaction with hydroxide. The result is eventual formation of inorganic species (CO_2, H_2O, and chloride ions) and small organic fragments.

At environmental concentrations, outdoor loss of PCP is largely complete within a few hours (Figure 3), although it is significantly slower in seawater due to competition of chloride (0.548 M)

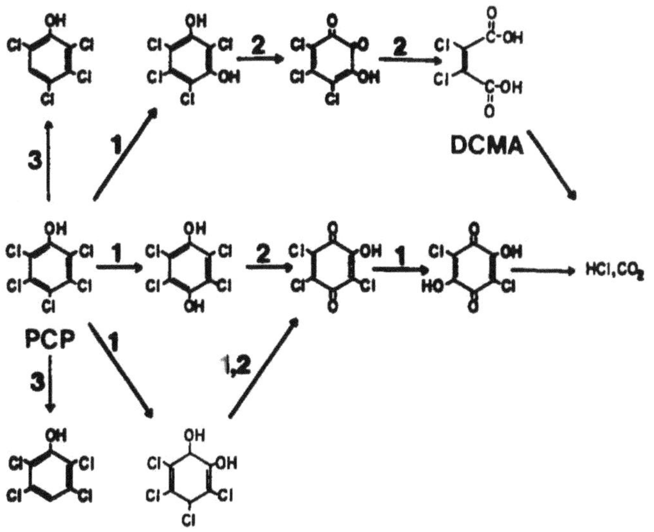

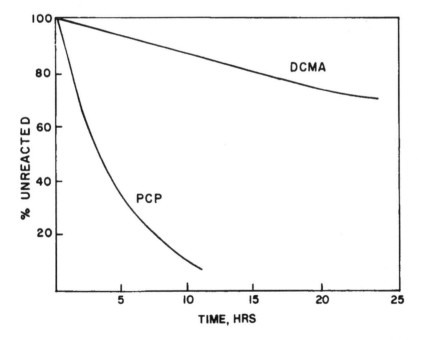

Figure 2. Photodegradation of pentachlorophenol in water. Initial concentration, 100 mg/l, initial pH, 7.3. (Adapted from Wong, A. S. and Crosby, D. G., *J. Agric. Food Chem.*, 29, 125–130, 1981.)

Figure 3. Rates of photodegradation of pentachlorophenol (PCP) and 2,3-dichloromaleic acid (DCMA) in water. (Adapted from Wong, A. S. and Crosby, D. G., *J. Agric. Food Chem.*, 29, 125–130, 1981.)

with hydroxide (10^{-6} M) for the available chlorines of PCP.[5] This relatively rapid lowering of PCP concentrations significantly affects its bioavailability, but raises another issue: the major photodegradation product, 2,3-dichloromaleic acid (DCMA), is only slowly photodegraded (Figure 3) and so tends to build up and, itself, may become bioavailable. We may only have exchanged one problem for another.

The rice herbicide, molinate, provides a different example. In field water, it is photodegraded to a variety of oxidation and hydrolysis products (Figure 4).[6] However, as it does not absorb UV energy from sunlight, its solutions in distilled water are stable toward UV irradiation in the

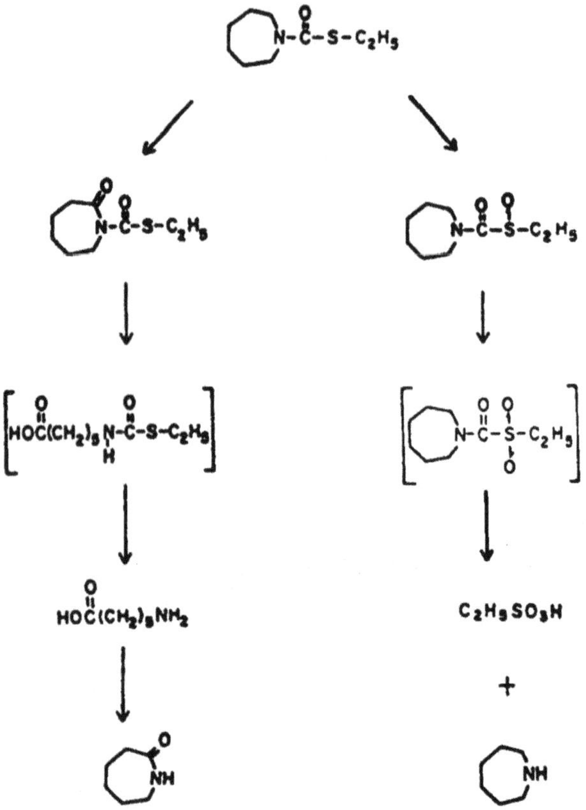

Figure 4. Photodegradation of molinate (S-ethylhexahydroazepine-1-carbothioate) in field water by two simultaneous pathways. (Adapted from Soderquist, C. J., Bowers, J. B., and Crosby, D. G., *J. Agric. Food Chem.*, 25, 940–946, 1977.)

laboratory. Indirect photodegradation provides the explanation. The action of UV radiation on a number of common constituents of natural water, including H_2O_2, NO_2^-, O_3, and humic substances leads to generation of hydroxyl radicals (HO·), a powerful oxidizing species. Hydroxyl then reacts with the pesticide as well as with many other natural and man-made solutes; the reaction products of the rice herbicide, thiobencarb (Figure 5), are typical of those from reactions of aliphatic and aromatic systems with hydroxyl (ring hydroxylation, N-dealkylation, C-oxidation, S-oxidation).[7] Indirect photooxidation via hydroxyl is probably the most significant of all environmental photoreactions in that the rates are high (k ≈ $10^{10}M^{-1}S^1$), reactions are broadly general, and hydroxyl sources are ubiquitous.

Another source of hydroxyl radical is the so-called Fenton reaction:

$$Fe^{2+} + H_2O_2 \rightarrow Fe^{3+} + HO· + HO^-$$

in which ferrous ion (and other divalent transition metal cations) reacts in aqueous solution with hydrogen peroxide; both iron and H_2O_2 are common constituents of natural waters,[8,9] but the environmental significance of the Fenton reaction has not been established. Similarly, the photoreaction of ferric complex ions such as Fe^{3+} (OH^-) to generate ferrous ion and hydroxyl radical[10] has not been investigated under environmental conditions. However, such reactions alternatively can be viewed as potential contributors to the metal redox chemistry of natural waters, much of which has only recently been clarified. For example, dissolved cupric complexes containing a variety of organic ligands are known to undergo photodegradation, with consequent reduction of Cu^{2+} to Cu^+ and oxidative disruption of the ligand.[11] Complex ions of copper (e.g., $CuCl_4^{2-}$),

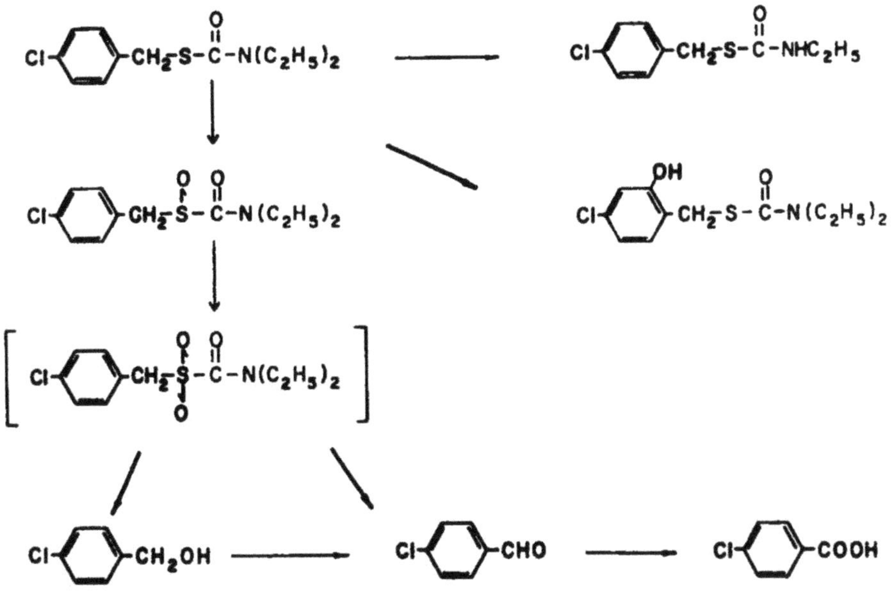

Figure 5. Photodegradation of thiobencarb (S-4-chlorobenzyl-N,N-diethylthiolcarbamate) in field water.

which predominate in seawater, likewise undergo photolysis to the cuprous state,[12] Cu^+ being rapidly reoxidized by molecular oxygen.

Photosensitized reactions also occur in the dissolved state. For example, the natural water constituent, riboflavin, strongly absorbs energy from sunlight and then passes it on to other molecules such as dissolved oxygen.[13] Among the most important aquatic photosensitizers are the soluble humic and fulvic acids, the high molecular weight polymers which cause dark coloration in both fresh and sea water. While Khan and Schnitzer[14] reported that such humic substances often inhibit photodegradation of other molecules by competitive absorption of UV energy, Zepp and co-workers[15,16] found that they also photosensitized reactions such as the *cis-trans* isomerism of 1,3-pentadiene and generated singlet oxygen which could then oxidize organic substrates such as aniline.

SUSPENDED PARTICLES

Suspended particulate matter often represents an important feature of the water column. While the most usual concentrations are in the range of a few milligrams per liter in surface waters to several hundred milligrams per liter in eutrophic lakes, a turbid river or estuary might contain up to 5×10^4 mg/l.[17] The particles in the photic zone are as much as 90% organic (bacteria, algae, and zooplankton, living and dead); the remaining mineral portion, primarily carried by rivers, is largely aluminosilicate, while oxides and carbonates generally are precipitated *in situ.* Colloidal oxides may be only a nanometer in diameter, but the more usual size range is 1 to 100 μm.[17]

Considerable attention has been given to the inorganic photoreactions of immediate concern to the bioavailability of metals. UV energy can cause the dissolution of otherwise insoluble Fe and Mn oxides,[18,19] sulfides,[20,21] and copper carbonate,[11] among other minerals. Hydrogen peroxide or DOM often may be required as reductants.[19] The subject has been reviewed by Stone and Morgan[22] and, recently, by Sulzberger.[23] A particularly pertinent example comes from the laboratory irradiation of suspended mercuric sulfide, HgS—one of the least water soluble and least bioavailable substances known—to provide soluble mercuric ion and sulfur.[20] Moreover, if the reaction mixture also contains acetate or acetic acid, the photoproducts are methylmercuric acetate and dimethylmercury, which then are bioavailable.

This reaction (and most other solubilizations) occurs because these minerals are semiconductors; that is, absorption of UV energy from sunlight promotes a valence electron into a conduction band, with subsequent movement of the electron to the particle surface where it reduces adsorbed oxygen to superoxide, while the remaining "hole" can oxidize hydroxide ion to hydroxyl radical:

$$ZnO \xrightarrow{380nm} e^- + ZnO^+$$
$$ZnO^+ + HO^- \longrightarrow HO\cdot + ZnO$$
$$e^- + O_2 \longrightarrow \cdot O_2^-$$

As an additional "benefit", these oxidizing species may then react with dissolved organic and inorganic compounds. The solubilization of insoluble minerals to release metal ions, the possible methylation to lipid-soluble form, and the cogeneration of powerful oxidants all indicate that sunlight irradiation of suspended inorganic particles can drastically alter bioavailability.

Organic solutes adsorbed to mineral surfaces also undergo photodegradation. For example, polycyclic aromatic hydrocarbons (PAH) became oxidized when irradiated on the surface of kaolinite or calcium carbonate suspensions,[24] and PCBs were dechlorinated on the surface of the semiconductor, titanium dioxide.[25] Although too dense a suspension simply resulted in exclusion of UV radiation and no reaction,[26] lighter suspensions allowed the photoreduction of both DDE and 3,4-dichloroaniline.[27] There appears to be little information about photodegradation reactions on organic particles, although Zepp and Schlotzhauer[28] reported that the indirect photodegradation of substituted anilines and phosphorothioates on the surface of suspended algae was almost independent of whether the cells were living or dead. Actually, even most suspended mineral surfaces appear to be coated with a film of natural organic matter,[29] so that most photochemical reactions on particles may be occurring with chemicals "dissolved" in an organic matrix.

BOTTOM SEDIMENTS

There are obviously several major difficulties with the investigation of photodegradation reactions on bottom sediments:

1. Such sediments are quite heterogeneous, depending upon the water composition, location, and history.
2. Sediment irradiation occurs only in relatively shallow water, with its constantly shifting surface.
3. There is difficulty in distinguishing the processes occurring on sediment from those taking place on adjacent suspended particles and in the water column.

Not surprisingly, little definitive work has been reported on sediment photochemistry.

Considerable photochemical investigation has been focused on terrestrial soil surfaces;[30] there is no doubt that solar irradiation is important to the fate of chemicals there, although photodegradation necessarily is limited to about the top millimeter due to UV screening.[30] Reactions with oxidizing species such as singlet oxygen are especially important.[31] Although precipitated humic substances could be the source of similar reactions on bottom sediments,[16] apparently mineral surfaces are not; Oliver and co-workers did not observe either photochemical oxidation or reduction on natural sediments,[32] possibly because of the influence of equilibration time.[33]

One possible sediment-pollutant interaction which could prove to be important is the spectral shift which can occur in sorbed species. Most existing evidence comes from artificial systems in which UV spectra of adsorbed chemicals are measured on the surface of silica gel suspended in cyclohexane and compared to those in cyclohexane solutions alone.[34,35] For example, the 2,6-dichlorophenol absorption maximum was shifted from 284 to 300 nm and that of p-nitrophenol from 286 to 315 nm, clearly from below the 290 nm sunlight UV cutoff to well above it. On the other hand, the 3,4-dichloroaniline maximum was shifted from 302 to 290 nm and that of *N*-methylaniline from 294 to 276 nm, in exactly the opposite direction. There is some evidence that similar shifts also can take place on natural clay surfaces,[36,37] but further evidence relating to photodegradation on actual sediments is sorely needed.

SURFACE MICROLAYER

Most natural surface waters are coated with a very thin (<1 mm) layer of hydrophobic organic film.[38,39] The upper surface of this layer (<1 μm thick) appears to be primarily lipid (natural hydrocarbons, fatty acids and alcohols, wax esters), below which there exists a hydrophilic polysaccharide-protein zone which is in equilibrium with the water (Figure 6). However, most surface sampling methods scrape or lift off a 50 to 150 μm slice at best, so the microlayers examined are diluted by a large proportion of water. In some locations, natural or anthropogenic petroleum may also contribute to the film, as do inorganic and plastic particles.

Not surprisingly, hydrophobic contaminants such as chlorinated hydrocarbons and PAH are concentrated into the microlayer from both the aqueous and atmospheric sides. Typically, PAH have been found there at concentrations up to 10^6-fold those in the adjacent water, PCBs at a 1000-fold enrichment, and DDT at >2600-fold. Metal derivatives also collect there in both particulate and complexed form: Pb at a 6-fold enrichment, Cu up to 36-fold, and Ni up to 50-fold. However, as the hydrophobic layer is diluted by sampling, actual enrichment may be greater by a factor of as much as 1000. Sn and Hg might be expected to concentrate in the lipid layer as their alkyl derivatives, while Cu might tend to be complexed into the subsurface protein layer. On the other hand, many more-polar substances show little or no enrichment.

Although this vast organic mixture, diurnally irradiated with sunlight and available to the reagents in both water and atmosphere, might seem to offer great potential as a photochemical reaction medium, about all there is to go on is speculation.[40] Several authors, such as Hansen,[41] describe the photodegradation of petroleum surface films which results in oxidation to aliphatic and aromatic carboxylic acids, alcohols, and phenols, but otherwise there appears to be little information. This may be due, in part, to some practical difficulties:

1. The capture cross section (light-path) is very thin, so that reaction rates may be low (typical PAH concentrations still are at best only <0.5 μg/l);
2. Representative microlayers are technically difficult to obtain and fractionate; and
3. The composition of the microlayer varies widely from place to place and time to time.

Still, the microlayer is of great significance to contaminant bioavailability because it represents home to countless bacteria, phytoplankton, and zooplankton; it is the depository for the eggs of both vertebrates and invertebrates and a feeding place which forms the basis of aquatic food chains (Figure 6). The surface microlayer acts as a contaminant reservoir that continuously supplies xenobiotics to a microcosm less than a millimeter deep and up to thousands of square kilometers in area, filled with countless tiny organisms—the neuston. To the (unknown) extent that solar UV radiation destroys or alters microlayer chemical components, bioavailability is directly and immediately affected.

SUMMARY AND CONCLUSIONS

Photodegradation by solar UV radiation takes place in both fresh and salt water, in the surface microlayer, on suspended particles, and almost certainly on the surface of bottom sediments. Water is unexpectedly transparent to UV, and, although molar concentrations of solutes generally are low (10^{-8} to $10^{-10}M$), the light path (depth) can be on the order of meters. On the other hand,

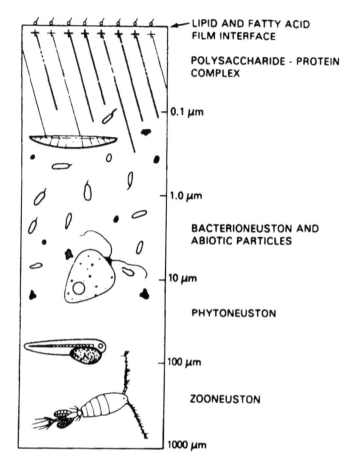

LIPID AND FATTY ACID
FILM INTERFACE

POLYSACCHARIDE - PROTEIN
COMPLEX

0.1 μm

1.0 μm

BACTERIONEUSTON AND
ABIOTIC PARTICLES

10 μm

PHYTONEUSTON

100 μm

ZOONEUSTON

1000 μm

Figure 6. Schematic structure of a typical surface microlayer; note the logarithmic scale. (From Hardy, J. T., *Prog. Oceanogr.*, 11, 307–328, 1983. With permission.)

although concentrations on sediments and in the microlayer can be thousands of times greater than in the adjacent water, the light path is short (on the order of micrometers).

In relation to bioavailability, photodegradation indeed can lower environmental concentrations of both contaminants and natural solutes, sometimes rapidly, and some photoproducts can be both persistent and toxic. Photodegradation also alters lipophilicity and, consequently, bioconcentration; photoproducts generally are more polar, less readily concentrated, and less toxic than the parent. On the other hand, UV radiation can convert unreactive and normally unavailable minerals into soluble and bioavailable form. Further, certain compounds, such as PAH, readily bioavailable but nontoxic to aquatic organisms, are absorbed by them and *then* photodegraded *in vivo* with lethal results.[42]

At present, we know the most about photoreactions in bulk water where oxidation, reduction, and hydrolysis reactions predominate. Xenobiotics and their degradation products are bioavailable via direct uptake through respiration, food, and skin absorption; almost certainly, water-column concentrations are regulated in large part by direct and indirect photolysis in shallow waters. Suspended sediments have received a modicum of photochemical attention and, in artificial systems at least, irradiation results in extensive degradation on many mineral and biological surfaces; dilution, shading, and surface coatings doubtless reduce the efficiency under natural conditions. Photodegradation on shallow-water bottom sediments and in the surface microlayer remains almost unexplored despite its probable significance for xenobiotic bioavailability to benthic organisms and neuston, respectively.

Quite apparently, major research needs include investigation of the degree and types of photodegradation reactions occurring on natural shallow-water sediments and concentration and photodegradation in surface microlayers. The entire area of photodegradation in the marine environment and its significance also need more attention, as does *direct* evidence for the influence of UV radiation on contaminant bioavailability. Any such research must consider the likelihood that, as our protective ozone layer thins, the high-energy UV intensity will increase, inevitably bringing greater penetration of radiation and higher reaction rates to aquatic environments.

REFERENCES

1. Baker, K.S. and Smith, R.C., Spectral irradiance penetration of natural waters, in Calkins, J., Ed., *The Role of Solar Ultraviolet Radiation in Marine Ecosystems*, Plenum Press, New York, 1982, 233–246.
2. Jerlöv, N.G., *Optical Oceanography*, Elsevier, New York, 1968.
3. Crosby, D.G., The photodecomposition of pesticides in water, in Gould, R.F., Ed., *Fate of Organic Pesticides in the Aquatic Environment*, Adv. Chem. Ser., Vol. 111, American Chemical Society, Washington, D.C., 1972, 173–188.
4. Wong, A.S. and Crosby, D.G., Photodecomposition of pentachlorophenol in water, *J. Agric. Food Chem.*, 29, 125–130, 1981.
5. Miille, M.J. and Crosby, D.G., Pentachlorophenol and 3,4-dichloroaniline as models for photochemical reactions in seawater, *Mar. Chem.*, 14, 111–120, 1983.
6. Soderquist, C.J., Bowers, J.B., and Crosby, D.G.,. Dissipation of molinate in a rice field, *J. Agric. Food Chem.*, 25, 940–946, 1977.
7. Draper, W.M. and Crosby, D.G., Hydrogen peroxide and hydroxyl radical. Intermediates in indirect photolysis reactions in water, *J. Agric. Food Chem.*, 29, 699–702, 1981.
8. Cooper, W.J., Zika, R.G., Petasne, R.G., and Plane, J.M.C., Photochemical formation of H_2O_2 in natural waters, exposed to sunlight, *Environ. Sci. Technol.*, 22, 1156–1160, 1988.
9. Draper, W.M. and Crosby, D.G., The photochemical generation of hydrogen peroxide in natural waters, *Arch. Environ. Contam. Toxicol.*, 12, 121–126, 1983.
10. Bates, H.G.C. and Uris, N., Oxidation of aromatic compounds in aqueous solution by free radicals produced by photo-excited electron transfer in iron complexes, *J. Am. Chem. Soc.*, 75, 2754–2759, 1953.
11. Moffett, J.W. and Zika, R.G., Photochemistry of copper complexes in sea water, in Zika, R.G. and Cooper, W.J., Eds., *Photochemistry of Environmental Aquatic Systems*, ACS Symp. Ser. 327, American Chemical Society, Washington, D.C., 1987, 116–130.
12. Ferraudi, G. and Muralidharan, S., Photochemical properties of copper complexes, *Coord. Chem. Rev.*, 36, 45–88, 1981.
13. Mopper, K. and Zika, R.G., Natural photosensitizers in sea water. Riboflavin and its breakdown products, in Zika, R.G. and Cooper, W.J., Eds., *Photochemistry of Environmental Aquatic Systems*, ACS Symp. Ser. 327, American Chemical Society, Washington, D.C., 1987, 174–190.
14. Khan, S.U. and Schnitzer, M., UV irradiation of atrazine in aqueous fulvic acid, *J. Environ. Sci. Health*, B13, 299–310, 1978.
15. Zepp, R.G., Baughman, G.L., and Schlotzhauer, P.F., Comparison of photochemical behavior of various humic substances in water. I. Sunlight induced reactions of aquatic pollutants photosensitized by humic substances, *Chemosphere*, 10, 109–117, 1981.
16. Zepp, R.G., Schlotzhauer, P., and Sink, R.M., Photosensitized transformations involving energy transfer in natural waters. Role of humic substances, *Environ. Sci. Technol.*, 19, 74–81, 1985.
17. Morel, F.M.M., *Principles of Aquatic Chemistry*, John Wiley & Sons, New York, 1983, chap. 8.
18. Waite, T.D. and Morel, F.M.M., Photoreductive dissolution of colloidal iron oxides in natural waters, *Environ. Sci. Technol.*, 18, 860–868, 1984.
19. Sunda, W.G., Huntsman, S.A., and Harvey, G.R., Photoreduction of manganese oxides in seawater and its geochemical and biological implications, *Nature*, 301, 234–236, 1983.
20. Akagi, H., Fujita, Y., and Takabatake, E., Methylmercury: photochemical transformation of mercuric sulfide into methylmercury in aqueous solution, *Photochem. Photobiol.*, 26, 363–370, 1977.
21. Hsieh, Y.H. and Huang, C.P., The dissolution of lead sulfide in dilute aqueous solutions, *J. Colloid Interface Sci.*, 131, 537–549, 1989.

22. Stone, A.T. and Morgan, J.J., Reductive dissolution of metal oxides, in Stumm, W., Ed., *Aquatic Surface Chemistry*, John Wiley & Sons, New York, 1987, 221–254.

23. Sulzberger, B., Photoredox reactions at hydrous metal oxide surfaces. A surface coordination chemistry approach, in Stumm, W., Ed., *Aquatic Chemical Kinetics*, John Wiley & Sons, New York, 1990, 401–429.

24. Andelman, J.B. and Suess, M.J., The photodecomposition of 3,4-benzpyrene sorbed on calcium carbonate, in Faust, S.J. and Hunter, J.V., Eds., *Organic Compounds in Aquatic Environments*, Marcel Dekker, New York, 1971, 439–468.

25. Carey, J.H., Lawrence, J., and Tosine, H.M., Photodechlorination of PCB's in the presence of titanium dioxide in aqueous suspension, *Bull. Environ. Contam. Toxicol.*, 16, 697–701, 1976.

26. Miller G.C. and Zepp, R.G., Effects of suspended sediments on photolysis rates of dissolved pollutants, *Water Res.*, 13, 453–459, 1979.

27. Miller, G.C. and Zepp, R.G., Photoreactivity of aquatic pollutants sorbed on suspended sediments, *Environ. Sci. Technol.*, 13, 860–863, 1979.

28. Zepp, R.G. and Schlotzhauer, P.F., Influence of algae on photolysis rates of chemicals in water, *Environ. Sci. Technol.*, 17, 462–468, 1983.

29. Davis, J.A., Adsorption of natural dissolved organic matter at the oxide-water interface, *Geochim. Cosmochim. Acta*, 46, 2381–2393, 1982.

30. Miller, G.C., Hebert, V.R., and Miller, W.W., Effect of sunlight on organic contaminants at the atmosphere-soil interface, in Sawhney, B.L. and Brown, K., Eds., *Reactions and Movement of Organic Chemicals in Soil*, SSSA Spec. Publ. 22, Soil Science Society of America, Madison, WI, 1989, 99–110.

31. Gohre, K. and Miller, G.C., Singlet oxygen generation on soil surfaces, *J. Agric. Food Chem.*, 31, 1104–1108, 1983.

32. Oliver, B.G., Cosgrove, E.G., and Carey, J.H., Effect of suspended sediments on the photolysis of organics in water, *Environ. Sci. Technol.*, 13, 1075–1077, 1979.

33. Zepp, R.G. and Schlotzhauer, P.F., Effects of equilibration time on photoreactivity of the pollutant DDE sorbed on natural sediments, *Chemosphere*, 10, 453–460, 1981.

34. Robin, M. and Trueblood, K.N., The ultraviolet absorption spectra of aromatic compounds adsorbed on silicic acid, *J. Am. Chem. Soc.*, 79, 5138–5142, 1957.

35. Plimmer, J.R., Photochemistry of pesticides. A discussion of the influence of some environmental factors, in Tahori, A.S., Ed., *Pesticide Chemistry: Fate of Pesticides in Environment*, Vol. VI, Gordon and Breach Science Publishers, New York, 1972, 47–76.

36. Bailey, G.W. and Karickhoff, S.W., Ultraviolet-visible spectroscopy in the characterization of clay mineral surfaces, *Anal. Lett.*, 6, 43–49, 1973.

37. Della Guardia, R.A. and Thomas, J.K., Photoprocesses on colloidal clay systems. Tris(2,2'-bipyridine)ruthenium(II) bound to colloidal kaolin and montmorillonite, *J. Phys. Chem.*, 87, 990–998, 1983.

38. Hardy, J.T., The sea surface microlayer: biology, chemistry, and anthropogenic enrichment, *Prog. Oceanogr.*, 11, 307–328, 1982.

39. Liss, P.S., Chemistry of the sea surface microlayer, in Riley, J.P. and Skirrow, G., Eds., *Chemical Oceanography*, 2nd. ed., Vol. 2, Academic Press, New York, 1975, 193–243.

40. Zafiriou, O.C., Photochemistry and the sea-surface microlayer. Natural processes and potential as a technique, in Burton, J.D., Brewer, P.G., and Chesselet, R., Eds., *Dynamic Processes in the Chemistry of the Upper Ocean*, Plenum Press, New York, 1986, 129–135.

41. Hansen, H.P., Photochemical degradation of petroleum hydrocarbon surface films on seawater, *Mar. Chem.*, 3, 183–195, 1975.

42. Bowling, J.W., Leversee, G.J., Landrum, P.F., and Giesy, J.P., Acute mortality of anthracene contaminated fish exposed to sunlight, *Aquat. Toxicol.*, 3, 79–90, 1983.

Chapter 2

Effects of Redox Processes on Acid-Producing Potential and Metal Mobility in Sediments

Jihua Hong, Ulrich Förstner, and Wolfgang Calmano

INTRODUCTION

Both pH and redox potential in sediment/water systems are significant parameters for mobilization and transformation of metals.[1,2] Many investigations have shown that pH decreases during oxidation of sediments.[3-7] Recently, considerable attention has been paid to the environmental impact on acid-sensitive waters and soils caused by redox processes.[8-14]

The concepts of acid-producing potential (APP) and potential acidity has been used in the prediction and calculation of acid mine drainage and waste tailing management.[15-17] Ferguson and Erickson[18] have summarized studies on acid mine drainage in North America. Recently, this concept has been applied to the oxidation research of anoxic sediments and other solid earth materials.[10,19]

Acidification in sediment/water systems occurs when hydrogen ions are generated during oxidation.[5,10,20] The acidification sensitivity depends on the acid neutralizing capacity (ANC) of the system.[20,21] When APP is addressed, ANC also should be considered, especially relative to its role in regulating the acid-base equilibrium of the system.[22-24]

There is increasing interest in the effects of periodical redox changes on APP and subsequent mobility of metals,[10,25,26] which has led to the development of several new concepts.[3,19] This paper will review the following processes related to metal availability:

1. Sources of acids in redox processes and APP in natural sediment/water systems;
2. Acid buffering capacity in the systems;
3. Periodical redox processes and their effects on APP;
4. Effects of redox processes on the mobilization of metals in sediment.

THE SOURCES OF ACIDS IN REDOX PROCESSES AND THE ACID-PRODUCING POTENTIAL (APP) IN SEDIMENT/WATER SYSTEMS

Main Oxidation Reactions of Acid Producers

In the presence of molecular oxygen, SO_4^{2-} and NO^{3-} are the thermodynamically stable forms of S and N. Compounds containing Fe, N, and S in lower oxidation states are ultimately oxidized to Fe^{3+}, $FeOOH$, NO^{3-}, and SO_4^{2-}, thereby releasing equivalent amounts of H^+ into the environment. Most of these oxidation reactions are microbially mediated.[24] Proton transfer processes which are important in the oxidation of sediment/water systems are summarized in Table 1.

S, Fe, and N are the most important elements in redox processes of a sediment/water system. This is not only due to their chemical reactivity, but also to their abundance in natural waters and sediments. For example, in tidal marsh sediments, pyrite contents (FeS_2) are often on the order of 1 to 5% on a mass basis, or higher.[28,30]

If chemical components in the sediment are known, APP can be calculated. The molar ratio of the reducing species to the produced hydrogen ions is listed in Table 1. We call this ratio the

1-56670-086-8/94/$0.00+$.50

Table 1.   Main Oxidation Reactions in Sediment/Water Systems

Elements oxidized	Reaction equations	f[a]	Ref.
Inorganic			
S	$H_2S + 2 O_2 = SO_4^{2-} + 2 H^+$	2	27
S	$S^\circ + 3/2 O_2 + H_2O = SO_4^{2-} + 2 H^+$	2	28
S, Fe	$FeS + 9/4 O_2 + 3/2 H_2O = FeOOH + SO_4^{2-} + 2 H^+$	2	27[b]
S, Fe	$FeS_2 + 15/4 O_2 + 5/2 H_2O = FeOOH + 2 SO_4^{2-} + 4 H^+$	4	29
Fe	$Fe^{2+} + 1/4 O_2 + 5/2 H_2O = Fe(OH)_3 + 2 H^+$	2	27
N	$NH_4^+ + 2 O_2 = NO_3^- + H_2O + 2 H^+$	2	27
N	$NO_x + 1/4(5-2x) O_2 + 1/2 H_2O = NO_3^- + H^+$	1	27
Organic			
N	$R-NH_2 + 2 O_2 = R-OH + NO_3^- + H^+$	1	27
S	$R-SH + H_2O + 2 O_2 = R-OH + SO_4^{2-} + 2 H^+$	2	27[b]

[a] f = Acid-producing coefficient.
[b] The coefficients have been revised by the authors of this paper.

acid-producing coefficient f. For example, 1 mol of FeS_2 can produce 4 mol of hydrogen ions. Because of its high f and its abundance in anoxic sediments, this reaction has been extensively discussed.[29-34]

Acid Production from the Oxidation of Inorganic Species

Oxidation of Sulfides

The oxidation of sulfides is a typical example of an acid source from inorganic species. A number of iron sulfide minerals are present in sedimentary environments: amorphous FeS, mackinawite (tetragonal FeS), greigite (cubic Fe_3S_4), and pyrite (cubic FeS_2).[32] In most cases, pyrite is the most important because of its relative abundance; hence, the oxidation of pyrite plays a significant role in acidification of sediments, soils, and other solid materials.[10,29,31]

Pyrite is stable only under strongly reduced conditions and changes into iron(II) sulfate at low pH and medium Eh, into iron-(hydr)oxides under near-neutral oxidizing and moderately reducing conditions (Table 1), and into jarosite ($KFe_3(SO_4)_2(OH)_6$) at low pH and high Eh.[29] The processes involved in oxidation of pyrite have been extensively studied in relation to acid mine drainage and were summarized by Nordstrom[31] and Ferguson and Erickson.[18] Numerous investigations[33,34] have shown that pyrite oxidation proceeds much faster in the presence of the chemoautotrophic micorbial genus, *Thiobacillus*.

Oxidation of Chalcopyrite

Oxidation of waste sulfidic ores in some special sites, particularly in the aquatic environment of mine tailing drainage areas, is one of the acid-producing sources. S^{2-} is completely oxidized to SO_4^{2-} and Fe^{2+} to Fe^{3+}, which tends to hydrolyze and precipitate as jarosite, thus releasing acid. The overall reaction is a net acid production:[35]

$$CuFeS_2 + 17/4 O_2 + 7/6 H_2O = 1/3 Fe_3(SO_4)_2(OH)_5 \qquad (1)$$
$$+ 4/3 SO_4^{2-} + Cu^{2+} + 2/3 H^+$$

The relationship between acid production and ore oxidation in the environment has been summarized by Ferguson and Erickson.[18]

Oxidation of Ammonium

Inorganic N is formed by the decomposition of organic matter and industrial sources in waters. When the pH value of a system is below about 9, NH_3 consumes H^+ from solution to form the weak acid ammonium (NH_4^+). The alkalinity associated with NH_4^+, in turn, is transferred to the solution or perhaps to the exchange complex.[36] When ammonium is oxidized (nitrification) by autotrophic microbes which obtain energy from the reaction, the acidity associated with ammonium is released and the Lewis acidity of the oxygen is transferred to nitric acid:[36]

$$NH_4^+ + 2\,O_2 = NO_3^- + H_2O + 2\,H^+ \tag{2a}$$

or

$$NH_3 + 2\,O_2 = NO_3^- + H_2O + H^+ \tag{2b}$$

Acid Production from the Oxidation of Organic Matter

Oxidation of Organic Nitrogen

The sediments in oceans, estuaries, and rivers can contain significant amounts of organic matter resulting from the decomposition of organisms.[37] Besides C and O, the elements most involved in redox cycles are those that are abundant in living matter: N and S. A stoichiometric description of the oxidation of protoplasm (as represented by the average proportions of the major elements in algal biomass) in natural waters is provided by the following reactions:[24,28]

$$\text{protoplasm } (CH_2O)_{106}(NH_3)_{16}(H_3PO_4) + 138\,O_2 \tag{3}$$
$$= 106\,CO_2 + 16\,NO_3^- + H_2PO_4^- + 122\,H_2O + 17\,H^+$$

When NO_3^- is the product of the oxidation, the alkalinity decreases by 0.16 ($= 17/106$) equivalent per mole of C fixed. When there is not enough oxygen for oxidizing all reducing compounds to their oxidized states, NH_4^+ will be produced; which consumes H^+. However, subsequent oxidation of NH_4^+ produces more H^+; summarized by Equation 3.

Oxidation of Organic Sulfur

Organic matter in sediments and soils contains many sulfur groups. For example, the amino acids cysteine[37] and methionine, and derivatives such as methionine sulfoxide,[39] methionine sulfone,[40] and cysteic acid[41] have been obtained from soil hydrolyzates. Clearly, the decomposition products of amino acids will include SO_4^{2-} and H^+ resulting from the oxidation of organic S:[27]

$$R\text{-}SH + H_2O + 2\,O_2 = R\text{-}OH + SO_4^{2-} + 2\,H^+ \tag{4}$$

Some other organic compounds such as oxalic acid, acetic acid, and hydroquinone can also produce hydrogen ions when they are oxidized.[42]

Calculations of Acid-Producing Potential (APP)

Acid-Producing Potential

APP can be defined as the largest amount of H^+ produced by the oxidation of the components per unit weight of sediments (or other solid materials), per unit volume of water (or other liquid substances), or in the whole sediment/water system. After the components of a sample are identified, APP can be easily calculated by using the following equations:

$$APP(s) = f \cdot \frac{C(s)}{M} \tag{5}$$

$$APP(aq) = f \cdot \frac{C(aq)}{M} \tag{6}$$

Table 2. Average Acid-Producing Potential (APP) of Some Samples

Parameter	Average content (mg/g)	C/M (mmol/g)	f	In suspension (1:10)		
				APP (mmol/g)	[H$^+$] (mmol/l)	pH^a
Suspension I						
Organic carbon^b	36					
Org-N^b	2.7	0.19	1	0.19	19	1.72
Org-S^b	0.49	0.015	2	0.030	3.0	2.52
Whole system				0.22	22	1.66
Suspension II						
FeS$_2$^{c}	1–5	0.083–0.417	4	0.332–1.668	33.2–166.8	1.48–0.78

^{a}Values calculated theoretically from Equations 5 and 7.
^{b}From Reference 44, average values of 14 samples.
^{c}From Reference 28, no other reducing species data are available for the calculation.

where C(s) and C(aq) are the contents (in grams per kilogram or grams per liter) of the reducing species (oxidized compounds) in 1 g of solid material and in 1 l of liquid, respectively. M is the molecular weight of the oxidized compound in grams per mole. The units of APP(s) and APP(aq) are moles per gram and moles per liter, respectively.

Total APP(s) is the sum of the APP(s) of each oxidized species (i):

$$T\text{-}APP(s) = \Sigma\ APP(s)(i) \tag{7}$$

and analogous for T-APP(aq):

$$T\text{-}APP(aq) = \Sigma\ APP(aq)(i) \tag{8}$$

So we have

$$T\text{-}APP = \Sigma\ APP(i) \tag{9}$$

Some Examples of APP

Experimental approaches for calculating APP for sulfidic mining residues have been summarized.[18] A test described by Sobek et al.[43] involves the analysis of total pyritic sulfur, whereas Bruynesteyn and Hackl[15] calculated APP from total sulfur analysis. However, the APP of sediments is more complex than that in sulfidic ores because the APP from organic matter also must be considered.

The N/C ratio in protoplasm is basically stable (16:106). If no directly measured org-N data are available, the APP(s) of org-N originating from the oxidation of N in protoplasm can approximately be calculated from a known carbon content in the sediment:

$$APP(s)\ (org\text{-}N) = 16/106 \times f \times C/12 \tag{10}$$

where C is the carbon content in the sediment.

The APP(s) of 14 samples were calculated by Swift,[44] according to the average content of each reducing species. We calculated the concentration of H$^+$ in a corresponding solid/liquid suspension of 1:10, disregarding the buffering and the oxidation kinetics. The results indicate that oxidation of either org-N or org-S can lead to a high concentration of H$^+$ and a low pH value in such systems (Table 2).

Apparent Acid-Producing Potential

In practice, actual APP is more important than the theoretically calculated values: how many hydrogen ions actually will be produced after overcoming buffering of the systems, and how

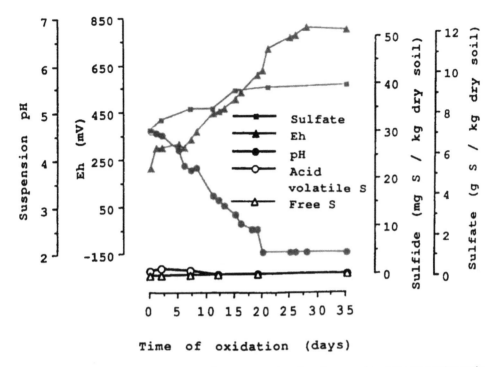

Figure 1. Changes in Eh, pH, sulfide, and sulfate concentrations in soil suspensions following continuous incubation for 35 days under aerobic conditions. (From Charoenchamratcheep, C., Smith, C.J., Stawathananont, S., and Patrick, W.H., Jr., *Soil Sci. Soc. Am. J.*, 51, 630, 1987. With permission.)

much is the pH value actually decreased during the oxidation under certain conditions? This leads to the development of the concept of "apparent APP." An acidification efficiency (β) related to the apparent APP has been developed:

$$\beta = H(a)/H(c) \tag{11}$$

with $\beta \leq 1$

where H(a) and H(c) are the amount or concentration of H^+ produced at given (a) and calculated (c) conditions, according to the reactions and the components of the sample, respectively.

Decreases in pH during oxidation have been found by many researchers.[4-7] Calmano et al.[6] observed the continuous decrease after pumping air in a poorly buffered sediment/water system. Charoenchamratcheep et al.[4] studied the changes in pH during oxidation and reduction in a soil/water suspension system. The decrease in pH was accompanied by an increase of SO_4^{2-} concentration in the solution. The pH values of all suspensions studied decreased below 4 (Figure 1). We have observed pH values of about 3 in a sediment/water experimental device after undergoing periodical redox processes.

BUFFERING PROCESSES DURING OXIDATION IN SEDIMENT/WATER SYSTEMS

When reducing species are oxidized, strong acidity may be produced (Table 2). However, buffering substances (or alkalinity)[21,24] also exist in most solid/water systems. In the buffering process the system restrains the pH change while H^+ is produced (input) or consumed. Many reactions participate in the buffering process (Figure 2 and Table 3).

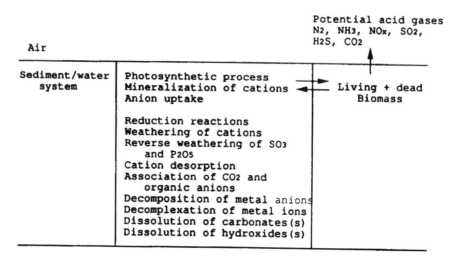

Figure 2. Conceptual model of H$^+$ consumption. (Modified from Driscoll, C.T. and Likens, G.E., *Tellus*, 34, 283, 1982.)

Table 3.   Proton Consuming Processes in Terrestrial and Aquatic Ecosystems

Mineralization of cations: (EQ. #)

$$(R\!-\!O)_nM(org) + n\ H^+(aq) \rightarrow M^{n+}(aq) + n\ R\!-\!OH(org)$$ (12)

Assimilation of anions:

$$n\ R\!-\!OH + A^{n-}(aq) + n\ H^+(aq) \rightarrow n\ H_2O + R_n\!-\!A(org)$$ (13)

Protonation of anions:

$$A^{n-}(aq) + n\ H^+(aq) \rightarrow H_nA(aq)$$ (14)

Reductions:

$$OX^{n-}(aq, s, g) + r\ H^+(aq) \rightarrow Red(aq, s, g) + m\ O_2 + n\ H_2O$$ (15)

Weathering of metal oxide components:

$$1/2\ n\ M_{2/n}O(s) + n\ H^+(aq) \rightarrow M^{n+}(aq) + 1/2\ n\ H_2O$$ (16)

Reverse weathering of anions:

$$NO^{2m-}_{(2m+n)}(aq) + 2\ m\ H^+(aq) \rightarrow NO_{(m+n)}(s) + m\ H_2O$$ (17–1)

Dissolution of carbonate minerals:

$$M_nCO_3 + 2\ H^+ \rightarrow CO_2 + H_2O + n\ M^{2/n}$$ (17–2)

Note: Quantitatively important cations (M^{n+}) include Ca^{2+}, Mg^{2+}, Na$^+$, K$^+$ (eq. 12, 16), NH$_4^+$ (eq. 12) and Al^{3+} (eq. 16). Important anions (A^{n-}) are H$_2$PO$_4^-$ (eq. 13, 17–1), NO$_3^-$ (eq. 13) and SO$_4^{2-}$ (eq. 13, 17–1). Mineralization and assimilation of SO$_4^{2-}$ and NO$_3^-$ involves oxidation and reduction processes but the net reactions are equivalent to eq. 13. Important reduction (red)-oxidation (OX)couples in gas/water/sediment (soil) systems include NH$_4^+$/NO$_3^-$, N$_2$/NO$_3^-$, H$_2$S/SO$_4^{2-}$, and Fe^{2+}/Fe$_2$O$_3$. Reduction/oxidation reactions (15) have been coupled to the O$_2$-H$_2$O system. In fact, organic matter (CH$_2$O) is usually the electron donor in reduction reactions (14); CH$_2$O oxidation is implicitly expressed in this reaction because organic matter is oxidized during aerobic respiration (O$_2$ + CH$_2$O $\leftrightarrow$ H$_2$O + CO$_2$). Cation and anion exchange reactions can be represented by reactions (16) and (17–1).

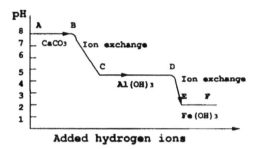

Figure 3. Acid buffering processes in sediments. (From Prenzel, J., *Z. D. Geol. Ges.*, 136, 293, 1985. With permission.)

Acidification/buffering interactions often have been investigated by adding strong acid into a sediment (or soil)/water system. A typical curve is given in Figure 3.[46] The curve combines two factors: the horizontal axis usually refers to the "capacity" factor, which shows the amount of strong acid added, whereas the vertical axis indicates the "intensity" factor (e.g., the pH or the degree of base saturation of the exchange complex).

Figure 3 summarizes many important aspects of acidification in a sediment/water system. Protons initially added to the system will associate with $CaCO_3$, resulting in mineral dissolution. As long as $CaCO_3$ is present, added protons will be consumed by the dissolution reaction and the pH of the system will remain constant (A → B in Figure 3). When the $CaCO_3$ is depleted by addition of protons, the system pH will decrease until the added protons associate with the next available proton energy level (B → C).

Between B and C the exchangeable bases on the exchange sites of clay are displaced by hydrogen ions. In this stage, sorbed metal ions will be released into the water phase. Ultimately, this process leads to the dissolution of clay minerals, which also results in buffering (C → D). Due to the abundance of Al in clay minerals, the buffering capacity in this process is high. When the reservoir of clay minerals is exhausted, further addition of protons results in an increase in dissolved $[H^+]$ and decrease in solution pH, respectively (D → E). The dissolution of ferric oxide or hydroxide will consume hydrogen ions and resist the pH decrease (E → F).[47] More acidic phenomena are rarely observed in natural environments.

Buffering Factors

For long-term effects, Binkley et al.[36] indicated that there are only four major processes consuming H^+ in ecosystems. They include:

1. Release of basic cations by decomposition of organic matter;
2. Specific anion adsorption;
3. Mineral weathering; and
4. Unbalanced reduction of oxidized compounds.

The release of basic cations from oxidized organic matter produces alkalinity in sediments, thus neutralizing the production of acidity. In sediments, SO_4^{2-} and PO_4^{3-} may be specifically adsorbed (enter into the hydration sphere) by aluminum and iron sesquioxides.[48] This adsorption transfers alkalinity associated with SO_4^{2-} to the water solution by consumption of 2 H^+. Of course, the acidity is actually retained in the sediment or soil at the sesquioxide surface, and subsequent desorption would restore the alkalinity to the SO_4^{2-} and the acidity to the solution. Because the oxidation of pyrite in sediments can produce SO_4^{2-}, this species becomes an important component in the sediment/water system. This problem will be discussed in greater detail later in this paper relative to its effects on the change in APP.

The weathering of minerals can consume or release H^+, but in most situations H^+ consumption predominates. For example, the weathering of aluminosilicate minerals consumes 1 mol of

H^+ per mole of charge of cation released. The aluminum released is considered to be an acidic cation because it behaves as an acid by H^+ donoring. As pH decreases, aluminum consumes H^+ and acts as a base. However, as the pH increases aluminum hydrolyzes, releasing H^+ to the solution and behaving as an acid. Similarly, Fe^{3+} will form different complex compounds consuming or releasing H^+.[49]

To date, the consumption of H^+ by photosynthetic processes has usually been neglected.[10,24] However, photosynthetic processes can consume significant amounts of hydrogen ions[24] (the reverse of Equation 3). The same net effect of acidity consumption occurs if nitrate is used as terminal electron acceptor (denitrification or dissimilatory nitrate reduction).[38]

Acid Neutralizing Capacity (ANC) of Sediment/Water Systems

The amount of H^+ required to reduce the pH in a system to a reference pH value is denoted ANC.[27] For example, a sediment with high carbonate content has higher ANC than one with low carbonate content. The ANC of a sediment/water or soil/water system consists of contributions by the solid phase ANC(s) and the aqueous phase ANC(aq):[10]

$$ANC = ANC(s) + ANC(aq) \qquad (18)$$

In aqueous solutions, ANC(aq) is defined as the base equivalence less the strong acid equivalence of the system, and is determined by strong acid titration to a reference pH. In most natural environments, ANC(aq) is mainly attributed to the bicarbonate concentration. However, at extreme pH or in the presence of additional weak bases such as natural organic anions (A^-), other species may contribute to ANC(aq):[27]

$$ANC(aq) = [HCO_3^-] + 2\,[CO_3^{2-}] + [OH^-] + [A^-] - [H^+] \qquad (19)$$

Other researchers[22] have indicated that most common species contributing to ANC in freshwater are

$$ANC(aq) = [HCO_3^-] + 2\,[CO_3^{2-}] + 2\,[S^{2-}] + [HS^-] + [NH_3] - [H^+] \qquad (20)$$

Silicates, borates, phosphates, and organic bases may also contribute to ANC, but usually only to a very small degree. Breemen[10] indicated that in natural waters ANC(aq) is quantitatively negligible compared to ANC(s).

The ANC of most sediments or soils is associated with silicate minerals, which have very slow dissolution kinetics. In practice, the best way to estimate the ANC(s) is by determination of component composition:[10]

$$\begin{aligned} ANC(s) = {} & 6\,[Al_2O_3] + 2\,[CaO] + 2\,[MgO] + 2\,[Na_2O] + 2\,[K_2O] \\ & + 4\,[MnO_2] + 2\,[MnO] + 6\,[Fe_2O_3] \\ & + 2\,[FeO] - 2\,[SO_4^{2-}] - 2\,[P_2O_5] - [HCl] \end{aligned} \qquad (21)$$

Above pH 5, the dissolution of Fe_2O_3, Al_2O_3 and MnO_2 is negligible. At a reference pH of 3, Al_2O_3 must be included as a basic component because of the appreciable solubility of aluminum at low pH.[10] Under extremely low pH conditions, e.g., in acid leaching procedures for contaminated sediments, significant dissolution of ferric iron and manganese(IV) will occur.

Acidification sensitivity is strongly associated with ANC. In fact, it is an index to overcome the acid buffering of the sediment/water system. Acidification sensitivity expresses the degree of pH change after H^+ are produced in the system or put into the system.

Four parameters have been defined as important in estimating soil or sediment sensitivity to acid water:

1. The total buffer capacity or cation exchange capacity (CEC), provided primarily by the clay minerals and organic matter;
2. The base saturation of that exchange capacity, which can be estimated by the pH of the sediment;
3. Anthropogenic influence; and
4. The presence or absence of carbonates in the sample.[50]

Experimental Determination of Buffer Capacity

Sobek et al.[43] studied the neutralization potential of mine tailings. Neutralization potential was obtained by adding a known amount of HCl, heating the sample, and titrating with standardized NaOH to pH 7. The method of Bruynesteyn and Hackl,[15] the so-called "determination of acid-consuming ability", is obtained by titration with standardized sulfuric acid to pH 3.5.

Calmano[51] has suggested a simple procedure to test the buffer intensity of a sludge: 10% sludge suspensions in distilled water (pH_o) and in 0.1 N acid (pH_x) are shaken for 1 h and the difference for the obtained pH values is calculated:

$$pH_{diff} = pH_o - pH_x \qquad (22)$$

Three categories of values can be established, ranging from $pH_{diff} < 2$ (strongly buffered), $pH_{diff} = 2$ to 4 (intermediate), to $pH_{diff} > 4$ (poorly buffered). These criteria should be used to decide if a dredged mud has to be stabilized for deposition, e.g., by addition of lime or limestone. It is also helpful to use this method for the judgement of the acidification or acidification sensitivity analysis of a sediment/water system.

EFFECTS OF PERIODIC REDOX PROCESSES ON ACID-PRODUCING POTENTIALS

Periodic redox processes can cause an increase or decrease in APP or pH in a sediment/water system. In a *closed* system, periodical redox processes can lead to the change or transfer between APP(s) and APP(aq), but the total APP of the system does not change. Conversely, in an *open* system, the total APP of the system will change, depending on the properties of the system and the reaction processes. Some processes are irreversible, e.g., when the components producing or consuming H^+ leave the system and cause the change in APP(s), APP(aq), and permanent ANC or BNC (base neutralizing capacity).

Change in APP By "Ferrolysis" in Periodical Redox Processes

Ferrolysis ("dissolution by iron") in redox processes was proposed by Brinkman.[52] When a system becomes reduced, part of the Fe^{2+} formed becomes exchangeable and displaces other cations, such as Ca^{2+} and Mg^{2+}. The displaced cations, together with the anions that appear simultaneously with dissolved Fe^{2+} (mainly HCO_3^- and organic anions), can be removed by percolation or by diffusion into the surface water followed by lateral flow. If the supply of cations by flood water, ground water, or mineral weathering is negligible, the surface complex may eventually become depleted of bases. The depletion is not immediately apparent in the reduced stage when the pH is high. During aeration of the system, however, exchangeable Fe^{2+} is oxidized to essentially insoluble Fe^{3+} oxide, and H^+ takes the place of adsorbed Fe^{2+} to such an extent that formerly adsorbed base cations are leached.[53]

In ferrolysis, exchangeable ferrous iron takes the place of ferrous sulfide as the immobile potentially acid substance formed during reduction, while exchangeable H^+ is the acidic product formed after oxidation of exchangeable ferrous iron. Breemen[10] suggested that ferrolysis is typical for the sediments or soils of older river or marine terraces in monsoon climates, which have a

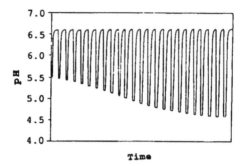

Figure 4. Temporal variation in pH by ferrolysis in periodical redox processes. (From Breemen, van, N., *Neth. J. Agric. Sci.*, 35, 271, 1987. With permission.)

seasonally alternate water table caused by submergence with rain water. In those conditions, hydrology favors either lateral or vertical drainage and, hence, removal of bases liberated from the exchange sites by ferrous iron. Over a number of redox cycles, the permanent decrease in ANC(s) is reflected by a decline in the pH of the aerobic system (Figure 4).[10]

Change in APP By the "Split" of Sulfate

Permanent acidification in alternating aerobic and anaerobic systems can occur in two stages.[27] The first stage is characterized by the reduction of SO_4^{2-}, particularly in tidal flats or sea bottom sediments. Most of the sulfide formed is fixed in the sediment as FeS or FeS_2. It leads to the increase in APP(s), while ANC(aq) (HCO_3^-) formed during sulfate reduction is removed by tidal turbulence or by diffusion into the overlying water. As a result, mobile ANC(aq) (HCO_3^-) and immobile potential acidity (FeS_2) are separated or "split". The increase in APP(s) leads to a permanent decrease in the ANC(aq) after the next aeration and oxidation cycle, and results in extreme acidification of the system.

Change in APP By Fe^{3+} Reduction and Some Displacement Reactions

The alkalinization by Fe^{3+} reduction and the desorption or displacement of SO_4^{2-} from solid surfaces has been reported.[27] The $FeSO_4$ concentration gradient caused by these processes promotes further transport of $FeSO_4$ to the sediment or soil surface. These processes will lead to a decrease in ANC(aq) and a permanent increase in ANC(s) by the adsorption of OH^- displacing SO_4^{2-}.

Change in APP By Volatilization of H_2S

A sediment/water system containing sulfate becomes alkaline after volatilization of H_2S from periodic sulfate reduction and retention of $M(HCO_3)_2$ in a saline alkaline lake or marsh. If the system reoxidizes, it will only reacidify partly, i.e., to the extent that reducing sulfur has been retained in the system (e.g., as FeS) and is available for the formation of H_2SO_4. A similar process has been described by Kelly et al.[22] and Kilham[23] in lakes receiving runoff from areas affected by acid atmospheric deposition. Thus, by sulfate reduction and volatilization of H_2S, such waters (lakes) could become more alkaline.

Change in APP By Denitrification

In periodical redox processes, APP(s)(org-N) will decrease due to the decomposition of solid organic matter. If denitrification does not occur, the total APP will not change but the APP(s)(org-N) transfers into APP(aq)(NH_3). On the other hand, if denitrification takes place, both the APP(s) and total APP in the system will decrease.[10,24]

$$(CH_2O)_{106}(NH_3)_{16}(H_3PO_4) + 84.8\ HNO_3 = \tag{23}$$
$$106\ CO_2 + 42.4\ N_2 + 16\ NH_3 + H_3PO_4 + 148.4\ H_2O$$

Because of the escape of N_2, the acidity from nitric acid before denitrification can not be reproduced by the next oxidation cycle, thus leading to a decline of APP in the system. Under certain conditions, including low pH and high NO_3^- concentrations, denitrification sometimes results in the formation of N_2O rather than N_2.[54]

EFFECTS OF REDOX AND pH VARIATIONS ON THE MOBILITY OF METALS

Typically, for systems involving solution/solid interactions, "mobility" reflects the flux of metal species in a certain medium, which contains both accelerating and inhibiting factors and processes.[55] The former influences comprise effects of pH lowering, redox changes, inorganic and organic complexation, and microbially mediated species transformations such as biomethylation. Among the spectrum of "barriers",[56] physical processes include adsorption, sedimentation, and filtration; chemical barriers comprise mechanisms such as complexation and precipitation; biological barriers are often associated with membrane processes which can limit translocation of metals (e.g., from plant roots to the shoots and fruits). "Complexation" in its various forms can both inhibit and accelerate metal flux,[55] particularly in biological systems consisting of different types of membranes.

Pore Water Chemistry Reflecting Metal/Solid Interactions

The composition of interstitial water is the most sensitive indicator of the types and the extent of reactions that take place between chemicals on sediment particles and the aqueous phase which contacts them. Particularly for fine-grained material, the large surface area related to the small volume of its entrapped interstitial water ensures that minor reactions with the solid phase will be indicated by major changes in the composition of the aqueous phase. Salomons[57] emphasized that it is important to know whether the concentrations of different components in pore waters are determined by adsorption/desorption processes or by precipitation/dissolution processes. If the latter is the case, pollutant concentrations in pore waters are somewhat independent from their respective concentrations in the solid phase.

In sediment, precipitation of sulfides is considered the dominant mechanism limiting the solubility of many trace elements.[58] Sulfide coordination is particularly strong for metals exhibiting so-called "B-character", such as Cu(I), Ag, Hg, Cd, Pb, and Zn; it also is an important mechanism for transition elements in decreasing order of the Irving-Williams series Cu(II) > Ni(II) > Co(II) > Fe(II) > Mn(II). There is strong direct[59,60] and indirect[61] evidence that the concentrations of Cu, Zn, and Cd in sulfidic pore waters are determined by precipitation/dissolution processes, whereas the concentrations of As and Cr in anoxic pore waters are probably controlled by adsorption/desorption processes.[57]

According to Davies-Colley et al.,[62] two situations can be distinguished in natural systems: the existence of a certain sulfide precipitation capacity (SPC), or (when exceeding the SPC) the accumulation of free sulfide (as H_2S or HS^-) in the aqueous phase. Concentrations of SO_4^{2-} and reducible Fe seem to be the principle factors influencing different evolutionary sequences in anoxic fresh and marine waters.[63] At excess S^{2-} concentrations (e.g., in Fe-poor environments), solubility of some metals can be increased by the formation of thio complexes. However, the stability of thio complexes is still questionable, possibly resulting in underestimates in most equilibrium models.[64]

There is some agreement that complexation by natural organic ligands is not important for most metals except Cu and Pb, due mainly to the competition by ions such as Ca^{2+}, Mg^{2+}, Fe^{2+},

Table 4.　Relative Mobilities of Elements in Solids and Sediments as a Function of Eh and pH

Relative mobility	Electron activity		Proton activity	
	Reducing	Oxidizing	Neutral-alkal.	Acid
Very low mobility	Al,Cr,Mo,V, U,Se,S,B,Hg, Cu,Cd,Pb	AL,Cr,Fe,Mn	Al,Cr,Hg,Cu, Ni,Co	Si
Low mobility	Si,K,P,Ni	Si,K,P,Pb	Si,K,P,Pb,Fe, Zn,Cd	K,Fe(III)
Medium mobility	Mn	Co,Ni,Hg,Cu, Zn,Cd	Mn	Al,Pb,Cu, Cr,V
High mobility	Ca,Na, Mg,Sr	Ca,Na,Mg,Sr, Mo,V,U,Se	Ca,Na,Mg,Cr	Ca,Na,Mg, Zn,Cd,Hg, Co,(Mn)
Very high mobility	Cl,I,Br	Cl,I,Br,B	Cl,I,Br,S,B Mo,V,U,Se	Cl,I,Br,B

From Plant, J.A. and Raiswell, R., *Applied Environmental Geochemistry*, Thornton, I., Ed., Academic Press, London, 1983. With permission.

and Mn^{2+}.[63-65] Even at high concentrations of organic substances (and high complexing capacity) from digested sewage sludge, only Cu and Pb seem to be slightly competitive with Ca and Mg ions.[66]

Typical curves for the adsorption of metals onto inorganic substrates, such as iron oxyhydrate, increase from almost nothing to near 100% as pH increases through a critical range of 1 to 2 units wide.[67] It is important to note that the location of the pH adsorption "edge" depends on adsorbent concentration. In the few cases where kinetics of sorption have been investigated, surface reactions were not found to be a single-step reaction.[68] Experiments performed by Benjamin and Leckie[69] showed an initial, rapid and almost complete metal uptake process perhaps lasting no more than a few minutes to hours, followed by a second, slower uptake process requiring from a few days to a few months. The first effect was thought to be true adsorption, and the second to be slow adsorbate diffusion into the solid substrate or coagulation of colloidal to filterable particles.[70]

For systems rich in organic matter, metal adsorption curves cover a wider pH range than is observed for inorganic substrates. Typically, a reduced reversibility of metal sorption has been observed in these organic systems.[71] Such effects may be important restrictions to using distribution coefficients in the assessment of metal mobility in rapidly changing environments, such as rivers, where equilibria between the solution and the solid phase often can not be achieved completely due to short residence times. In practice, applicability of distribution coefficients may find further limitations due to methodological problems. Sample pretreatment (e.g., dry or wet condition), solid/liquid separation technique (filtration or centrifugation), and grain size distribution of solid material all strongly affect K_D factors of metals.[72,73]

Redox Reactions Influencing Metal Mobility in Sediments

Regarding the potential release of metals from sediments, changes in the pH and redox conditions are of prime importance (Table 4). It can be expected that changes from reducing to oxidizing conditions, which involve transformations of sulfides and a shift to more acid conditions, increase the mobility of typical "B" or "chalcophilic" elements, such as Hg, Zn, Pb, Cu, and Cd. On the other hand, the mobility is characteristically lowered for Mn and Fe under oxidizing conditions. Elements exhibiting anionic species, such as S, As, Se, Cr, and Mo are solubilized, for example, from fly ash sluicing/ponding systems at neutral to alkaline pH conditions.[75,76]

Processes affecting metal mobilization from sulfide oxidation are highly complex. Factors involved are not only the extent of protonation, but also exchange processes involving interactions with Fe^{3+} and alkaline earths as well as the buffering effect of organic substances. Organic matter

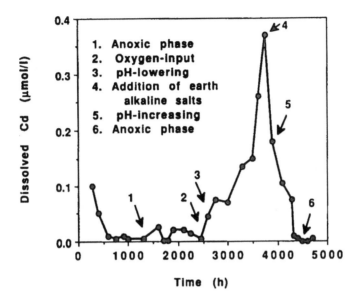

Figure 5. Experiments on cadmium release from solid waste material. (From Peiffer, S., Dissertation, Universitat Bayreuth, München, 1989. With permission.)

plays a major role as a buffer for acidity. Either acid-producing Fe^{3+} is directly taken up, or there is a permanent titration of the organic matter with H^+ from the hydroxide precipitation of $Fe(OH)_3$. This titration leads to the release of metals, and the increase in concentration of released metals should equal to the increase in SO_4^{2-} concentration. During resuspension/oxidation of harbor sediments, Eh of the system increased and concentrations of trace metals were highly correlated ($r = 0.98, 0.98, 0.93,$ and 0.92 for Cu, Zn, Cd, and Pb, respectively).[77] These results demonstrate that metal sulfide oxidation provides an important source of metal release.

The solubility of sulfide minerals is quite low. In an estuarine environment before the sediment is suspended, sulfide ions in the anoxic environment will "titrate" heavy metal ions,[78] resulting in their removal from pore waters and precipitation as solid phase. Moreover, pyrite formed during the "titration" can adsorb or coprecipitate heavy metal ions.[79] The results from the sequential extraction for chemical forms of heavy metals indicated that the metals in sulfidic/organic fraction can reach a high percentage of the total. The percentage of this fraction for heavy metals is typically between 30 to 90%.[6,80]

Using scanning electron microscopy with energy dispersive X-ray analysis on sediments of Newark Bay, New Jersey, Luther et al.[59] detected zinc sulfide minerals. This was the first report of sulfide minerals which contain a cation, other than Fe^{2+}, as the major constituent in recent estuarine sediments. Similarly, using an electron beam microprobe, Lee and Kittrick[60] found that 89% of the Cd and 83% of the Zn were associated with S in an anoxic harbor sediment heavily contaminated with trace metals.

Iron is one of the major components of most sediments. Under an anoxic diagenetic environment, it will probably be precipitated as pyrite. Kornicker and Morse[79] reported pyrite adsorption for divalent cations of heavy metals increased in the order: $Co^{2+} < Cd^{2+} < Mn^{2+} < Ni^{2+} < Zn^{2+}$. This adsorption or coprecipitation is often observed in pyrite samples.[81] If the pyrite is subsequently oxidized, heavy metals adsorbed on the pyrite can be released to solution.

Experimental investigations by Peiffer[64] on the long-term development of organic-rich solid matter provide detailed insight into the sequence of processes taking place in the postmethanogenic stage of such deposits (Figure 5). During an initial phase, anoxic cadmium is bound to sulfides, resulting in very low metal concentrations in solution (phase 1). Aeration by addition of dissolved oxygen initiates release of cadmium from the solid substrate (phase 2); this process is enhanced by the production of acidity, which lowers the pH from 6.7 to approximately 6.4 (phase 3). An even stronger effect is observed from the addition of alkaline earth ions (phase 4); such competitive desorption of trace metals has been widely overlooked so far, but seems to be a characteristic factor in these complex interactions. The pH increase, which may be induced from buffering components within the system, leads to a decrease in the dissolved cadmium concentration (phase 5). Formation of new sulfide ions from the degradation of organic matter brings the concentrations of dissolved cadmium back to its original, extreme low level (phase 6). The observed pH decrease seems to indicate that zinc and cadmium are being exchanged for protons, whereas lead and copper, because of their much stronger bonding to the solid substrate, are not.

Metal Mobilization from Aquatic Sediments

Release of potentially toxic metals from contaminated sediments poses problems both in aquatic systems and in subsequent land deposition of dredged materials. Effects of resuspension/ oxidation on metal release have been intensively investigated.[5,6,8,77,82,83] Khalid et al.[8] and Wall-mann[7] found an increase in dissolved Cu, Cd, and Pb with time during the oxidation of an anoxic estuarine sediment in seawater and freshwater. In these experiments, the pH was not controlled and it progressively decreased. Förstner et al.[84] reported 1.3% of total Cu was released from the dredged mud of Hamburg harbor when treated with seawater in a 3-week suspension experiment. Calmano et al.[85] observed that 9.1% of the total Zn in dredged mud was released during a suspension experiment when treated with seawater.

Some other examples illustrating the major factors, processes, and rates of metal mobilization in aquatic sediments follow:

- Field evidence for changing cadmium mobilities was reported by Holmes et al.[86] from Corpus Christi Bay harbor. During the summer when the harbor water was stagnant, Cd precipitated as CdS at the sediment/water interface; in the winter months, however, the increased flow of O_2-rich water into the bay resulted in a release of the precipitated metal.
- In the St. Lawrence estuary, Gendron et al.[87] found evidence for different release mechanisms near the sediment/water interface. The profiles for Co resemble those for Mn and Fe with increased concentration with depth, suggesting a mobilization of these elements in the reducing zone and a reprecipitation at the surface of the sediment profile. On the other hand, Cd appeared to be released at the surface, probably as a result of the aerobic remobilization of organically bound Cd.
- Biological activities are typically involved in these processes. Remobilization of trace metals has been explained by the removal of S^{2-} from pore waters via ventilation of the upper sediment layer with oxic overlying water by the biota, allowing the enrichment of dissolved Cd that would otherwise exhibit very low concentrations due to the formation of insoluble sulfides in reduced, H_2S-containing sediments. Emerson et al.[88] suggest a significant enhancement of metal fluxes to the bottom waters may occur by these mechanisms. As evidenced by Hines et al.,[89] biological activity in surface sediments greatly enhances remobilization of metals by the input of oxidized water; these processes are more effective during spring and summer than during the winter months.
- Prause et al.[83] studied the release of Pb and Cd from contaminated dredged material after dumping in a harbor environment. During an observation period of up to 24 h, no significant Cd or Pb release could be found from the dredged sludge used in the experiments. But during long-term experiments, extensive Cd remobilization of 1 to 2 mg/kg solids occurred.
- Cores were taken from tidal Elbe River sediments, where diurnal inundation of the fine-grained fluvial deposits takes place.[90] In the upper part of the sediment column, total particulate Cd content was approximately 10 mg/kg, whereas in the deeper anoxic zone 60 to 80% of the Cd was associated with the sulfidic/organic fraction. In the upper zone (oxic and transition) the association of Cd in the carbonatic and exchangeable fractions simultaneously increased up to 40% of total Cd. This distribution suggests that the release of metals from particulate phases into the pore water, and further transfer into biota, is controlled by the frequent downward flux of oxygenated surface water.[90]

Typical early diagenetic geochemical changes and subsequent element mobilization via pore water result from dredging activities. A study performed by Darby et al.[91] in a man-made estuarine

Table 5.   Metal Mobilization from Dredged Material after Land Deposition

	Reducing water	Oxidizing water
NH_4^+	125 mg/l	< 3 mg/l
Fe	80 mg/L	<3 mg/L
NO_3^-	<3 mg/L	120 mg/L
Zn	<10 mg/L	5000 µg/L
Cd	<0.5 µg/L	80 µg/l

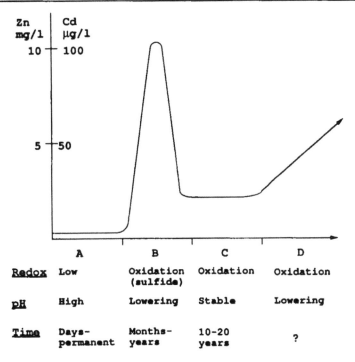

Figure 6. Schematic diagram illustrating different phases of metal release from land-disposed dredged material. (From Maass, B. and Miehlich, G., *Mitt. Dtsch. Bodenkunde Ges.*, 56, 289, 1988. With permission.)

marsh demonstrates the characteristic effects of oxidation. Compared to the river water concentration, the channel sediment pore water was enriched by a factor of 200 for Fe and Mn, 30 to 50-fold for Ni and Pb, approximately 10-fold for Cd and Hg, and 2 to 3-fold for Cu and Zn. Following hydraulic dredging, the expected concentration of metals which had been calculated based on a ratio of pore water to river water of about 1:4, were compared with the actual measurements at the pipe exiting the dredging device. Negative deviations were found for Fe and Mn, suggesting reprecipitation of Fe/Mn oxide minerals; the positive deviations of Zn (factor of 80), Cu, Pb, and Cd (factors of 7 to 8) indicate that during dumping of the sludge-water mixture significant proportions of these elements were mobilized and transferred into the effluent water.

Pore water data from dredged material from Hamburg harbor indicate typical differences in the kinetics of proton release from organic and sulfidic sources (Table 5). Recent deposits are characterized by low concentrations of NO_3^-, Cd, and Zn; when these low-buffered sediments are oxidized during a time period of a few months to years, the concentrations of NH_4^+ and Fe in the pore water typically decrease, whereas those of Cd and Zn increase. The different steps are schematically shown in Figure 6. Oxidation of sulfides during stage B strongly increases the concentrations of Cd and Zn relatively fast. When acidity is consumed by buffer reactions (phase C), Cd and Zn concentrations decrease, but are still higher than in the original sulfidic system. In phase D, oxidation of organic matter again lowers pH values and can induce a long-term mobilization of Zn and Cd.

Metal Transfer Between Inorganic and Organic Substrates

With respect to modeling metal partitioning between dissolved and particulate phases in a natural system, e.g., for estuarine sediments, the following requirements have been listed by Luoma and Davis:[94]

- Binding intensities and capacities for important sediment components,
- Relative abundance of these components,
- Assessment of the effect of particle coatings and of multicomponent aggregation on binding capacity of each substrate,
- Consideration of the effect of major competitors (Ca^{2+}, Mg^{2+}, Na^+, Cl^-),
- Evaluation of the kinetics of metal redistribution among sediment components.

It seems that thermodynamic models are still restricted because:

1. Adsorption characteristics are related not only to the system conditions (e.g., solid types, concentrations, and adsorbing species), but also to changes in the net system surface properties resulting from particle/particle interactions such as coagulation;
2. Influences of organic ligands in the aqueous phase can rarely be predicted as yet;
3. Effects of competition differ between various sorption sites; and
4. Reaction kinetics of the individual constituents cannot be evaluated in a mixture of sedimentary components.

At present, experimental studies on the dissolved/solid interactions in such complex systems seem to be more useful. One approach employs a six-chamber device, where the individual components are separated by membranes which still permit phase interactions via solute transport of the elements.[85] In this way, exchange reactions and biological uptake rates can be studied for individual phases under the influence of pH, redox, ionic strength, solid and/or solute concentration, and other parameters.

This system is made of a central chamber connected with six external chambers and separated by membranes of 0.45-μm pore diameter (Figure 7). The volume of the central chamber is 6 l and each of the external chambers contains 250 ml. Either a solution or a suspension can be inserted into the central chamber; in each external chamber the individual solid components are kept in suspension by magnetic stirring. Redox, pH, and other parameters may be controlled and adjusted in each chamber. In an experimental series on the effect on salinity by disposal of anoxic dredged mud into sea water, quantities of model components were chosen in analogy to an average sediment composition; i.e., 0.5 g algal cell wall (5%), 3 g bentonite (30%), 0.2 g manganese oxide (2%), 0.5 g goethite (5%), and 5 g quartz powder (50%). In the central chamber, 100 g of anoxic mud from Hamburg harbor was inserted; salts were added corresponding to the composition of sea water. After 3 weeks, solid samples and filtered water samples were collected from each chamber and analyzed.

The effect of salinity on metal remobilization from contaminated sediments is different for the individual elements. While approximately 16 and 9% of Cd and Zn, respectively, in the dredged mud from Hamburg harbor is released; for metals such as Cu, a salinity increase seems to be less important in the transfer, both among sediment substrates and to aquatic biota. This is, however, not true, as can be demonstrated from a mass balance for the Cu in Figure 7B: only 1.3% of the inventory of Cu of the sludge sample was released when treated with seawater. Only one third stays in solution, equivalent to approximately 40 μg l^{-1}, and there is no significant difference from conditions before salt addition; two thirds of the released Cu is reabsorbed at different affinities to the model substrates. Slight enrichment of Cu is observed in the iron hydroxide (approximately 80 mg/kg) and manganese oxide (100 mg/kg), whereas the cell walls— a minor component in the model sediment—accumulated nearly 300 mg/kg of Cu.

The dominant role of organic substrates in the binding of metals such as Cd and Cu is of particular relevance for the transfer of these elements into biological systems. It can be expected

A

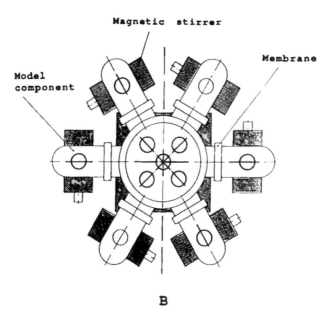

B

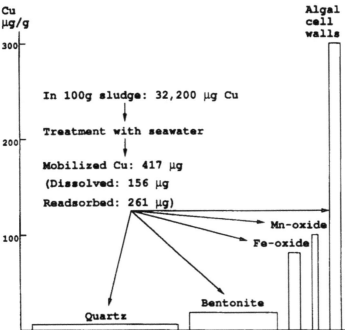

Figure 7. Metal transfer between sedimentary components. 7A: Schematic view of the multichamber device. 7B: Transfer of copper from anoxic harbor mud into different model substrates after treatment with artificial seawater. (From Calmano, W., Ahlf, W., and Förstner, U., *Environ. Geol. Water Sci.*, 11, 77, 1988. With permission.)

Table 6.   Some Examples from *In Situ* Studies for Prediction of Trace Metal Availability to Benthic Organisms from Sediment Characteristics

Organism	Metal	Best predictor in the sediment	Ref.
Scrobicularia plana	Pb	[Pb]/[Fe] extracted with 1 *N* HCl	98
Scrobicularia plana	As	[As]/[Fe] extracted with 1 *N* HCl	99
Scrobicularia plana and *Macoma balthica*	Hg	[Hg] extracted with HNO₃/organic content (%)	100
Anadonta grandis and *Elliptio complanata*	Cu	[Cu]/[Fe] extracted with NH₂OH·HCl	101
			102

From Tessier, A. and Campbell, P.G.C., *Hydrobiologia*, 149, 43, 1987. With permission.

that even at relatively small percentages of organic substrates, these materials are primarily involved in metabolic processes, and thus may constitute the major carriers by which metals are transferred within the food chain.

Bioavailability of Sediment-Bound Metal Species

Estimation of the remobilization of metals under changing environmental conditions and of the potential uptake by biota are two major objectives of species differentiation on particle-bound trace metals.[95] However, many authors have shown that with respect to bioavailability, as distinct from geochemical mobility, the present state of knowledge on solid matter speciation of metals is still unsatisfactory. For example, the leachable fraction does not necessarily correspond to the amount available to biota.[96] This handicap is primarily due to a lack of information about the specific mechanism by which organisms actively translocate trace-element species. In the case of plant root activities, interactions with soil and sediment components include redox changes, pH alterations, and organic complexing processes.

Studies on the prediction of the trace-metal levels in benthic organisms have shown that the prognostic value of extraction data is improved when the trace-metal concentrations are normalized with respect to the Fe (hydrous oxide) and/or organic content of the sediments[97] (Table 6).

It has been inferred by Tessier and Campbell[97] that a strong dependence of trace metal accumulation upon sediment characteristics does not imply that the main route of entry of trace metals is necessarily via ingestion of particulate metals. It can be explained by a control through adsorption reactions of the dissolved trace metal concentrations in the solution to which the organisms are exposed, such as in the case of filter-feeders, where high levels of trace metals were found associated with the gills and mantle.[102] For the latter pathway, study of the intermediate water phase (e.g., pore water) and the different forms and availabilities of metals in this medium seems to be particularly promising.[103]

Recently, there have been a number of studies focused on evaluating the bioavailability to benthos of heavy metals in anoxic sediments relative to the ratio of acid volatile sulfide (AVS) to those metals.[103–106] Metal ions in pore water can displace iron from iron monosulfide to form metal sulfide:

$$Me^{2+} + FeS(s) \leftrightarrow Fe^{2+} + MeS(s). \tag{24}$$

In this process, Me^{2+} forms essentially insoluble precipitates and is scavenged from the pore water of the sediment. If no other strong complexing ligand is present, the metal activity will approximate the metal in excess of the AVS. Hence, the concentration of AVS determines the boundary between low metal activity and potentially high metal activity. In the experiment carried out by Di Toro et al.,[107] if [SEM (simultaneously extracted metal concentration)]/[AVS] < 1, no acute toxicity (mortality > 50%) has been found in any sediment for any benthic test organism. The mortality of sensitive species increases in the ratio range of 1.5 to 2.5.[107]

SUMMARY

Redox processes can significantly affect the acid-producing potential and metal bioavailability in sediments. Many oxidation reactions producing acid can occur in natural aquatic sediment systems and human-affected environments. S, Fe, and N are the most important elements involved in redox processes of sediment/water systems. A calculation method developed in this paper can be used to assess APP in such systems. The actual extent of acidification in a sediment/water system depends not only on APP, but also on the acid neutralizing capacity (ANC) of the system.

Periodic redox processes leading to "ferrolysis", "split" of sulfate, and the volatilization of H_2S can cause changes in APP, and then affect metal transformation and bioavailability. The effects of redox and pH variations on the mobility of metals has been successfully assessed by pore water chemistry, which reflects the interactions of metals in aqueous and solid systems under changing redox environments with depth of sediments. Redox reactions, especially reoxidation of anoxic sediments, can lead to metal release from contaminated sediments. The bioavailability of metals in sediments depends on metal species, especially metal sulfide species.

A lack of understanding of the kinetic processes controlling acid production and about the accurate determination of ANC quantitatively limits our ability to predict apparent APP. Heavy metal mobility following oxidation of anoxic sediments has been observed. Although acid volatile sulfide (AVS) may control the toxicity of heavy metals in anoxic sediments, under changing redox conditions AVS is oxidized and dissolved metal sulfates are released. Furthermore, the acidity which follows sulfide oxidation may accelerate metal release. Thus, assessment of the relationship between metal sulfides and their bioavailability in periodic redox environments needs much more research.

REFERENCES

1. Stumm, W. and Morgan, J.J., *Aquatic Chemistry*, 2nd ed., John Wiley & Sons, New York, 1981.
2. Förstner, U. and Wittmann, G.T.W., *Metal Pollution in the Aquatic Environment*, Springer-Verlag, New York, 1979.
3. Calmano, W., *Schwermetalle in kontaminierten Feststoffen*, Verlag TÜV Rheinland, Köln, 1989, 237.
4. Charoenchamratcheep, C., Smith, C.J., Stawathananont, S., and Patrick, W.H., Jr., Reduction and oxidation of acid sulfate soils of Thailand, *Soil Sci. Soc. Am. J.*, 51, 630–634, 1987.
5. Hong, J., Calmano, W., Wallmann, K., Petersen, W., Schröder, F., Knauth, H.-D., and Förstner, U., Change in pH and release of heavy metals in the polluted sediments of Hamburg-Harburg and the downstream Elbe during oxidation, in Farmer, J. G., Ed., *Proc. Int. Conf. Heavy Metals in the Environment*, CEP Consultants, Edinburgh, 1991, 330–333.
6. Calmano, W., Hong, J., and Förstner, U., Einfluss von pH-Wert und Redoxpotential auf die Bindung und Mobilisierung von Schwermetallen in kontaminierten Sedimenten, *Vom Wasser*, 78, 77–84, 1992.
7. Wallmann, K., Die Frühdiagenese und ihr Einfluss auf die Mobilität der Spurenelemente As, Cd, Co, Cu, Ni, Pb und Zn in Sediment- und Schwebstoff-Suspensionen, Doctoral Dissertation, Technische Universität Hamburg-Harburg, Germany, 1990.
8. Khalid, R.A., Patrick, W.H., Jr., and Gambrell, R.P., Effect of dissolved oxygen on chemical transformations of heavy metals, phosphorus and nitrogen in an estuarine sediment, *Estuarine Coastal Mar. Sci.*, 6, 21–35, 1978.
9. Drever, J.I., *The Geochemistry of Natural Waters*, Prentice-Hall, Englewood Cliffs, N.J., 1982, 61–63.
10. Breemen, van, N. Effects of redox processes on soil acidity, *Neth. J. Agric. Sci.*, 35, 271–279, 1987.
11. Warren, C.J. and Dudas, M.J., Acidification adjacent to an elemental sulfur stockpile: I. Mineral weathering, *Can. J. Soil Sci.*, 72, 113–127, 1992.
12. Warren, C.J. and Dudas, M.J., Acidification adjacent to an elemental sulfur stockpile. II. Trace element redistribution, *Can. J. Soil Sci.*, 72, 127–135, 1992.
13. Schnoor, J.L. and Stumm, W., Acidification of aquatic and terrestrial systems, in Stumm, W., Ed., *Chemical Processes in Lakes*, John Wiley & Sons, New York, 1985, 311–339.

14. Nelson, W.O. and Campbell, P.G.C., The effects of acidification on the geochemistry of Al, Cd, Pb, and Hg in fresh water environments: a literature review, *Environ. Pollut.,* 71,91–130, 1991.

15. Bruynesteyn, A. and Hackl, R.P., Evaluation of acid production potential of mining waste materials, *Miner. Environ.,* 4, 5–8, 1984.

16. Breeman, van, N., Genesis and Soil Solution Chemistry of Acid Sulfate Soils in Thailand, Agric. Res. Rep. 848, Purdoc, Wageningen, 1976, 263.

17. Rahn, P.H., A method to mitigate acid-mine drainage in the Shamokin area, Pennsylvania, U.S.A., *Environ. Geol. Water Sci.,* 19, 47–54, 1992.

18. Ferguson, K.D. and Erickson, P.M., Pre-mine prediction of acid mine drainage, in Salomons, W. and Förstner, U., Eds., *Environmental Management of Solid Waste,* Springer-Verlag, New York, 1989, 24–43.

19. Förstner, U., Calmano, W., Ahlf, W., and Kersten, M., Sediment criteria development, in Heling, D., Rothe, R., Förstner, U., and Stoffers, P., Eds., *Sediments and Environmental Geochemistry,* Springer-Verlag, New York, 1990, 311–338.

20. Breemen, van, N., Driscoll, C.T., and Mulder, J., Acidic deposition and internal proton sources in acidification of soils and waters, *Nature,* 307, 599–604, 1984.

21. Reuss, J.O. and Johnson, D.W., *Acid Deposition and the Acidification of Soil and Waters,* Springer-Verlag, 1986, New York, 119.

22. Kelly, C.A., Rudd, W.M., Cook, R.B., and Schindler, D.W., The potential importance of bacterial processes in regulating rate of lake acidification, *Limnol. Oceanogr.,* 27, 868–882, 1982.

23. Kilham, P. Acid precipitation. Its role in the alkalization of a lake in Michigan, *Limnol. Oceanogr.,* 27, 856–867, 1982.

24. Morel, F.M.M., *Principles of Aquatic Chemistry,* John Wiley & Sons, New York, 1983, 446.

25. Kersten, M. and Kerner, M., Transformation of heavy metals and plant nutrients in a tidal freshwater flat sediment of the Elbe estuary as affected by Eh and tidal cycle, in Lekkas, T.D., Ed., *Proc. Int. Conf. on Heavy Metals in the Environment,* Vol. 1, CEP Consultants, Edinburgh, 1985, 533–535.

26. Reddy, K.R. and Patrick, W.H., Jr., Effect of alternate aerobic and anaerobic conditions on redox potential, organic matter decomposition and nitrogen loss in a flooded soil, *Soil Biol. Biochem.,* 7, 87–94, 1975.

27. Breemen, van, N., Mulder, J., and Driscoll, C.J., Acidification and alkalinization of soils, *Plant Soil,* 75, 283–308, 1983.

28. Breemen, van, N., Soil forming processes in acid sulphate soil, in Dost, H., Ed., Acid Sulphate Soils, IL RI Publ. 18, Wageningen, The Netherlands, 1973, 66–130.

29. Breemen, van, N., Redox processes of iron and sulfur involved in the formation of acid sulfate soils, in Stucki, J.W., et al., Eds., *Iron in Soil and Clay Minerals,* Reidel Publishing, Dordrecht, The Netherlands, 1988, 825–841.

30. Postma, D., Pyrite and siderite oxidation in swamp sediments, *J. Soil Sci.,* 34, 163–182, 1983.

31. Nordstrom, D.K., Aqueous pyrite oxidation and the consequent formation of secondary iron minerals, in Hossner, L.R., Ed., *Acid Sulfate Weathering,* SSSA Spec. Publ. 10, Soil Science Society of America, Madison, WI, 1982, 37–56.

32. Berner, R., Thermodynamic stability of sedimentary iron sulfides, *Am. J. Sci.,* 265, 773–785, 1967.

33. Sullivan, P.J., Yelton, J.L., and Reddy, K.J., Iron sulfide oxidation and the chemistry of acid generation, *Environ. Geol. Water Sci.,* 11, 289–295, 1988.

34. Lorenz, W.C., Progress in Controlling Acid Mine Water. A Literature Review, I.C. 8080, U.S. Bureau of Mines, Washington, D.C., 1962, 40.

35. Blancarte-Zurita, M.A., Microbiological leaching of chalcopyrite concentrates by *Thiobacillus ferrooxidans.* A comparative study of a conventional process and a catalyzed process, in Norris, P.R. and Kelley, D.P., Eds., *Biohydrometallurgy: Proc. Int. Symp.,* Warwick, 1987, Science and Technology Letters, Kew Survey, U.K., pp. 273–285.

36. Binkley, D., Driscoll, C.T., Allen, H.L., Schoeneberger, P., and McAvoy, D., *Acidic Deposition and Forest Soils,* Springer-Verlag, New York, 1989, 146.

37. Bremner, J.M., The amino-acid composition of the protein material in soil, *Biochem. J.,* 47, 538–543, 1950.

38. Froelich, P.N., Klinkhammer, G.P., Bender, M.L., Ludetke, N.A., Heath, G.R., Cullen, D., Dauphin, P., Hammond, D., Hartman, B., and Maynard, V., Early oxidation of organic matter in pelagic sediments of the eastern equatorial Atlantic. Suboxic diagenesis, *Geochim. Cosmochim. Acta*, 43, 1075–1090, 1979.

39. Sowden, F. J., Estimation of amino acids in soil hydrolyzates by the Moore and Stein method, *Soil Sci.*, 80, 181–186, 1955.

40. Sowden, F.J., The forms of nitrogen in the organic matter of different horizons of soil profiles, *Can. J. Soil Sci.*, 38, 147–153, 1958.

41. Stevenson, F.J., Effect of some long-time rotations on the amino-acid composition of the soil, *Soil Sci. Soc. Am. Proc.*, 20, 204–209, 1956.

42. Stone, A.T. and Morgan, J.J., Reaction and dissolution of manganese (III) and the manganese (IV) oxides by organics. II. Survey of the reactivity of organics, *Environ. Sci. Technol.*, 18, 617–624, 1984.

43. Sobek, A.A., Schuller, W.A., Freeman, J.R., and Smith, R.M. Field and Laboratory Methods Applicable to Overburden and Mine Soils, EPA-600/2–78–054, U.S. Environmental Protection Agency, Washington, D.C., 1978.

44. Swift, R.S., *Soil Organic Matter Studies*, International Atomic Energy Agency, Vienna, 1977, 275–281.

45. Driscoll, C.T. and Likens, G.E., Hydrogen ion budget of an aggrading forested ecosystem, *Tellus*, 34, 283–292, 1982.

46. Prenzel, J., Verlauf und Ursache der Bodenversauerung, *Z. Dtsch. Geol. Ges.*, 136, 293–302, 1985.

47. Fiedler, H.J., Bodenchemische Gesichtspunktezur Wiederaufforstung in Immissionsgebieten der Mittelgebirge, *Wiss. Z. Tech. Universität Dresden*, 37, 205–210,1988.

48. Tabatabei, M.A., Physico-chemical fate of sulfate in soils, *J. Am. Phys. Chem. Assoc.*, 37, 34–38, 1987.

49. Sullivan, P.J., Yelton, J.L., and Reddy, K.J., Solubility relationships of aluminum and iron minerals associated with acid mine drainage, *Environ. Geol. Water Sci.*, 11, 283–287, 1988.

50. McFee, W.W., Sensitivity of soil regions to long term acid precipitation, in Shriner, D.S., Ed., *Atmospheric Sulfur Deposition: Environmental Impact and Health Effects*, Science Publishers, Ann Arbor, Michigan, 1980, 495–506.

51. Calmano, W., Stabilization of dredged mud, in Salomons, W., and Förstner, U., Eds., *Environmental Management of Solid Waste*, Springer-Verlag, Berlin, 1988, 80–98.

52. Brinkman, R., Ferrolysis, a hydromorphic soil forming process, *Geoderma*, 3, 199–206, 1970.

53. Breemen, van, N., Long-term chemical, mineralogical, and morphological effects of iron-redox processes in periodically flooded soils, in Stucki, J.W., Goodman, B.A., and Schwertmann, U., Eds., *Soils and Clay Minerals*, Dreidel, Dordrecht, 1988, 811–823.

54. Firestone, M.K., Firestone, R.B., and Tiedje, J.M., Nitrous oxide from soil denitrification: Factors controlling its biological production, *Science*, 208, 749–751, 1980.

55. Förstner, U., Ahlf, W., Calmano, W., Kersten, M., and Schoer, J., Assessment of metal mobility in sludges and solid wastes, in Broekaert, J.A.C., Gucer, S., Adams, F., Eds., *Metal Speciation in the Environment*, NATO AST Ser. Vol. G 23, Springer-Verlag, Berlin, 1990, 1–41.

56. Pele'man, A.I., *Geochemistry of Episenesis*, Plenum Press, New York, 1967, 213–234.

57. Salomons, W., Sediments and water quality, *Environ. Technol. Lett.*, 6, 315–368, 1985.

58. Boulegue, J., Lord, C.J., III, and Church, T.M., Sulfur speciation and associated trace metals (Fe, Cu) in the pore waters of Great Marsh, Delaware, *Geochim. Cosmochim. Acta*, 46, 453–464, 1982.

59. Luther, G.W., Meyerson, A.L., Krajewski, J.J., and Hires, R., Metal sulfides in estuarine sediments, *J. Sediment. Petrol.*, 50, 1117–1120, 1980.

60. Lee, F.Y. and Kittrick, J.A., Elements associated with the cadmium phase in a harbor sediment as determined with the electron beam microprobe, *J. Environ. Qual.*, 13, 337–340, 1984.

61. Lu, C.S.J. and Chen, K.Y., Migration of trace metals in interfaces of seawater and polluted surficial sediments, *Environ. Sci. Technol.*, 11, 174–182, 1977.

62. Davies-Colley, R.J., Nelson, P.O., and Williamson, K.J., Sulfide control of cadmium and copper concentrations in anaerobic estuarine sediments *Mar. Chem.*, 16, 173–186, 1985.

63. Salomons, W., De Rooij, N.M., Kerdijk, H., and Bril, J., Sediments as a source for contaminants? in Thomas, R.L. et al., Eds., Ecological Effects of In-Situ Sediment Contaminants, *Hydrobiologia*, 149, 13–30, 1987.

64. Peiffer, S., Biogeochemische Regulation der Spurenmetallöslichkeit während der anaeroben Zersetrung fester kommuler Abfalle, Dissertation Universität Bayreuth, München, 1989.

65. Stumm, W. and Bilinski, H., Trace metals in natural waters: difficulties of interpretation arising from our ignorance on their speciation, in Jenkins, S.H., Ed., *Proc. 8th Int. Conf. Water Pollution Research*, Pergamon Press, Oxford, 1973, 39–49.

66. Fletcher, P. and Beckett, P.H.T., The chemistry of heavy metals in digested sewage sludge. II. Heavy metal complexation with soluble organic matter, *Water Res.*, 21, 1163–1172, 1987.

67. Benjamin, M.M., Hayes, K.L., and Leckie, J.O., Removal of toxic metals from power-generated waste streams by adsorption and coprecipitation, *J. Water Pollut. Control Fed.*, 54, 1472–1481, 1982.

68. Anderson, M.A., Kinetics and equilibrium control of interfacial reactions involving inorganic ionic solutes and hydrous oxide solids, in Brinckman, F.E. and Fish, R.H., Eds., Environmental Speciation and Monitoring Needs for Trace Metal-Containing Substances from Energy-related Processes, U.S. Deptartment of Commerce, Washington D.C., 1981, 146–162.

69. Benjamin, M.M. and Leckie, J.O., Multiple-site adsorption of Cd, Cu, Zn, and Pb on amorphous iron oxyhydroxide, *J. Colloid Interface Sci.*, 79, 209–211, 1981.

70. Santschi, P.H. and Honeyman, B.D., Radioisotopes as tracers for the interactions between trace metals, colloids and particles in natural waters, in Vernet, J.P., Ed., *Proc. Int. Conf. Heavy Metals in the Environment*, Vol. 1, CEP Consultants, Edinburgh, 1989, 243–252.

71. Lion, L.W., Altman, R.S., and Leckie, J.O., Trace-metal adsorption characteristics of estuarine particulate matter: evaluation of contribution of Fe/Mn oxide and organic surface coatings, *Environ. Sci. Technol.*, 16, 660–666, 1982.

72. Calmano, W., Untersuchungen uber das Verhalten von Spurenelementen an Rhein- und Mainschwebstoffen mit Hilfe radioanalytischer Methoden, Doctoral Dissertation, TH Darmstadt, 1979.

73. Dursma, E.K., Problems of sediment sampling and conservation for radionuclide accumulation studies, in *Sediments and Pollution in Waterways*, IAEA-TecDoc-302, International Atomic Energy Agency, Vienna, 1984.

74. Plant, J.A. and Raiswell, R., Principles of environmental geochemistry, in Thornton, I., Ed., *Applied Environmental Geochemistry*, Academic Press, London, 1983, 1–39.

75. Dreesen, D. R., Gladney, E.S., Owens, J.W., Perkins, B.L., Wienke, C.L., and Wangen, L.E., Comparison of levels found in effluent waters from a coal-fired power plant, *Environ. Sci. Technol.*, 11, 1017–1019, 1977.

76. Turner, R.R., Lowry, P., Levin, M., Lindberg, S.E., and Tamura, T., Leachability and Aqueous Speciation of Selected Trace Constituents of Coal Fly Ash. Final Report, Research Project 1061EA-2588, Electric Power Research Institute, Palo Alto, CA, 1982.

77. Calmano, W., Förstner, U., and Hong, J., Mobilization and scavenging of heavy metals following resuspension of anoxic sediments from the Elbe River, in *Environmental Geochemistry of Sulfide Oxidation*, Alpers, C.N. and Blowers, D.W., Eds., American Chemical Society, Washington, D.C., 1994, 298–321.

78. Berner, R.A., *Early Diagenesis—A Theoretical Approach*, Princeton University Press, Princeton, NJ, 1980, 241.

79. Kornicker, W.K. and Morse, J.W., Interactions of divalent cations with the surface of pyrite, *Geochim. Cosmochim. Acta*, 55, 2159–2171, 1991.

80. Kersten, M. and Förstner, U., Chemical fractionation of heavy metals in anoxic estuarine and coastal sediments, *Water Sci. Technol.*, 13, 121–130, 1986.

81. Nichoson, R.V., Gillham, R.W., and Reardon, E.J., Pyrite oxidation in carbonate-buffered solution. 1. Experimental kinetics, *Geochim. Cosmochim. Acta*, 52, 1077–1085, 1988.

82. Gambrell, R.P., Khalid, R.A., and Patrick, W.H., Jr., Chemical availability of mercury, lead and zinc in Mobile Bay sediment suspensions as affected by pH and oxidation-reduction conditions, *Environ. Sci. Technol.*, 14, 431–436, 1980.

83. Prause, B., Rehm, E., and Schulz-Baldes, M., The remobilization of Pb and Cd from contaminated dredged spoil after dumping in the marine environment, *Environ. Technol. Lett.*, 6, 261–266, 1985.

84. Förstner, U., Ahlf, W., and Calmano, W., Studies on the transfer of heavy metals between sedimentary phases with a multi-chamber device: combined effects of salinity and redox variation, *Mar. Chem.*, 28, 145–158, 1989.

85. Calmano, W., Ahlf, W., and Förstner, U., Study of metal sorption/desorption processes on competing sediment components with a multi-chamber device, *Environ. Geol. Water Sci.*, 11, 77–84, 1988.

86. Holmes, C.M., Slade, E.A., and Mclerran, C.J., Migration and redistribution of zinc and cadmium in marine estuarine systems, *Environ. Sci. Technol.*, 8, 255–259, 1974.

87. Gendron, A., Silverberg, N., Sundby, B., and Lebel, V., Early diagenesis of cadmium and cobalt in Laurentian Trough sediments, *Geochim. Cosmochim. Acta*, 50, 741–747, 1986.

88. Emerson, S., Jahnke, R., and Heggie, D., Sediment-water exchange in shallow water estuarine sediments, *J. Mar. Res.*, 42, 709–730, 1984.

89. Hines, M.E., Lyons, B.W.M., Armstrong, P.B., Orem, W.H., Spencer, M.J., and Gaudette, H.E., Seasonal metal remobilization in the sediments of Great Bay, New Hampshire, *Mar. Chem.*, 15, 173–187, 1984.

90. Kersten, M., Mechanisms und Bilanz der Schwermetallfreisetzung aus einem Susswasswatt der Elbe, Doctoral Dissertation, Technische Universität Hamburg-Harburg, Germany, 1989.

91. Darby, D.A., Adams, D.D., and Nivens, W.T., Early sediment changes and element mobilization in a man-made estuarine marsh, in Sly, P.G., Ed., *Sediment and Water Interactions*, Springer-Verlag, New York, 1986, 343–351.

92. Maass, B., Miehlich, G., and Grongroft, A., Untersuchungen zur Grundwassergefahrdung durch Hafenschlick-Spulfelder. II. Inhaltsstoffe in Spulfeldsedimenten und Porenwassern, *Mitt. Dtsch. Bodenkundl. Ges.*, 43/1, 253–258, 1985.

93. Maass, B. and Miehlich, G., Die Wirkung des Redoxpotentials auf die Zusammensetzung der Porenlosung in Hafenschlickspulfendern, *Mitt. Dtsch. Bodenkundl. Ges.*, 56, 289–294, 1988.

94. Luoma, S.N. and Davis, J.A., Requirements for modeling trace metal partioning in oxidized estuarine sediments, *Mar. Chem.*, 12, 159–181, 1983.

95. Kersten, M. and Förstner, U., Speciation of trace elements in sediment, in Batley, G.E., Ed., *Trace Element Speciation Analytical Methods and Problems*, CRC Press, Boca Raton, FL, 1989, 245–318.

96. Pickering, W.F., Selective chemical extraction of soil components and bound metal species, *Crit. Rev. Anal. Chem.*, 1981, 233–266.

97. Tessier, A. and Campbell, P.G.C., Partitioning of trace metals in sediments: relationships with bioavailability, in Thomas, R.L. et al., Eds., *Ecological Effects of In-Situ Sediment Contaminants*, *Hydrobiologia*, 149, 43–52, 1987.

98. Luoma, S. M. and Bryan, G.W., Factors controlling the availability of sediment-bound lead to the estuarine bivalve *Scrobicularia plana*, *J. Mar. Biol. Assoc. U.K.*, 58, 793–802, 1978.

99. Langston, W.J., Arsenic in U.K. estuarine sediments and its availability to deposit-feeding bivalves, *J. Mar. Biol. Assoc. U.K.*, 60, 869–881, 1980.

100. Langston, W.J., Distribution of mercury in British estuarine sediments and its availability to deposit-feeding bivalves, *J. Mar. Biol. Assoc. U.K.*, 62, 667–684, 1982.

101. Tessier, A., Campbell, P.G.C., and Auclair, J.C., Relationships between trace metal partitioning in sediments and their bioaccumulation in freshwater pelecypods, in Müller, G., Ed., *Proc. Int. Conf. Heavy Metals in the Environment*, CEP Consultants, Edinburgh, 1983, 1086–1089.

102. Tessier, A., Campbell, P.G.C., Auclair, J.C., and Bissn, M., Relationships between the partitioning of trace metals in sediments and their accumulation in the tissues of the freshwater mollusc *Elliptio complanata* in a mining area, *Can. J. Fish Aquat. Sci.*, 41, 1463–1472, 1984.

103. Kersten, M. and Förstner, U., Assessment of metal mobility in dredged material and mine waste by pore water chemistry and solid speciation, in Salomons, W. and Förstner, U., Eds., *Chemistry and Biology of Solid Waste—Dredged Material and Mine Tailings*, Springer-Verlag, Berlin, 1988, 213–237.

104. Carlson, A.G., Phipps, G.L., Mattson, V.K., Kosian, P.A., and Cotter, A.M., The role of acid-volatile sulfide in determining cadmium bioavailability and toxicity in fresh water sediments, *Environ. Toxicol. Chem.*, 10, 1309–1319, 1990.

105. Di Toro, D.M., Mahony, J.D., Hansen, J.D., Scott, M.B., Hicks, M.B., Mayr, S.M., and Redmond, M.S., Toxicology of cadmium in sediments. The role of acid volatile sulfide, *Environ. Toxicol. Chem.*, 9, 1487–1502, 1990.

106. Ankley, G.T., Phipps, G.L., Leonard, E.N., Benoit, D.A., Mettson, V.R., Kosian, P.A., Cotter, A.M., Dierkes, J.R., Hansen. D.J., and Mahony, J.D., Acid-volatile sulfide as a factor mediating cadmium and nickel bioavailability in contaminated sediments, *Environ. Toxicol. Chem.*, 10, 1299–1307, 1991.

107. Di Toro, D.M., Mahony, J.D., Hansen, D.J., Scott, K.J., Carlson, A.R., and Ankley, G.T., Acid volatile sulfide predicts the acute toxicity of cadmium and nickel sediments, *Environ. Sci. Technol.*, 26, 96–101, 1992.

Chapter 3

Sediment-Water Exchange Processes

Bjorn Sundby

INTRODUCTION

There are many ways to think about the sediment on the bottom of lakes and seas. A physicist who is interested in describing the circulation of water masses might, for example, think about the bottom sediment as the walls of a container that encloses a body of water; the environmental manager who is looking for ways to describe the quality of the sedimentary environment might think of it as an unstructured layer of mud. The modern view of the bottom sediment takes into account that the sediment is an integral part of the aquatic environment, with transfer of mass and momentum taking place across the water-sediment boundary. Because it is an open system, processes within the sediment depend directly on processes in the water column, particularly those near the sediment-water interface. The state of the sediment is influenced by variations in the sedimentation rate and the composition of the sedimenting particulate matter, by variations in the composition of the overlying water column, and by variations in near-bottom flow and turbulence.

The box corer and the benthic flux chamber are the most popular tools for studying the sediment. With a box corer one can collect undisturbed sediment cores for vertical profiling of the concentrations of dissolved and particulate sediment components, and for sediment-water exchange studies. The benthic flux chamber encloses a small area of bottom sediment and a volume of the overlying water either *in situ* or in the laboratory (Figure 1). The chamber allows one to measure the exchange of various solutes across the sediment-water interface by monitoring the composition of the enclosed water column. By its very nature, the benthic chamber interferes with the transfer of mass and momentum across the sediment-water interface since it creates a closed sediment-water system. Furthermore, since the composition of the enclosed water column changes from the moment the chamber is put in place, the chamber imposes a set of continuously changing conditions at the sediment-water boundary. Because of these interferences, the results of benthic chamber experiments can be difficult to interpret in terms of what is actually taking place at the sediment-water interface. However, when interpreted in terms of how changes in the water column lead to changes within the sediment, benthic chamber studies can be very useful. It is in this context that they will be considered in this brief overview of sediment-water exchange processes.

THE STRUCTURE AND DYNAMICS OF THE SEDIMENTARY ENVIRONMENT

If one takes an undisturbed core of sediment and examines the vertical distribution of the sediment composition from the surface down, one finds a series of layers or zones of progressively decreasing redox potential. The thickness of these zones can be variable, ranging from millimeters in organic-rich coastal muds to tens of centimeters in organic-poor deep-sea sediments. Each of these zones is a particular chemical environment, created by the activity of microorganisms. The microorganisms oxidize organic matter, using a series of electron acceptors that are consumed in a sequence determined by the free energy yield of the particular reaction involved. Starting at the surface of the sediment, these reactions correspond to oxygen reduction, nitrate reduction followed by manganese and iron oxide reduction, sulfate reduction and, finally, carbon dioxide reduction.[1] In addition to these biochemical shifts, the sediment also contains physical structures—animal

1-56670-086-8/94/$0.00+$.50
© 1994 by CRC Press, Inc.

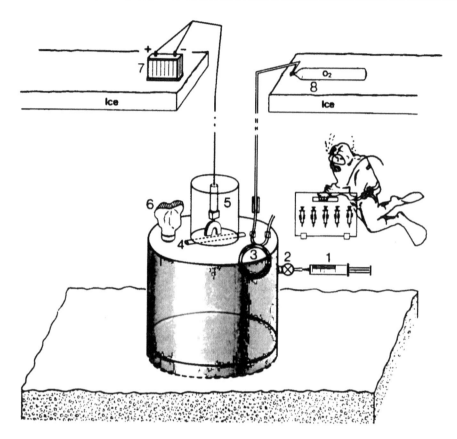

Figure 1. A benthic chamber for *in situ* studies of sediment-water exchange processes 1: syringe for withdrawing samples; 2: Teflon® valve; 3: coil of Teflon® tubing for diffusing oxygen into the chamber; 4: magnetic stirring bar; 5: electric motor; 6: soft membrane allowing compensation of volume withdrawn during sampling; 7: battery; 8: oxygen supply. (From Hall, P., Anderson, L., Rutgers van der Loeff, M., Sundby, B., and Westerlund, S., *Limnol. Oceanogr.*, 34, 734, 1989. With permission.)

burrows—that are created by tube- and gallery-building benthic organisms. These structures communicate with the water above—and the composition of the water within them, although variable, is often closer to the composition of the overlying bottom water than to that of bulk interstitial water. For this reason, the composition of the burrow wall is usually closer to the composition of the sediment surface than to the bulk sediment, and burrow walls are often regarded as extensions of the sediment surface.[2]

The redox zonation created by microorganisms gives rise to concentration gradients in the sediment pore water, which are the driving force for diffusion. Solutes, for example nutrients and trace metals, diffuse along the gradients from zones of high concentrations to zones of low concentrations. This mass flux in the pore water can develop zones of different solid-phase composition, i.e., layers with relatively lower and relatively higher concentrations of certain constituents. When benthic organisms mix the solid phases across the layers, these constituents are transported in the opposite direction of the solute flux. In this way, some sediment constituents cycle between regions of low redox potential and regions of higher redox potential.

A well-documented example of this is the case of manganese. Manganese dissolves in the anoxic reducing subsurface zone of a sediment, diffuses towards the sediment surface, and precipitates in the oxic surface layer, creating manganese enrichment in this layer (Figure 2). When benthic organisms mix particles from the manganese-enriched surface layer with manganese-poor particles from the deeper layers, the net result is a downward flux of particulate manganese. When this manganese dissolves, the pore water concentration of dissolved manganese increases and the

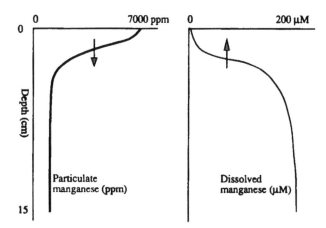

Figure 2. The distribution of total and dissolved manganese in a coastal marine sediment.

cycle starts over again. The fluxes may be confined to the interior of the sediment or include the sediment-water interface, depending on the efficiency of the precipitation (Figure 3). The cycling can be quite rapid—the manganese-rich layer in some coastal marine sediments goes through a complete cycle of downward mixing, dissolution, upward diffusion, and precipitation several times per year.[3]

The sediment is thus a very dynamic environment where dissolved and particulate material constantly migrate along concentration gradients. Microorganisms create the redox zonation and the concentration gradients that drive molecular diffusion, and macro organisms mediate the transport by mixing across the gradients.

BENTHIC RESPIRATION: OXYGEN AND pH VARIATIONS NEAR THE SEDIMENT-WATER INTERFACE

The oxygen concentration in the pore water of the sediment, and the depth to which oxygen penetrates into the sediment, determine the redox conditions within the upper layer of a bottom sediment. The upper layer in this context also includes the walls of the ventilated burrows of benthic organisms. Oxygen is therefore one of the most important variables to consider in studies of speciation and bioavailability of trace metals. This section describes some examples of what one can learn about oxygen with benthic chamber experiments.

Figure 4 shows an example of how the oxygen concentration in a water column enclosed in a benthic chamber evolves over time. These data come from a benthic chamber experiment that was carried out *in situ* at a 6-m-deep site in the Gullmarsfjord in Sweden.[4] In the beginning of the experiment, the oxygen concentration decreased at an approximately constant rate, but once the oxygen concentration had reached about 100 µM the curvature became quite pronounced. This pattern is typical of such experiments, and reflects the gradual decrease of the concentration gradient across the diffusive boundary layer at the sediment-water interface. The diffusive boundary layer is the region near the sediment surface where solute transport takes place via molecular diffusion rather than via turbulent diffusion. The thickness of this region is on the order of millimeters or a fraction of a millimeter.[5]

There are two reasons why the oxygen flux into the sediment decreases during an incubation experiment. One is that the concentration gradient across the boundary layer decreases with the decreasing oxygen concentration in the water column; the other is that the oxygen-containing layer of the sediment decreases in thickness, thereby reducing the number of microorganisms that respire with oxygen.[6]

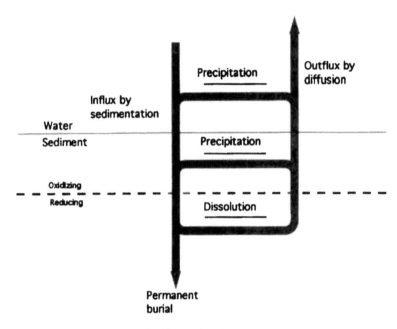

Figure 3. The manganese cycle in coastal marine sediments.

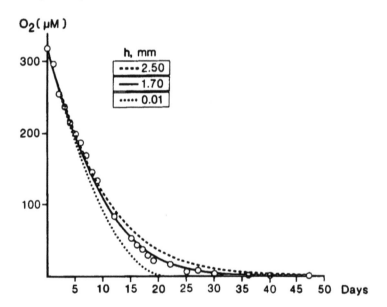

Figure 4. Oxygen uptake from a benthic chamber during an *in situ* experiment. The circled points are the measurements; the lines are calculated from a model of the oxygen flux across diffusive boundary layers of different thicknesses, assuming zero order oxygen uptake kinetics within the sediment. (From Hall, P., Anderson, L., Rutgers van der Loeff, M., Sundby, B., and Westerlund, S., *Limnol. Oceanogr.,* 34, 734, 1989. With permission.)

In shallow locations, oxygen is produced at the sediment-water interface by photosynthesis by epipelagic algae during the day, and is consumed by respiration during the night.[7] This creates a cycle in the oxygen distribution within the sediment, whereby both the concentration and the penetration depth of oxygen fluctuate diurnally.

In shallow sediments one can also expect a diurnal cycle in the pH of the pore water near the sediment-water interface, with pH varying between high values during the day when CO_2 is

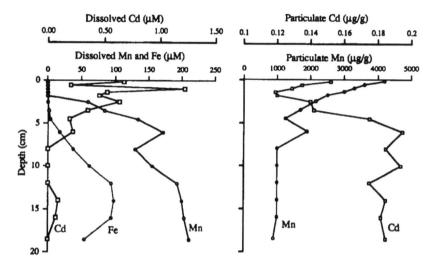

Figure 5. The distribution of manganese, iron, and cadmium three cores of fine grained sediment from the St. Lawrence Estuary. Left: distribution in the solid phase sediment. Right: distribution in the pore water. (From Gobeil, C., Silverberg, N., Sundby, B., and Cossa, D., *Geochim. Cosmochim. Acta*, 51, 589, 1987. With permission.)

consumed by photosynthesis, and low values during the night when CO_2 is produced as organic carbon is oxidized.[8] In environments where there is no production of oxygen, the production rate of CO_2 may at times be sufficiently high to depress the pH near the sediment-water interface.[9,10]

These experiments illustrate the sensitivity of the oxygen concentration in the pore water of sediments to changes in the boundary layer thickness, the oxygen concentration at the sediment-water interface, and the rate of organic carbon oxidation. One can therefore expect that the oxygen distribution and the pH in the pore water of sediments will vary, not only in response to seasonal or longer-term phenomena, but also to short time-scale phenomena such as diurnal variations in photosynthesis and respiration, and tidal-induced or storm-induced variations in bottom currents and oxygen concentrations.

TRACE METALS AND THE OXYGEN REGIME

The concentration of trace metals in the pore water of sediments is controlled by oxidation and reduction, dissolution and precipitation, and adsorption and desorption onto surfaces. Since these reactions depend directly or indirectly on the presence or absence of oxygen in the sediment pore water, as well as on the pH, any change induced in the distribution of oxygen and pH in the sediment pore water will affect the concentration and fluxes of trace metals. The pH changes will also influence the speciation of trace metals in the pore water.

When fine-grained sediments are sampled with a high degree of vertical resolution, it is often possible to observe how the presence or absence of oxygen affects the distribution of trace metals. Figure 5a is an example of how differently the two metals, manganese and cadmium, are influenced by the absence or presence of oxygen. Within the oxygen-containing surface layer of the sediment, one observes at the same time the lowest manganese concentrations and the highest cadmium concentrations. Deeper in the sediment column, where there is no oxygen, the concentration of dissolved manganese is high whereas dissolved cadmium can barely be detected. The distributions of manganese and cadmium are equally contrasting in the solid phase of the sediment (Figure 5b). In the surface layer, manganese is enriched and cadmium is depleted; in the deeper layer, manganese is depleted and cadmium is enriched.[11]

The manganese and cadmium profiles illustrate several important points. Some metal ions including manganese, iron, and cobalt precipitate in the presence of oxygen and dissolve in the

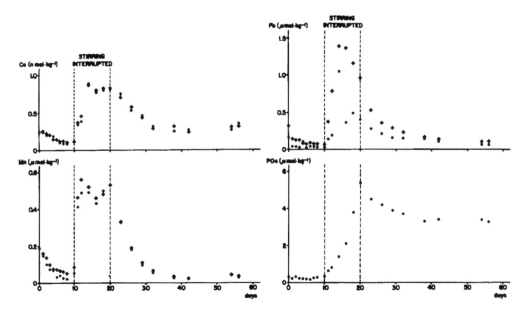

Figure 6. Release and uptake of manganese, iron, cobalt, and phosphate during an *in situ* experiment with a benthic chamber. The crosses represent the total concentration of an element in the water column inside the chamber, and the dots represent the dissolved concentration. The oxygen concentration in the chamber was kept constant during the experiment, but the stirring was interrupted between day 10 and day 20. During this period, the water column stratified and anoxia developed at the sediment-water interface. (From Sundby, B., Anderson, L., Hall, P., Iverfeldt, A., Rutgers van der Loeff, M., and Westerlund, S., *Geochim. Cosmochim. Acta*, 50, 1281, 1986. With permission.)

absence of oxygen. The reason for this is the different solubilities of their reduced and oxidized forms. Other metal ions including cadmium, zinc, and to some extent also copper and nickel, are soluble in the presence of oxygen and precipitate, probably as sulfides, in the sediment when oxygen is absent. The different solubilities create strong concentration gradients along which the ions migrate. In the example given, dissolved manganese migrates upward into the oxygen-containing layer, whereas cadmium migrates downward into the oxygen depleted layer as well as upward into the water column. The reactions that cause the dissolution of cadmium and other trace metals within the oxic sediment layer are not well understood. It may possibly be attributed to the accumulation of metal-containing biogenic material at the sediment surface and the release of metals when this material decomposes. The presence of oxygen appears to be essential for the release of these metals to the water column.[12] The postdepositional migration of dissolved metal ions is responsible for creating the zonation observed in the solid phase of the sediment. The zonation does not need to be vertical: if benthic organisms create ventilated burrows, the zonation radiates out from the burrow walls.[13]

The strong influence of redox potential and pH on the concentration of dissolved trace metals means that the distribution of trace metals is sensitive to variations in the oxygen regime. This sensitivity can be illustrated with results from the benthic chamber experiment that was referred to above. In this experiment, the oxygen concentration in the enclosed water was maintained approximately constant by adding oxygen through a diffuser to replace the oxygen consumed by respiration (Figure 1). Then, after 10 days, the stirring was interrupted, allowing the water column to stratify and become anoxic near the sediment surface. After another 10 days, the stirring was resumed, and oxygenated water was brought down again to the sediment-water interface. All this time the chamber was kept dark to avoid benthic photosynthesis.

Consider first what happened to manganese, iron, and cobalt (Figure 6).[14] As long as the oxygen concentration in the water was maintained near ambient levels and the water column was kept well mixed, none of these metals was released from the sediment. Instead, there was a small flux of all three metals from the water column into the sediment. Interrupting the stirring on day 10 resulted in a sudden release of dissolved iron, manganese, and cobalt from the sediment. When the stirring resumed and oxygenated water was brought down to the sediment-water interface again, all three metals gradually disappeared from the water column back into the sediment. The disappearance can be described with first-order reaction kinetics, indicating that the actual removal took place at the bottom of the diffusive sublayer where the concentration of the dissolved ions was kept low by continuous precipitation of insoluble oxidized forms.

Consider now what happened to the other trace metals in this experiment (Figure 7).[12] During the initial 10-day period, cadmium, zinc, copper, and nickel were all released to the water column and at nearly constant rates. When the stirring was interrupted, the fluxes were reversed and all four metals were taken up by the sediment until the stirring resumed. From then on, cadmium, zinc, and copper were once again released from the sediment to the water column. Nickel was not released during this phase, and the release rates for the three other trace metals were significantly lower than during the first phase of the experiment.

The effects of the changes in the oxygen regime on the trace-metal fluxes are entirely consistent with what we learned from the steady-state trace-metal distributions in sediment and the pore water. Manganese, iron, and cobalt, which precipitate in the presence of oxygen, were not released from the sediment when the water column was kept oxygenated and well stirred. In fact, they were taken up by the sediment. Cadmium, zinc, copper, and nickel, all of which dissolve in oxygenated pore water, were all released when the water column was kept oxygenated, but were taken up again by the sediment when the bottom of the water column went anoxic, perhaps because of the formation of sulfides at the sediment-water interface. The rapidity with which the fluxes changed direction after the events that changed the oxygen regime is remarkable. It shows that both the distribution of trace metals in the pore water of sediments, and the fluxes across the sediment-water interface, will respond quickly to changes in the ambient oxygen regime. Through the influence of dissolved oxygen, trace metals in sediments are influenced by short-term and long-term variations in photosynthesis and respiration. They can therefore be expected to vary in response to tidal-induced or storm-induced variations in bottom currents and oxygen concentrations.

THE PHOSPHORUS CYCLE AND THE OXYGEN REGIME IN SEDIMENTS

Our approach to trace metals in sediments has been greatly influenced by work done on the phosphorus cycle, and it may be useful to consider some notions that relate to phosphate exchange at the sediment-water interface. The importance of oxygen for the release and uptake of phosphate by sediments is well documented, dating back to the work of Einsele[15] and Mortimer.[16] Briefly, the idea is that iron oxides in the sediment surface layer adsorb phosphate diffusing up from the deeper sediment layers and immobilize it until the bottom water becomes anoxic, at which time the iron oxides in the sediment are reduced and the adsorbed phosphate is released. This basic idea has recently been refined by introducing the notion that sorption equilibria between dissolved and adsorbed phosphate buffer the phosphate concentration in the sediment pore water.[17] The fact that the concentration of phosphate in the pore water is buffered at an equilibrium concentration (which has a value of about 6 μM in the pore water of the sediments in the St. Lawrence Estuary) constrains the concentration gradient across the sediment-water interface and places upper limits on the instantaneous flux into or out of the sediment. Whether the concentration of trace metals that participate in sorption reactions with iron oxides is constrained in some way by sorption equilibria is not known, but it is tempting to suggest that the exchange of arsenic, whose chemistry in sediments in many ways parallels phosphorus,[18] might be subject to similar constraints.

The phosphorus cycle is also a good illustration of what a dynamic and complex environment the sediment really is. Consider the sedimentation of phosphorus as a mixture of inorganic and

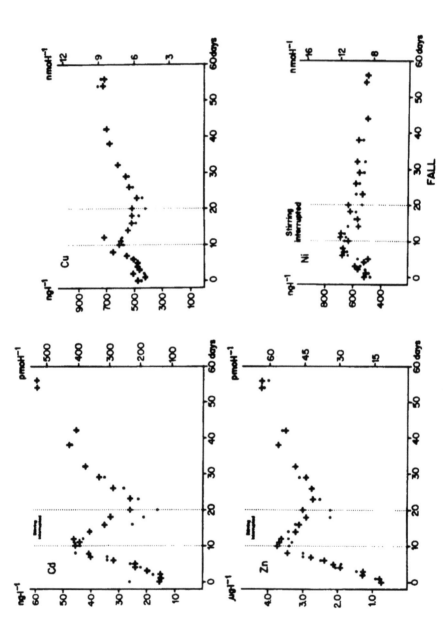

Figure 7. Release and uptake of cadmium, zinc, copper, and nickel during an *in situ* experiment with a benthic chamber. The crosses represent the total concentration of an element in the water column inside the chamber, and the dots represent the dissolved concentration. The oxygen concentration was kept constant during the experiment, but stirring was interrupted between day 10 and day 20. During this period, the water column stratified and anoxia developed at the sediment-water interface. (From Westerlund, S., Anderson, L., Hall, P., Iverfeldt, A., Rutgers van der Loeff, M., and Sundby, B., *Geochim. Cosmochim. Acta*, 50, 1289, 1986. With permission.)

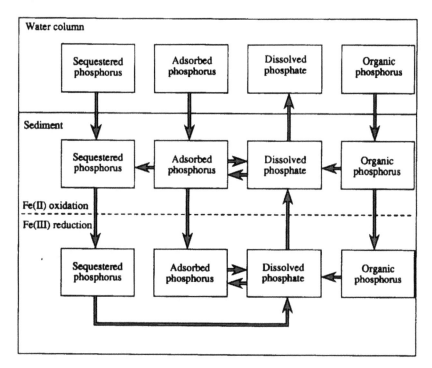

Figure 8. The phosphorus cycle and phosphorus-iron interactions at the sediment-water interface. "Sequestered" phosphorus is defined as phosphorus that is coprecipitated with iron oxides or has diffused into the interior of the iron oxide matrix. Sequestered phosphorus is released when the iron oxide matrix dissolves. (From Sundby, B., Gobeil, C., Silverberg, N., and Mucci, N., *Limnol. Oceanogr.*, 37, 1129, 1992. With permission.)

organic phosphorus, with the latter being both adsorbed to the surface of iron oxides and coprecipitated with iron oxides or otherwise sequestered in the interior of the iron oxide matrix. In the oxidizing surface sediment, a major portion of the sedimentation flux of organic phosphorus is mineralized and the released phosphate is partitioned between the pore water and surface adsorption sites. Surface-adsorbed phosphate is released to the pore water as needed to maintain the equilibrium concentration and replace the dissolved phosphate that escapes to the overlying water. More phosphate is released deeper in the sediment column from iron oxides undergoing reduction. This phosphate is now free to participate in exchange reactions with remaining adsorption sites and to migrate upward and out of the sediment, which it could not do as long as it was sequestered. Sedimentation and biological mixing transport adsorbed, sequestered, and organic phosphorus downward into the reducing region of the sediment, and bioenhanced diffusion transports dissolved phosphate upward towards the sediment surface. Mobilized phosphate, by being reabsorbed in the oxidizing layer, can be recycled several times across the redox boundary for iron, before escaping the sediment. This schematic model, which is consistent with the known chemistry of phosphorus/iron interactions, is summarized in Figure 8.

The buffer capacity of the sediment, which determines how much phosphate can be adsorbed or desorbed before the concentration of dissolved phosphate changes from the equilibrium value, is determined by the concentration of iron oxides which, in turn, are dependent on oxygen. If so much phosphate has been adsorbed on the iron oxides that the buffer capacity is close to being saturated, a small decrease in the iron oxide concentration, brought about by reducing the supply of oxygen, would have a large effect on the phosphate concentration and flux. Otherwise, an abundance of iron oxides would buffer the iron/phosphate system against short-term fluctuations in oxygen. An indication that this mechanism can occur could be seen in the benthic chamber experiment (Figure 6). Iron was rapidly released from the sediment when the water at the interface went anoxic; the release of phosphate was much more gradual.

In shallow-water sediments another mechanism, involving the diurnal cycle of oxygen production and consumption, may be important in controlling phosphate uptake and release. Within the photic zone, epipelagic algae mediate the exchange of phosphate between sediment and water by creating a thin oxygen-containing layer during the day that disappears during the night because of respiration. Carlton and Wetzel[8] have shown that phosphate, diffusing up from deeper layers, is taken up in the oxic microzone during the period of illumination and released again during darkness. This process is attributed to rapid uptake or release of dissolved phosphate by sediment microorganisms in response to oxic or anoxic conditions, respectively.

SUMMARY

This brief overview of sediment-water exchange processes has attempted to demonstrate that the bottom sediment is a highly dynamic environment which reflects the conditions that exist at the sediment-water interface. The upper layer of the sediment is very sensitive to changes in the variables that affect the flux of oxygen into the sediment. The distribution of oxygen and many other solutes in the pore water of sediments, including the trace metals, will adjust quickly to fluctuations in bottom water currents and oxygen concentrations, and consequently there can be rapid changes in the fluxes across the sediment-water interface.

REFERENCES

1. Froelich, P.N., Klinkhammer, G.P., Bender, M.L., Luedtke, N.A., Heath, G.R., Cullen, D., Dauphin, P., Hammond, D., Hartman, B., and Maynard, V., Early oxidation of organic matter in pelagic sediments of the eastern equatorial Atlantic: suboxic diagenesis, *Geochim. Cosmochim. Acta,* 43, 1075–1090, 1979.
2. Aller, R.C. and Yingst, J.Y., Biogeochemistry of tube-dwellings: a study of the sedentary polychaete *Amphitrite ornata* (Leidy), *J. Mar. Res.,* 36, 201–254, 1978.
3. Sundby, B. and Silverberg, N., Manganese fluxes in the benthic boundary layer, *Limnol. Oceanogr.,* 30, 374–382, 1985.
4. Hall, P., Anderson, L., Rutgers van der Loeff, M., Sundby, B., and Westerlund, S., Oxygen uptake kinetics at the sediment-water interface, *Limnol. Oceanogr.,* 34, 734–746, 1989.
5. Boudreau, B.P. and Guinasso, N.L., The influence of a diffusive sublayer on accretion, dissolution and diagenesis at the sea floor, in Fanning, K.A. and Manheim, F.T., Ed., *The Dynamic Environment of the Sea Floor,* Lexington Books, Toronto, 1982, 116–145.
6. Rasmussen, H. and Jørgensen, B.B., Microelectrode studies of seasonal oxygen uptake in a coastal sediment: role of molecular diffusion, *Mar. Ecol. Prog. Ser.,* 81, 289–303, 1992.
7. Revsbech, N.P., Jørgensen, B.B., Blackburn, T.H., and Cohen, Y., Microelectrode studies of photosynthesis and O_2, H_2S, and pH profiles of a microbial mat, *Limnol. Oceanogr.,* 28, 1062–1074, 1983.
8. Carlton, R.G. and Wetzel, R.G., Phosphorus flux from lake sediments. Effect of epipelic algal oxygen production, *Limnol. Oceanogr.,* 33, 562–570, 1988.
9. Howarth, R.W. and Jørgensen, B.B., Formation of 35S-labelled elemental sulfur and pyrite in coastal marine sediments during short-term 35SO42-reduction measurements, *Geochim. Cosmochim. Acta,* 48, 1807–1818, 1984.
10. Anderson, L., Hall, P., Iverfeldt, A., Rutgers van der Loeff, M., Sundby, B., and Westerlund, S., Benthic respiration measured by total carbonate production, *Limnol. Oceanogr.,* 31, 319–329, 1986.
11. Gobeil, C., Silverberg, N., Sundby B., and Cossa, D., Cadmium diagenesis in Laurentian Trough sediments, *Geochim. Cosmochim. Acta,* 51, 589–596, 1987.
12. Westerlund, S., Anderson, L., Hall, P., Iverfeldt, A., Rutgers van der Loeff, M., and Sundby, B., Benthic fluxes of cadmium, copper, nickel, zinc, and lead in the coastal environment, *Geochim. Cosmochim. Acta,* 50, 1289–1296, 1986.

13. Aller, R.C., The effects of macrobenthos on chemical properties of marine sediment and overlying water, in McCall, P.L. and Tevesz, M.J.S., Eds., *Animal-Sediment Relations*, Plenum Press, New York, 1982, 53–102.

14. Sundby, B., Anderson, L., Hall, P., Iverfeldt, A., Rutgers van der Loeff, M., and Westerlund, S., The effect of oxygen on release and uptake of iron, manganese, cobalt, and phosphate at the sediment-water interface, *Geochim. Cosmochim. Acta*, 50, 1281–1288, 1986.

15. Einsele, W., Über Die Beziehugen des Eisenkreislaufs zum Phosphatkreislauf im eutrophen See, *Arch. Hydrobiol.*, 29, 664–686, 1936.

16. Mortimer, C.H., The exchange of dissolved substances between mud and water in lakes. I and II, *J. Ecol.*, 30, 280–329, 1941.

17. Sundby, B., Gobeil, C., Silverberg, N., and Mucci, N., The phosphorus cycle in coastal marine sediments, *Limnol. Oceanogr.*, 37, 1129–1145, 1992.

18. Belzile, N., The fate of arsenic in sediments of the Laurentian Trough, *Geochim. Cosmochim. Acta*, 52, 2293–2302, 1988.

Chapter 4

Synopsis of Discussion Session: The Effects of Variable Redox Potentials, pH, and Light on Bioavailability in Dynamic Water-Sediment Environments

Joseph S. Meyer (Chair), William Davison, Bjorn Sundby, James T. Oris, Darrel J. Laurén, Ulrich Förstner, Jihua Hong, and Donald G. Crosby

INTRODUCTION

In aquatic environments, a predictable sequence of electron acceptors is used during the microbial degradation of organic matter. These reactions determine the redox potential and may alter the pH; they also determine, directly or indirectly, the bioavailability of many inorganic and organic chemicals.

The concept of sequential electron acceptors dates back to the publications of Einsele[1] and Mortimer[2] about annual cycles of redox-related processes affecting the availability of Fe, P, and Si in eutrophic lakes. Subsequent research refined and expanded the concept (See Reference 3) and demonstrated that cycles of C, O, N, P, Si, S, Fe, and Mn are interrelated.[4] Figure 1 shows the set of sequential redox reactions and dissociation equilibria of the classical biogeochemical paradigm. Organic matter (C_{org}) is synthesized when CO_2 (or HCO_3^-) is reduced by plants and bacteria during photosynthesis or chemosynthesis, thus storing energy in an organic-chemical "battery" that is then slowly "discharged" as the organic matter is processed and decomposed by the aquatic community.

pH can be directly affected by such anabolic and catabolic reactions. In systems that are not at equilibrium with the atmosphere (e.g., most sediments and subsurface waters), a decrease in H_2CO_3 consumes protons (H^+); thus, in insufficiently buffered systems, pH increases as a result of photosynthesis and chemosynthesis. In freshwater systems, the rate of pH change depends on the rates of the photosynthetic and chemosynthetic reactions and on the capacity of the bicarbonate-carbonate buffer system (depicted by the H_2CO_3 dissociation and $CaCO_3$ dissolution equilibria in Figure 1); in marine systems, pH remains relatively constant at about 8.2, due to the large buffering capacity of the combined borate and bicarbonate-carbonate systems. pH does not change as CO_2 is consumed in an open system that is at equilibrium with the atmosphere.

A net release of CO_2 occurs as a result of most oxidations of C_{org}, thus providing a driving force to decrease the pH in systems that are not at equilibrium with the atmosphere. However, the natural buffer systems and the presence of acid-soluble minerals containing anions of weak acids (e.g., $CaCO_3$) will at least partly offset the associated increase in H^+ concentration when H_2CO_3 dissociates. Moreover, protons are also consumed during the reduction of most electron acceptors.

Each of the decomposition reactions in Figure 1 that occur under anoxic conditions consumes protons through the redox transformation of the paired inorganic electron acceptor; Davison[5] lists the net proton consumption of each reaction. Even when the production of H^+—a result of the partial dissociation of H_2CO_3—is considered, all of the coupled redox reactions depicted below NO_3^- respiration (Mn reduction through methanogenesis) will tend to consume protons and *increase* the pH (i.e., produce acid-neutralizing capacity).

Besides the direct detrimental effect of anoxia on most aquatic fauna, a major toxicity concern in the classical biogeochemical paradigm is the release of trace metals (e.g., Cd, Cu, Pb, Hg;

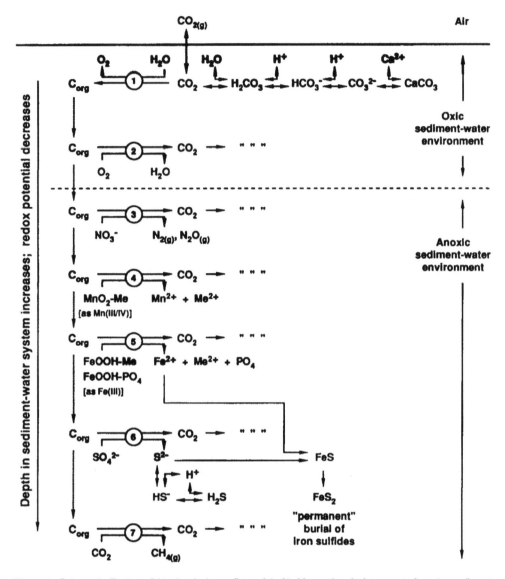

Figure 1. Schematic diagram of the classical paradigm related to biogeochemical processes in water-sediment systems, including metal ions (represented as "Me") and phosphate (PO_4) that associate with iron and manganese oxides. Circled numbers indicate the following reactions: 1—photosynthesis, 2—oxidative metabolism, 3—denitrification, 4—manganese reduction, 5—iron reduction, 6—sulfate reduction, and 7—methanogenesis.

symbolized by "Me" in Figure 1). Trace metals adsorb to manganese oxides (MnO_2) and iron oxyhydroxides (FeOOH), which exist in particulate and colloidal form in oxic water-sediment systems. When the redox potential decreases sufficiently, MnO_2 [as Mn(III/IV)] and FeOOH [as Fe(III)] are reduced to Mn(II) and Fe(II). Because these reduced forms (Mn^{2+} and Fe^{2+}) are relatively soluble in water, the MnO_2-Me and FeOOH-Me complexes dissolve, and the metal ions (Me^{2+}) may become bioavailable. When the redox potential decreases to the level where SO_4^{2-} is reduced to S^{2-}, Fe^{2+} reacts with the S^{2-} to form insoluble monosulfide precipitates (e.g., FeS) that may become "permanently" buried in the sediments as insoluble metal sulfides (e.g., pyrite, FeS_2). Conversely, if the water-sediment system is reoxidized, the classical paradigm predicts that FeOOH particles will form rapidly and readsorb trace metals. Thus, trace metals are depicted as cycling predominantly with Fe. Although geochemists have known for decades that trace metals also form metal-sulfide complexes that can be "permanently" buried in sediments,[6] the interaction between trace metals and sulfides only recently has become a major concern in aquatic toxicology.[7]

The sequence of redox reactions depicted in Figure 1 produces a vertical zonation in a water-sediment system, such as might occur in the sediment of a shallow, well-aerated lake or in the hypolimnion of a stratified, eutrophic lake. Under this scenario, reactants and products and related chemical species could diffuse and otherwise be transported vertically from one layer to another. However, the same reaction sequence also depicts the processes that can occur over time within a layer in a water-sediment system during a period of stratification, as the supplies of electron acceptors diminish. In this scenario, rather than representing increasing depth, the vertical arrow in Figure 1 represents increasing time.

Temporal and Spatial Scales

Periodicity is the rule rather than the exception in nature. Astronomic forces influence life cycles of plants and animals, and we routinely think about the world around us in terms of annual, monthly, and daily cycles. The redox properties of aqueous environments are also influenced by periodic forces, whether they are the annual turnover cycle of the water column or the periodic flushing of a burrow by a benthic organism.

A redox cycle can be thought of as a chain of chemical reactions that is initiated by bringing together reducing and oxidizing compounds, as shown in Figure 1, so they can react with each other. Given enough time, these reactions will eventually proceed to completion and reach a state of thermodynamic equilibrium. More often, though, the chain of redox reactions will be interrupted by some external event before equilibrium is reached; subsequently, the process will start over again. For example, consider the oxidation of organic matter at the sediment-water interface. In an open system with unlimited supply of O_2, C_{org} will be oxidized by bacteria using O_2 as the terminal electron acceptor. In a closed or transport-limited system, however, the reservoir of O_2 may be exhausted before all of the available C_{org} is oxidized. In that case, the oxidation will proceed using NO_3^-, followed by metal oxides, and then SO_4^{2-} as the successive electron acceptors (Figure 1). If the system is not disturbed, each redox pair will buffer the redox potential until the reservoir of the electron acceptor is exhausted (Figure 2a). However, if a mixing event transports oxygenated water down to the sediment surface before the O_2 reservoir is depleted, the use of other inorganic electron acceptors will not be required.

The *periodicity* of the redox cycle, therefore, depends on the periodicity of the forcing event, whereas the *amplitude* of the redox cycle depends on how far the reactions proceed towards equilibrium during each cycle. Because the reservoir of C_{org} in sediments is usually much greater than the reservoir of inorganic electron acceptors, the extent to which the electron acceptors are consumed depends on the amounts initially present, the rates at which they are consumed, and the rates at which they are resupplied. Thus, the amplitude of the redox cycle depends on the periodicity of the forcing event, the amounts of each electron acceptor initially present, the reaction rates, and the rates of resupply of the electron acceptors. Figure 2b shows conceptually the stepwise sequence of decreasing redox potentials that might be observed between two disturbances.

Consider the bottom water of a stratified, deep lake and the water inside the burrow of a benthic organism as two examples of this general principle. The bottom water of the lake is effectively isolated from the atmosphere for long periods of time, and O_2 is only significantly resupplied during the annual (or semiannual) turnover that follows the breakdown of thermal stratification. Between these events, there may be time for the O_2 reservoir to be consumed completely and for the reactions to proceed to SO_4^{2-} reduction and beyond. Therefore, the lake's bottom-water environment—especially at the sediment-water interface—has low forcing periodicity and high redox amplitude. In contrast, a burrow is flushed frequently, because low O_2

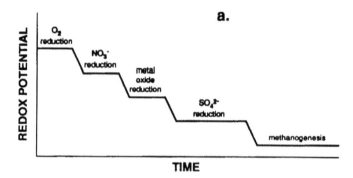

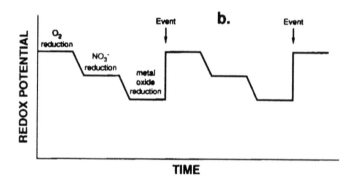

Figure 2. Stepwise decreases in redox potential over time in a sediment, as progressively less energetically favorable electron acceptors are used to oxidize organic matter: (a) in a closed system that proceeds to SO_4^{2-} reduction without disturbance, and (b) in a system that is reoxygenated periodically during the phase when a metal oxide is being used as the electron acceptor.

concentrations induce the burrow dweller to exchange the water in the burrow with the O_2-containing overlying water, thus resupplying its habitat with O_2 and evacuating metabolic by-products. Therefore, the burrow environment has high forcing periodicity and low redox amplitude.

Periodic physical, chemical, and biological phenomena linked to redox processes—with time scales of minutes to years—are listed in Table 1 in terms of driving forces, resulting processes, and effects on bioavailability. Although some of the processes that are related to annual cycles in water-sediment systems have been studied extensively, information about processes that have shorter time scales is relatively limited. Furthermore, little information exists about the effects of such periodic redox cycles on speciation and transport of potential toxic chemicals in sediments, and about the exposure of sediment-dwelling organisms to those chemicals.

Freshwater Vs. Marine Systems

A major difference between many freshwaters and marine environments is that seawater has a high capacity to buffer pH. This high buffering capacity tends to maintain constant conditions in seawater for reactions involving acid-base dissociations, ion exchanges involving protons, and pH-dependent dissolutions and precipitations. Freshwater and marine systems also differ because SO_4^{2-} concentrations generally are much higher in marine waters. Therefore, in marine environments, SO_4^{2-} reduction is relatively more important and methanogenesis usually begins later in

Table 1.   Periodic Disturbances That Affect Redox-Related Processes and Bioavailability of Chemicals

Time scale	Phenomenon	Driving force	Resulting process	Effect on bioavailability
1 year	Seasonal production	CO_2 consumption in photic zone	pH increase (up to pH 10) in photic zone during summer	Release of PO_4 from shallow-water sediments Hydrolysis/adsorption of trace metals*
1 year	Seasonal production	Increased supply of organic carbon	Trace-metal uptake by growing phytoplankton and/ or their exudates	Decrease of available trace-metal concentration in photic zone
1 year	Seasonal production	Increased supply of organic carbon	Increased supply of metals in biota, and increased reducing intensity in surface sediment (0–5 cm)	Chain of events initiated: 1. Initial release of trace metals, due to oxidative decomposition of biota 2. Subsequent binding of trace metals to sulfides, when SO_4^{2-} is reduced
1 year	Seasonal production	Increased supply of organic carbon	pH decrease (~0.5 pH unit) in hypolimnion and oxic upper layer of sediment, due to decomposition	Release of trace metals due to dissolution and exchange reactions*
1 year	Seasonal stratification of lakes	Diminished supply of O_2	Release of Fe^{2+}, Mn^{2+}, and S^{2-} in hypolimnion (minimum of several m deep)	Possible release of Co and Ni, but no effect on Cu, Pb, and Zn
1 month	Lunar tidal cycle (maximum current velocities are lunar-dependent)	Gravitational attraction of moon	Periodic inundation, flushing, and aeration of extreme intertidal-zone sediment	Resuspension and oxidation of reducing sediment; can include longer-term (years) change in sediment character*
1 day	Photosynthesis	Diurnal CO_2 consumption	Diurnal pH shift (usually < 0.5 pH unit)	Small diurnal release of PO_4 from shallow-water sediments?* Diurnal hydrolysis/ adsorption of metals?*
1 day	Photosynthesis	Diurnal organic carbon production	Diurnal uptake of metals by phytoplankton (probably undetectable)*	Diurnal changes in trace-metal availability (probably undetectable)*
1 day	Photoreduction	Light in photic zone	Reduction of Fe-, Mn-, and other metal oxides	Fe (essential nutrient) available for uptake by organisms as Fe^{2+}; more general availability of trace metals (e.g., Zn)?*

Table 1. (Continued)

Time scale	Phenomenon	Driving force	Resulting process	Effect on bioavailability
1 day	Photooxidation	Light in photic zone	Partial or complete oxidation of organic ligands (e.g., humics)	Release or uptake of inorganic and organic ions*
12 hours	Semi-diurnal tidal cycle (prime determinant of instantaneous current velocities)	Gravitational attraction of moon	Periodic inundation, flushing, and aeration of intertidal-zone sediment	Resuspension and oxidation of reducing sediment; can include longer-term (years) change in sediment character*
Minutes	Advection in rippled, sandy sediment	Surface gravity waves	Advection of water through surface sediment (0–5 cm)	Exchange flux of material at most active aquatic site— the sediment-water interface
Minutes	Pore-water flushing	Physical motion of organisms	Exchange of water between sediment environment and overlying water (0– 30 cm)	Perturbation of redox environment within sediment

Note: Asterisk (*) indicates a reasonable hypothesis, rather than an established fact.

the sequence of C_{org} oxidation.[8] In fact, methanogenesis may be absent when the rate of supply of C_{org} is relatively low compared to the rate of supply of SO_4^{2-}.

It has often been suggested that the pyrite (FeS_2) content of marine sediments is higher than the pyrite content of freshwater sediments because of the greater supply of S^{2-} in marine sediments.[9] However, the high FeS_2 content in many freshwater sediments caused Davison et al.[10] to question that generality. Formation of iron sulfides in sediments may depend as much on the rate of supply of electrons in the form of reactive C_{org} as on the rate of supply of S.[11]

Thus, it is not surprising that DiToro et al.[7] found that the ranges of reported concentrations of acid volatile sulfide (AVS, a measure of mobilizable sulfides that mainly comprises iron monosulfides) were similar in freshwater and marine sediments. Sulfide can be supplied by the decomposition of organic matter as well as by reduction of SO_4^{2-}. Consequently, the concentrations of AVS in sediments ultimately may be limited by the amount of available metal ions that can react with S^{2-}. Because the amounts of available metal ions are similar in freshwater and marine systems,[12] the two systems may have similar maximum possible concentrations of AVS. For example, DiToro et al.[7] compiled published AVS contents that ranged from 0 to 43 μmol/g in marine sediments and from 0.3 to 112 μmol/g in freshwater sediments. However, the observed AVS concentration will also depend on the rate of conversion of iron monosulfides to pyrite, which is not a component of AVS. Because the ranges presented in DiToro et al.[7] were so large, the overlapping ranges of AVS concentrations in freshwater and marine sediments may have little practical significance.

ORGANIC CHEMICALS

In our representation of the classical biogeochemical paradigm (Figure 1), the aerobic and anaerobic degradation of C_{org} to CO_2 is greatly simplified. In reality, the pathways of organic matter degradation are complex, especially when considering the flow of organic toxicants through an ecosystem. Microbial catabolism (primarily oxidation and hydrolysis) of organic matter is important, but a variety of other important reactions and processes—abiotic and biotic—also alter the structure or transport of organic chemicals in water-sediment systems.

Abiotic Factors

Solar radiation is an important abiotic factor affecting chemical degradation in the aquatic environment. Two important classes of abiotic reactions caused by solar radiation are photolysis (light-induced hydrolysis) and photoredox reactions (light-induced reductions and oxidations).

Photolysis reactions in aquatic systems have been investigated for several decades (See References 13 and 14), and these reactions are generally considered to produce compounds that are more water-soluble and less toxic than the parent compound. However, organic chemicals long known to cause photosensitization reactions in terrestrial organisms have also demonstrated light-mediated toxicity in aquatic organisms. For example, polycyclic aromatic hydrocarbons cause photo-induced toxicity in fish and invertebrates.[15]

Photooxidations and photoreductions may play an important role in altering chemicals, either increasing or decreasing bioavailability and toxicity. In some aquatic systems, photoredox reactions could dominate degradation of xenobiotics within the photic zone and the upper 1 mm of sediment, especially for pesticides in shallow aquatic systems.[16]

Biotic Factors

Organisms may either directly or indirectly affect the bioavailability of toxicants. Although this concept received little attention during the workshop, the direct and indirect effects of biotic factors may be significant.

Direct Effects of Biota

The metabolic detoxification systems (e.g., Phase I—cytochrome P450 enzymes; Phase II—conjugating enzymes) of many organisms are evolutionarily conserved.[17] The primary function of Phase I enzymes is to catalyze monooxygenation reactions, the net result of which is the addition of OH—imparting greater water solubility and, hence, expediting excretion of the metabolites. Water-soluble metabolites excreted by an organism generally are assumed to be less bioavailable to other organisms, although they are not necessarily less toxic. For example, the conversion of benzo[a]pyrene to its carcinogenic 7,8-dihydro-9,10-epoxide is a classic example of xenobiotic activation as a result of a monooxygenation reaction.

Another important reaction is the reductive dehalogenation of a variety of pesticides and other halogenated compounds. Crosby and Hamedmad[18] and Reinhard et al.[19] demonstrated that there is both a biotic and an abiotic component to this reaction. These reactions have been observed in and attributed to microbes, some insects, and plants; however, hydrophobic chemicals sorbed to particles seem to be protected from reductive biodegradation. The relative importance of these reactions may be related to the ecosystem's productivity.[20] For example, reductive dehalogenation of DDT to the less persistent DDD (and thence to MDE, etc.) presumably dominated in eutrophic Lake Erie, whereas oxidative dehalogenation to the stable end product and biologically active DDE appeared to dominate in the oligotrophic Great Lakes—Michigan, Huron, and Superior.[53]

Bioavailability of several metals (As, Hg, Pb, Se) may increase as a result of biomethylation. These forms are lipid soluble and can be highly toxic. D'Itri[21] reviewed the biotic and abiotic processes by which organometallic compounds of As, Hg, and Pb are formed in sediments, and Cooke and Bruland[22] recently presented evidence of the biotic formation of organo-Se compounds in aquatic systems.

Because most aquatic organisms are ectothermic, metabolic rates will vary seasonally with temperature, and community-level metabolic rates could play an important role in chemical speciation. However, little is known concerning this potential impact.

Indirect Effects of Biota

Macroorganisms that use sediments either as a habitat or as a source of food may disturb the sediments. This disturbance can suspend sediment particles and allow the desorption and release of toxic chemicals.[23,24] Activities of tube-dwelling and burrowing benthic organisms can increase the depth of the oxic zone, change the redox potential, move sequestered substances to the surface, and affect the transport of dissolved chemicals. It also appears that the tube itself alters the microenvironment of the organism, enhancing microbial activity, adsorbing neutral organics, and altering the uptake of anions.[54]

As a result of excreted metabolites and secondary decomposition products, concentrations of dissolved organic carbon (DOC) may closely parallel the seasonal patterns of primary productivity in aquatic systems. Additionally, seasonality in the intensity of solar irradiance will cause seasonality in the relative importance of photoredox reactions affecting the DOC.[25] Because DOC is known to alter bioavailability, seasonal cycles of DOC concentrations can create seasonal cycles in the response of aquatic communities to toxicants. For example, during early spring, aquatic invertebrate communities demonstrate much greater sensitivity to Cu than in summer or fall.[26] When combined with the possibility of other seasonal influences on the bioavailability of toxicants in early spring (e.g., melt-off of "acid snow" that reduces pH and increases the concentrations of available metals), this seasonal low of DOC concentration may become significant, even in areas not receiving point-source discharges.

METALS

According to conventional wisdom, all trace metals are viewed as cycling with Fe and Mn. When iron and manganese oxides are present, trace metals are thought to adsorb primarily to those solid phases. Because this is not entirely correct, we briefly review in the following subsections the current knowledge about (1) the abiotic and biotic reduction and dissolution of metal oxides, and (2) the associated uptake and release of trace metals from those metal oxides. Then we discuss the binding of trace metals to acid volatile sulfide (AVS) and the release of those trace metals from dredged sediments.

Mechanisms of Reduction and Dissolution of Metal Oxides

The reductions of Fe(III) to Fe(II) and of Mn(III/IV) to Mn(II) are influenced by complexation and by the mineral form of the oxyhydroxide. Complexing agents generally interact more strongly with Fe(III) than Fe(II); thus, with few exceptions, complexation shifts the thermodynamic equilibrium in favor of the reduced species.[27] However, in natural systems, the extent of oxide reduction is usually controlled by the kinetics of the reduction reactions, rather than by thermodynamic considerations.

Precise mechanisms for the reduction of Fe(III) oxyhydroxides in natural systems are still poorly defined.[28] Most dissolved organics resulting from the decomposition of natural organic matter are thermodynamically capable of reducing Fe(III), but they may be kinetically limited. Photoactivation enhances the reduction in surface waters. Fe(III) may also be reduced at the surfaces of phytoplankton by organic compounds that they exude. The rate of reduction generally increases with decreasing pH, and reduction is thermodynamically more favorable at low pH.

More is known about the mechanism of reduction by S^{2-}, which is likely to be a significant reductant of iron oxyhydroxides in strongly reducing sediments, at the sediment-water interface, and in some anoxic basins. The rate of reduction depends directly on the concentration of protonated surface sites and on the concentration of HS^- in solution,[29] resulting in a marked pH dependency with a pronounced maximum rate between pH 6 and 7. Such pH values are typical of anoxic waters, because the redox reactions that occur there actively buffer the pH to a value close to 7. Anoxic systems therefore exhibit a synergism: reduction reactions buffer pH to a value that favors reduction reactions.

In anoxic sediments, not all of the iron oxide is reduced. Black iron sulfide particles collected from the anoxic bottom water of a lake have been shown to be a mixture of intact iron oxide and

FeS.[30] Furthermore, Fe(III) oxyhydroxides are a substantial component of the Fe fraction in strongly reducing, sulfidic sediments.[28] Conversely, manganese oxyhydroxides are generally more readily reduced than their Fe counterparts. Natural organic material isolated from freshwater and sea water is able to reduce Mn(III/IV) to Mn(II) directly. In the presence of light, the reaction proceeds much more rapidly.[31]

Mechanisms of Uptake and Release of Trace Metals from Metal Oxides

Laboratory studies, which clearly show that trace metals are strongly adsorbed to iron and manganese oxides, are not supported by field evidence. For example, in a seasonally anoxic lake that had elevated concentrations (3 mg/l) of freshly formed iron oxides in the vicinity of the redox boundary in the water column, there was no accompanying diminution in the concentrations of dissolved Zn, Pb, or Cu.[32] Neither was there an accumulation of these elements in the bottom water, which might have indicated release from Fe or Mn particles as they were reduced in the deep waters or at the sediment-water interface. Furthermore, Gendron et al.[33] observed strongly contrasting distributions of Co and Cd in a marine sediment. Co appeared to precipitate within the thin, oxic upper layer of sediment (where iron and manganese oxides are abundant), whereas Cd remained dissolved.

Balistrieri et al.[34] also observed that vertical profiles of Cd, Cu, and Zn in the water column of a lake were not affected by the redox-driven cycling of Fe and Mn. However, concentrations of Co and Ni were elevated in bottom waters, and those concentrations correlated better with concentrations of reduced Mn than with Fe. Cr(VI) can be reduced to Cr(III) in anoxic lake waters,[34,35] but the Cr(III) is rapidly depleted, indicating that it is adsorbed to iron or manganese oxyhydroxides. Consequently, there is circumstantial evidence from water-column concentrations that Co, Ni, and Cr(III) adsorb to freshly formed manganese and iron oxyhydroxides in natural lake waters, but no evidence for uptake of Cd, Cu, or Zn. However, flux measurements using sediment traps have indicated that Cu and Zn may be transported by manganese oxides in the bottom waters of lakes.[36]

Selective extraction schemes also have indicated that substantial fractions of trace metals in sediments may be associated with the iron oxide fraction; hence, they presumably are available for release on reduction.[11] Binding constants derived from sediment data are also compatible with those for iron oxides,[37] although accurate comparisons are difficult due to the large range of values obtained by different workers as a result of variations in experimental conditions.

Reductive release of metals at the sediment-water interface, or within sediments, may not result in elevated concentrations in bottom waters or interstitial waters. The reducing conditions favor the formation of sulfides, which may react with the trace metals to form insoluble metal sulfides. In sediment-interstitial waters of productive lakes, the trace-metal concentrations are usually lower than the values at the sediment-water interface, presumably due to the formation of sulfides, either directly or by replacing Fe^{2+} or Mn^{2+} in their sulfides.[7] There is also the possibility that iron sulfides may be extremely reactive with respect to trace-metal adsorption,[30] but systematic adsorption experiments have not been conducted.

Uptake of trace metals by iron and manganese oxyhydroxides is known to increase dramatically as pH increases. Therefore, if the oxyhydroxides encounter a low pH environment, release may be possible. Reversibility of adsorption/desorption processes for natural manganese oxides has been demonstrated for Cu, Zn, and Cd by exposing a natural sample collected from a disused mine to a series of solutions in the pH range 1 to 7;[38] Pb and Co were less easily released. By contrast, the required pH for desorption of metals from a natural goethite sample was lower than would be expected from adsorption experiments, perhaps due to aging effects. Whatever the reason, the results indicated that adsorption experiments cannot be used to predict the pH-dependent release of metals from natural iron oxides.

In a different experiment,[39] Co, Zn and Mn were released when the pH of lake water was lowered from 6.5 to 4.8, indicating that Co and Zn may be associated with a reactive fraction of manganese oxyhydroxides or with reactive exchange sites. Concentrations of Sn, Se, Cr, Ag, and Hg were not affected by this treatment.

General scavenging of trace metals by natural particles has been demonstrated by several workers,[36,40] but organic compounds rather than metal oxides may be primarily responsible for this removal. For example, Fu et al.[41] recently reported that all of the binding of Cd that they added to three river sediments could be explained by the presence of humic acids associated with those sediments. In fact, it is likely that most naturally occurring metal oxide surfaces are coated with a layer of adsorbed organic compounds.[42] Yet relative to the large amount of research conducted in laboratory systems using "pure" metal oxides, little is known about the details of the associations between trace metal ions and iron and manganese oxides in natural systems. Trace metals may actually associate with organic matter that only incidentally is associated with the metal oxides.[43]

Binding of Trace Metals By Acid Volatile Sulfide

The use of the total concentration of a trace metal in sediment as a measure of its bioavailability is not supported by data, because different sediments exhibit different degrees of bioavailability for the same total quantity of metal.[7,44] These differences have recently been reconciled by relating organism response to the metal concentration in the sediment-interstitial water.[7,45] Presumably, the chemical activity of the metal—not the concentration—correlates best with toxicity (see other chapters of this volume for further discussion of concentration vs. activity as a measure of bioavailability).

One of the major chemical components that controls metal activities in the interstitial water of anoxic sediments is acid volatile sulfide (AVS). AVS is operationally defined as the sulfides that are liberated from a sediment sample to which acid has been added at room temperature under anoxic conditions. This operational definition includes most of the amorphous and moderately crystalline monosulfides (e.g., FeS) and lesser percentages of other sulfides.[16] A closely related term is simultaneously extractable metals (SEM), which can be operationally defined as "metals . . . [that] form less soluble sulfides than do Fe or Mn and [that] are at least partially soluble under [the same test conditions in which the AVS content of the sediment is determined]."[46]

The chemical basis for the primacy of the sediment sulfide phase for metal binding is assumed to be that, at equilibrium, S^{2-} successfully outcompetes all other common dissolved or particle-associated ligands for metal ions and forms insoluble metal sulfides.[7,47] Because the Fe content of many sediments is much higher than the content of other metals, iron sulfides usually dominate the metal-sulfide pool. Most trace metals (e.g., Ni^{2+}, Zn^{2+}, Cd^{2+}, Pb^{2+}, Cu^{2+}, Hg^{2+}, listed in increasing order of affinity)[7] have higher affinity for S^{2-} than does Fe^{2+} (i.e., the solubility products of the trace-metal monosulfides are lower than the solubility product of FeS). If trace-metal ions are added to the interstitial water, the following net replacement reaction with FeS is postulated to occur:[7]

$$Me^{2+} + FeS_{(s)} \rightarrow Fe^{2+} + MeS_{(s)} \tag{1}$$

The overall stoichiometry of the replacement reaction is 1 mol Me^{2+} removed per mole of FeS. Hence, if no other strong complexing ligand is present, the trace metal activity will reflect the molar amount of metal in excess of the molar amount of AVS in the sediment. It is likely that sediment quality criteria that are being developed in the U.S. for metals will take into account the amount of AVS present relative to the amount of SEM. Under this scenario, a ratio of $\Sigma SEM/AVS < 1$ could be assumed to indicate a potentially low metal activity (and, thus, potentially low metal toxicity), whereas a ratio of $\Sigma SEM/AVS > 1$ could indicate potentially high metal activity (and, thus, potentially high metal toxicity).

We have several concerns about using the ΣSEM/AVS ratio. First, the replacement reaction suggested by Equation 1 has not been unequivocally established to occur in nature. Although experiments are being conducted to address that question,[55] no one has yet reported in the published literature an equimolar increase of Fe^{2+} concentration as sediment that contains FeS is titrated with a trace metal ion that would replace the Fe^{2+}.

Second, results of chemical analyses of homogenized sediment collected with grab samplers do not necessarily indicate the bioavailability of trace metals in intact, in-place sediments. Natural sediments are stratified,[48] and only in certain circumstances will AVS be at the sediment surface. If the sediment is oxic or weakly reducing, SO_4^{2-} will not be reduced to S^{2-}, and most of the previously formed sulfides (if any were present) will be oxidized to SO_4^{2-}, allowing the trace metal ions to be released and possibly associate with other dissolved or particulate components. Thus, the ratio ΣSEM/AVS may not accurately indicate trace-metal activity at the sediment-water interface. Addressing this question, studies are currently being conducted to test whether toxicity of metals in intact, stratified sediments correlates well with the ΣSEM/AVS ratio.[56]

Third, AVS is not likely to be present at a uniform concentration in the sediment environment. An instantaneous measurement of AVS represents, at best, an ephemeral state that depends on the rates of supply of reactive C_{org}, SO_4^{2-}, and trace metals. These rates will be affected by periodic and episodic events. Addressing this question, studies are currently being conducted to determine temporal variation in sediment AVS concentrations in freshwater and marine systems.[56]

Fourth, and most importantly, in an established sediment the biota create microenvironments in which the chemistry differs from the bulk sediment. Burrowing organisms, which occur more commonly in marine systems than in freshwaters, pump oxic water into their burrows,[49] causing a localized high redox potential that affects the local concentrations of AVS and trace metals, regardless of the bulk content of AVS in the surrounding sediment. Additionally, pH may be altered by excretion of CO_2 and NH_3. Bioavailability of metals at the temporal and spatial scales relevant to such organisms has not yet been studied, and almost nothing is known about the exposure of sediment dwellers to trace metals (or, for that matter, to other inorganic and organic compounds) due to advection currents occurring in the vicinity of the sediment-water interface.

In summary, AVS appears to be important for removing trace metals from polluted waters (e.g., wastewater treatment wetlands) and for net burial of trace metals in sediments over geologically relevant time scales. The ratio ΣSEM/AVS also appears to be a useful concept to explain toxicity of organisms exposed to homogenized bulk sediments in laboratory test systems. However, it remains to be demonstrated whether that ratio is appropriate for predicting bioavailability of metals to organisms inhabiting oxic or partially reducing regions of in-place sediments. Tests with homogenized bulk sediments may be suitable to mimic the exposure conditions that might occur during a dredging operation, before the AVS is oxidized.[50] However, in that situation the potential toxicity of trace metals probably will be overshadowed by the catastrophic effects of the dredging operation on the physical habitat of the organisms.

Release of Trace Metals from Dredged Sediments

Regulatory concern about dredging sediments from harbors has prompted speculation about the potential release of trace metals when anoxic sediments that contain metal sulfides are exposed to oxygenated waters and/or to the air. It can be expected that trace metals will, at least temporarily, dissolve when the metal sulfides are oxidized. Although one might speculate that those trace metals would immediately adsorb onto iron oxyhydroxides when the iron in the resuspended sediments is oxidized (e.g., see Reference 51), long-term monitoring studies showing that trace metals can be released during the dredging operation[52] and from dredged sediments that are stored on land for decades[50,53] indicate that such speculation is not necessarily correct.

Therefore, we believe that the concept that *all* trace metals cycle with iron and manganese is false and should be discarded. Sundby[48] showed that the concept was not supported by the results of benthic flux chamber experiments; and Hong et al.[50] showed that a large percentage of Cu remobilized from harbor sediment can partition onto algal cell walls, and that smaller percentages will partition to iron and manganese oxides and to clay. In order to predict trace metal availability from sediments that are resuspended into the water column during dredging, additional research will be needed on (1) the associations of metals with iron oxides, clay particles, algal cells, etc., in natural waters; (2) pH-mediated exchange reactions; and (3) the range of environmental conditions under which these potential associations occur.

CONCLUSIONS AND RECOMMENDATIONS

In this chapter we have raised many questions about physical, chemical, and biological interactions that affect bioavailability in the water column and sediments, but we have offered few suggestions to improve the regulatory approach to protecting the quality of the sediment environment. The large number of questions and the limited number of suggestions reflect the limited knowledge about temporal and spatial scales of photoredox reactions occurring in the water column and on the sediment surface, bacterial processes and chemical reactions, and transport of chemicals within sediments and across the sediment-water interface.

We feel compelled to raise a concern that does not appear to be well appreciated by many environmental toxicologists. Contrary to popular belief, a sediment is not simply a pile of mud. A sediment is the habitat of living organisms, it is heterogeneous, and it is spatially and temporally structured. Although relatively little is known about the physical and chemical features of the microenvironments that sediment-dwellers inhabit, and even less about the bioavailability of chemicals to those organisms, the general concepts of sediment chemistry and the dynamic nature of the sediment environment should be incorporated into proposed sediment quality criteria.

The lack of appreciation of the structured nature of sediments, and of the way organisms inhabit and modify the sediment, has led to what we consider to be a "steam-shovel" approach to sampling and testing; i.e., collecting sediment with a grab-sampler and then homogenizing it for chemical analysis and toxicity testing. We believe that this is the equivalent of studying gill physiology by excising a gill arch from a fish with a butcher knife, homogenizing it in a blender, placing the homogenate in a beaker, and then adding a metal to determine uptake rates.

The structured nature of sediments, and the way organisms inhabit and modify the sediment, suggests that sediment toxicity experiments should be conducted on intact cores, rather than on homogenates. For certain purposes, homogenizing may be heuristic. However, a more realistic simulation of the actual situation is achieved using intact sediments. Technologies that are currently being developed for microscale (submillimeter) levels of resolution in physical and chemical analyses will be needed to advance the understanding of the distribution and transport of chemicals in water-sediment systems.[54]

In summary, we offer the following general conclusions about the current understanding of environmental factors that affect bioavailability in dynamic water-sediment environments. Additionally, we suggest new research needed to improve the scientific basis of regulatory activities.

1. Dynamic processes that affect bioavailability occur in sediments on time scales ranging from minutes to years. Generally, processes that occur at higher frequencies become increasingly important as depth of the water column decreases.

   Research needs:
   • Conduct basic research on short-, intermediate- and long-term, interrelated dynamic processes (physical, chemical and biological) occurring in sediment systems.

2. Most benthic organisms live in microenvironments in which the bioavailability of chemicals cannot be adequately studied by testing bulk sediment samples.

Research needs:
- Develop techniques to investigate the chemical conditions in structured sediments at submillimeter levels of resolution.

3. In most sediments, it will not be true that the organic fraction controls only the associations of organic chemicals, and the inorganic fraction controls only the associations of inorganic chemicals.

Research needs:
- Conduct basic and applied research into chemical-chemical and chemical-particle interactions in sediments under *realistic* conditions (i.e., not only with pure chemicals in the laboratory).

4. Microbial catabolism is not always the major process by which organic matter is degraded in a water-sediment system. Photoredox reactions may also directly and indirectly affect the bioavailability of organics (and metals) in the photic zone of the water column and in the upper 1 mm of sediments to which sufficient light penetrates.

Research needs:
- Conduct basic research on microbial ecology in sediments.
- Conduct research on microbial catabolism of xenobiotics in marine sediments.
- Continue research on photoredox reactions in natural waters and sediments, especially related to the influence of short-wavelength UV light.

5. Oxidation, reduction, and hydrolysis of organic chemicals affect the availability of those chemicals, but these reactions do not always lead to detoxification.

Research needs:
- Demonstrate a direct link between photodegradation and bioavailability.
- Conduct research on detoxification processes across a broad range of organisms.

6. Not all trace metals cycle with Fe and Mn. For example, Cd, Cu, and Zn (and possibly Ni) appear to be involved in other element cycles.

Research needs:
- Conduct more *in situ*, detailed studies of the behavior of trace metals in natural and contaminated sediments.

7. Trace metals can be released from resuspended and dredged sediments and become bioavailable, immediately after resuspension or dredging and after long storage times in oxic environments.

Research needs:
- Determine kinetics of desorption of metals from, and ion exchange with, components of natural sediments under anoxic conditions and during subsequent oxidation of the sediment.
- Determine kinetics of oxidation of sediment components.

8. The concentration of acid volatile sulfide (AVS) in freshwater and marine sediments depends on many factors, including the rates of supply of reactive organic matter, reactive metal ions, sulfide from the reduction of sulfate, and sulfide from decomposition of organic matter.

Research needs:
- Develop an expanded database on AVS in freshwater and marine sediments, especially related to vertical profiles of AVS.
- Conduct basic research on the formation and stability of metal sulfides in sediments, especially related to the effects of natural and artificial organic complexing agents.
- Conduct basic research on fluxes of S in sediments.

9. AVS may not directly control bioavailability of trace metals in oxic sediments.

Research needs:
  • Correlate the ratio ΣSEM/AVS with survival of organisms inhabiting *in-place* sediments.

ACKNOWLEDGMENTS

We thank Herb Allen and Lee Wolfe for providing timely information and stimulating discussions during the workshop, and Gary Ankley and Dominic DiToro for comments on the manuscript.

REFERENCES

1. Einsele, W., Ueber die Beziehungen des Eisenkreislaufes zum Phosphatkreislauf im eutrophen See, *Arch. Hydrobiol.*, 29, 664–686, 1936.
2. Mortimer, C.H., The exchange of dissolved substances between mud and water in lakes, *J. Ecol.*, 29, 280–329, 1941; 30, 147–201, 1942.
3. Davison, W. and Tipping, E., Treading in Mortimer's footsteps: the geochemical cycling of iron and manganese in Esthwaite Water, in *Fifty-second Annu. Rep. Freshwater Biological Association*, Freshwater Biological Association, Ambleside, U.K., 1984, 91–101.
4. Froelich, P.N., Klinkhammer, G.P., Bender, M.L., Luedtke, N.A., Heath, G.R., Cullen, D., Dauphin, P., Hammond, D., Hartman, B., and Maynard, V., Early oxidation of organic matter in pelagic sediments of the eastern equatorial Atlantic: suboxic diagenesis, *Geochim. Cosmochim. Acta*, 43, 1075–1090, 1979.
5. Davison, W., Internal elemental cycles affecting the long-term alkalinity status of lakes: implications for lake restoration, *Schweiz. Z. Hydrol.*, 49, 186–201, 1987.
6. Morse, J.W., Millero, F.J., Cornwell, J.C., and Rickard, D., The chemistry of hydrogen sulfide and iron sulfide systems in natural waters, *Earth-Sci. Rev.*, 24, 1–42, 1987.
7. DiToro, D.M., Mahony, J.D., Hansen, D.J., Scott, K.J., Hicks, M.B., Mayr, S.M., and Redmond, M.S., Toxicity of cadmium in sediments: the role of acid volatile sulfide, *Environ. Toxicol. Chem.*, 9, 1487–1502, 1990.
8. Mackenzie, F.T. and Wollast, R., Thermodynamic and kinetic control of global chemical cycles of the elements, *Phys. Chem. Res. Rep.*, 2, 45–59, 1977.
9. Berner, R.A. and Rainwell, R., C/S method for distinguishing freshwater from marine sedimentary rocks, *Geology*, 12, 365–368, 1984.
10. Davison, W., Lishman, J.P., and Hilton, J., Formation of pyrite in freshwater sediments: implications for C/S ratios, *Geochim. Cosmochim. Acta*, 49, 1615–1620, 1985.
11. Davison, W., Interactions of iron, carbon and sulphur in marine and lacustrine sediments, in Fleet, A.J., Kelts, K., and Talbot, M.R., Eds., *Lacustrine Petroleum Source Rocks*, Spec. Publ. 40, Geological Society of America, Boulder, CO, 1988, 131–137.
12. Salomons, W. and Förstner, U., *Metals in the Hydrocycle*, Springer-Verlag, Berlin, 1984.
13. Crosby, D.G., The photodecomposition of pesticides in water, in Gould, R.F., Ed., *Fate of Organic Pesticides in the Aquatic Environment*, Adv. Chem. Ser. III, American Chemical Society, Washington, DC, 1972, 173–188.
14. Zepp, R.G., Photochemical transformations induced by solar ultraviolet radiation in marine ecosystems, in Calkins, J., Ed., *The Role of Solar Ultraviolet Radiation in Marine Ecosystems*, Plenum Press, New York, 1982, 293–307.
15. Oris, J.T. and Giesy, J.P., Jr., The photoinduced toxicity of anthracene to juvenile bluegill sunfish (*Lepomis macrochirus* Rafinesque): photoperiod effects and predictive hazard evaluation, *Environ. Toxicol. Chem.*, 5, 761–768, 1986.
16. Crosby, D.G., Photochemical aspects of bioavailability, this volume, Section 5, Chapter 1, 1994.
17. Nebert, D.W., Nelson, D.R., and Feyereisn, R., Evolution of the cytochrome P450 genes, *Xenobiotica*, 19, 1149–1160, 1989.

18. Crosby, D.G. and Hamedmad, N., Photoreduction of pentachlorobenzenes, *J. Agric. Food Chem.*, 19, 1171–1174, 1971.

19. Reinhard, M., Curtis, G.P., and Kriegman, M., Abiotic Reductive Dechlorination of Carbon Tetra-chloride and Hexachloroethane By Environmental Reductants, Res. Dev. Project Summary No. EPA/600/S2–90/040, U.S. Environmental Protection Agency, Ada, OK, 1990.

20. Vanderford, M.J. and Hamelink, J.L., Influence of environmental factors on pesticide levels in sport fish, *Pestic. Monit. J.*, 11, 138–145, 1977.

21. D'Itri, F.M., The biomethylation and cycling of selected metals and metalloids in aquatic sediments, in Baudo, R., Giesy, J.P., and Muntau, H., Eds., *Sediments: Chemistry and Toxicity of In-place Pollutants*, Lewis Publishers, Ann Arbor, MI, 1990, 163–214.

22. Cooke, T.D. and Bruland, K.W., Aquatic chemistry of selenium: evidence of biomethylation, *Environ. Sci. Technol.*, 21, 1214–1219, 1987.

23. Keilty, T.J., White, D., and Landrum, P.F. Sublethal responses to endrin in sediment by *Limnodrilus hoffmeisteri* (Tubificidae), and in mixed culture with *Stylodrilus heringianus* (Lumbriculidae), *Aquat. Toxicol.*, 13, 227–250, 1988.

24. Clements, W.H., Oris, J.T., and Wissing, T.E., Transfer of polycyclic aromatic hydrocarbons from sediments to benthic macroinvertebrates and fish, *Bull. N. Am. Benthol. Soc.*, 7, 56, 1990.

25. Zepp, R.G., Schlotzhauer, P., and Sink, R.M., Photosensitized transformations involving energy transfer in natural waters: role of humic substances, *Environ. Sci. Technol.*, 19, 74–81, 1985.

26. Winner, R.W., Owen, H., and Moore, M.V., Seasonal variability in the sensitivity of freshwater lentic communities to a chronic copper stress, *Aquat. Toxicol.*, 17, 75–92, 1990.

27. Stumm, W., *Chemistry of the Solid-water Interface*, John Wiley & Sons, New York, NY, 1992.

28. Davison, W. and DeVitre, R., Iron particles in freshwater, in Buffle, J. and van Leeuwen, H.P., Eds., *Environmental Particles*, Vol. 1, Lewis Publishers, Boca Raton, FL, 1992, 315–335.

29. Peiffer, S., The reaction of H_2S with ferric oxides; some conceptual ideas on its significance for sediment-water interactions, *Adv. Chem. Ser.*, 237, 371–390, 1994.

30. Davison, W., Grime, G.W., and Woof, C., Characterization of lacustrine iron sulfide particles with proton-induced X-ray emission, *Limnol. Oceanogr.*, 37, 1770–1777, 1992.

31. DeVitre, R. and Davison, W., Manganese particles in freshwaters, in van Leeuwen, H.P. and Buffle, J., Eds., *Environmental Particles*, Vol. 2, Lewis Publishers, Boca Raton, FL, 1993, 317–353.

32. Morfett, K., Davison, W., and Hamilton-Taylor, J., Trace metal dynamics in a seasonally anoxic lake, *Environ. Geol. Water Sci.*, 11, 107–114, 1988.

33. Gendron, A., Silverberg, N., Sundby, B., and Lebel, J., Early diagenesis of cadmium and cobalt in Laurentian Trough sediments, *Geochim. Cosmochim. Acta*, 50, 741–747, 1986.

34. Balistrieri, L.S., Murray, J.W., and Paul, B., The biogeochemical cycling of trace metals in the water column of Lake Sammamish, WA: response to seasonally anoxic conditions, *Limnol. Oceanogr.*, 37, 529–548, 1992.

35. Johnson, C.A., Sigg, L., and Lindauer, U., The chromium cycle in a seasonally anoxic lake, *Limnol. Oceanogr.*, 37, 15–321, 1992.

36. Sigg, L., Sturm, M., and Kistler, D., Vertical transport of heavy metals by settling particles in Lake Zurich, *Limnol. Oceanogr.*, 32, 112–130, 1987.

37. Tessier, A., Sorption of trace elements on natural particles in oxic environments, in Buffle, J. and van Leeuwen, H.P., Eds., *Environmental Particles*, Vol. 1, Lewis Publishers, Boca Raton, FL, 1992, 425–453.

38. Tipping, E., Thompson, D.W., Ohnstad, M., and Hetherington, N.B., Effects of pH on the release of metals from naturally-occurring oxides of Mn and Fe, *Environ. Technol. Lett.*, 7, 109–114, 1988.

39. Santschi, P.H., Nyffeler, V.P., Anderson, R.F., Schiff, S.L., O'Hara, P., and Hesslein, R.H., Response of radioactive trace metals to acid-base titrations in controlled experimental ecosystems: evaluation of transport parameters for application to whole-lake radiotracer experiments, *Can. J. Fish. Aquat. Sci.*, 43, 60–77, 1986.

40. Jackson, T.A., Kipphut, G., Hesslein, R.H., and Schindler, D.W., Experimental study of trace metal chemistry in soft-water lakes at different pH levels, *Can. J. Fish. Aquat. Sci.*, 37, 387–402, 1980.

41. Fu, G., Allen, H.E., and Cao, Y., The importance of humic acids to proton and cadmium binding in sediments, *Environ. Toxicol. Chem.*, 11, 1363–1372, 1992.

42. Tipping, E., Some aspects of the interactions between particulate oxides and aquatic humic substances, *Mar. Chem.*, 18, 161–169, 1986.

43. Laxen, D.P.H., Trace metal adsorption/coprecipitation on hydrous ferric oxide under realistic conditions: the role of humic substances, *Water Res.*, 10, 1229–1236, 1985.

44. Luoma, S., Bioavailability of trace metals to aquatic organisms—a review, *Sci. Total Environ.*, 28, 1–22, 1983.

45. Adams, W.J., Kimerle, R.A., and Mosher, R.G., Aquatic safety assessment of chemicals sorbed to sediments, in Cardwell, R.D., Purdy, R., and Bahner, R.C., Eds., *Aquatic Toxicology and Hazard Assessment: Seventh Annu. Symp.*, American Society for Testing and Materials, Philadelphia, PA, 1985, 429–453.

46. Allen, H.E., Fu, G., and Deng, B., Analysis of acid volatile sulfide (AVS) and simultaneously extracted metals (SEM) for the estimation of potential toxicity in aquatic sediments, *Environ. Toxicol. Chem.*, 12, 1441–1453, 1993.

47. Emerson, S., Jacobs, L., and Tebo, B., The behavior of trace metals in marine anoxic waters: solubilities at the oxygen-hydrogen sulfide interface, in Wong, C.S., Boyle, E., Bruland, K.W., and Burton, J.D., Eds., *Trace Metals in Sea Water*, Plenum Press, New York, 1983, 579–608.

48. Sundby, B., Sediment-water exchange processes, this volume, Section 5, Chapter 3, 1994.

49. Aller, R.C., The effects of macrobenthos on chemical properties of marine sediment and overlying water, in McCall, P.L. and Tevesz, M.J.S., Eds., *Animal-Sediment Relations*, Plenum Press, New York, 1982, 53–102.

50. Hong, J., Calmano, W., and Förstner, U., Effects of redox processes on the acid-producing potential and metal mobility in sediments, this volume, Session 5, Chapter 2, 1994.

51. Lee, G.F. and Jones, R.A., Sediment Quality Criteria Development: Technical Difficulties with Current Approaches, and Suggested Alternatives, unpublished report, G. Fred Lee and Associates, El Macero, CA, 1992.

52. Darby, D.A., Adams, D.D., and Nivens, W.T., Early sediment changes and elemental mobilization in a man-made estuarine marsh, in Sly, P.G., Ed., *Sediment and Water Interactions*, Springer-Verlag, New York, 1986, 343–351.

53. Maass, B. and Miehlich, G., Die Wirkung des Redoxpotentials auf die Zusammensetzung der Porenlösung in Hafenschlickspülfeldern, *Mitt. Dtsch. Bodenkundl. Ges.*, 56, 289–294, 1988.

54. Davison, W., Grime, G., Morgan, A., and Clarke, C., Distribution of dissolved iron in sediment porewaters at submillimetre resolution, *Nature*, 352, 323–325, 1991.

55. Hamelink, J., Dow Corning Corporation, Midland, MI, personal communication.

56. Lee, H., U.S. Environmental Protection Agency, Newport, OR, personal communication.

57. DiToro, D., Manhattan College, The Bronx, NY, personal communication.

58. Ankley, G., U.S. Environmental Protection Agency, Duluth, MN, personal communication.

SESSION 6

KINETIC LIMITATIONS OR DISSOLUTION

Chapter 1

Unraveling the Choreography of Contaminant Kinetics: Approaches to Quantifying the Uptake of Chemicals by Organisms

Donald Mackay

INTRODUCTION

It is clearly important that we understand how kinetic factors influence bioavailability and bioconcentration. Although in many situations we can assume equilibrium to have been attained and kinetics may thus be irrelevant, there are many environmental conditions in which the state of a system is a reflection of how fast it is journeying towards some ultimate, unattainable state of equilibrium. Landrum et al.[1] have recently reviewed models used in toxicokinetic studies, to which the reader is referred for a more detailed discussion.

At the outset we should discriminate between the thermodynamic quantity of *equilibrium* and the phenomenon of the *steady-state* condition. Equilibrium implies that the system has reached its ultimate desired state where, for example, forward and reverse processes are occurring at equal rates. Steady state implies merely that the situation is unchanging with time. Dissolved oxygen in a river at 2 mg/l may be at *steady-state,* but it is not at *equilibrium* with the atmosphere, which may be 8 mg/l. A *steady-state* condition such as this is dictated by the rates of oxygen entry to the river being balanced by the rates of consumption. It is very much a kinetically controlled system.

When characterizing kinetics in bio-uptake situations it is usual to invoke some form of mass balance expression in the form of an equation stating the axiomatic relationship that the rate of inventory change equals the rate of input less the rate of output. There are several ways of expressing this equation which are ultimately equivalent. To illustrate these approaches we can consider the uptake of chemical from water by a particle. The particle may be biotic or abiotic. How long will it take to reach equilibrium? We first assume that the particle is homogeneous, is well mixed internally, and that the "resistance" to uptake lies in the diffusion of chemical through the water to the particle. The approaches exploit the concepts of (1) rate constants, (2) mass transfer, and (3) equilibrium or fugacity. Ultimately, as we show, they are algebraically identical, but each approach provides its own insights. Having established these basic ideas, we then examine some complications or deviations from this ideal picture of bio-uptake.

1-56670-086-8/94/$0.00+$.50
© 1994 by CRC Press, Inc.

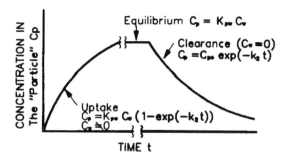

Figure 1. Uptake and Clearance regimes over time.

METHODS

The Rate Constant Approach

The differential equation is

$$dC_P/dt = k_1C_W - k_2C_P \tag{1}$$

where: C_P is the concentration of chemical in the particle, C_W is the concentration of chemical in the water, t is time, and k_1 and k_2 are uptake and clearance rate constants with dimensions of reciprocal time.

For uptake from water from a zero initial C_P and constant C_W we can solve this equation to obtain:

$$C_P/C_W = (k_1/k_2)C_W(1-exp(-k_2t)) = K_{PW}C_W(1-exp(-k_2t)) \tag{2}$$

where K_{PW} is k_1/k_2 and is C_P/C_W at equilibrium, i.e., where t is large and $exp(-k_2t)$ is zero. The group in the exponent can also be written t/τ where τ is $1/k_2$ and is a characteristic time, or as $0.693t/\tau_{1/2}$ where $\tau_{1/2}$ is the half time of $0.693/k_2$ where 0.693 is the natural log of 2. It is conceptually easier to think in terms of τ.

For clearance from an initial concentration C_{PO} the decay equation into chemical-free water is

$$C_P = C_{PO} exp(-k_2t) \tag{3}$$

These changes are illustrated in Figure 1. Interestingly, the uptake and clearance *times* τ are the same, and are $1/k_2$, the reciprocal *clearance* rate constant, not the uptake rate constant. The initial *rate* of uptake, however, is k_1C_W.

The reason for this apparent anomaly is that the uptake and clearance times $1/k_2$ are really K_{PW}/k_1. The time required for uptake is proportional to K_{PW}, the equilibrium concentration ratio. Obviously, it will take longer to reach higher concentrations implied by a large K_{PW}. It is also inversely proportional to k_1, because the faster the uptake, the shorter the uptake time. This is analogous to stating that the time of the journey is the distance to be traveled divided by the speed.

Of the three quantities k_1, k_2, and K_{PW}, only two can be defined independently because they are related, i.e., K_{PW} is k_1/k_2. Probably, k_1 and K_{PW} are the most meaningful and fundamental.

From this analysis it can be argued that it is misleading to assert that high bioconcentration in organisms (larger K_{PW}) is a result of fast uptake (a large k_1) and slow release (small k_2). In reality, bioconcentration is a manifestation of the value of the thermodynamic quantity K_{PW}, which is determined by the physical chemistry of the chemical in water and the particle or organism.

The uptake rate constant k_1 is determined by diffusion and flow phenomena, such as the gill ventilation rate in the case of a fish, and k_2 is simply k_1/K_{PW}. In short, bioconcentration is caused by the thermodynamic equilibrium quantity K_{PW}: the kinetics or time to achieve equilibrium follow. Kinetics do not cause bioconcentration. They are the result of it!

Experimentally, it is best to measure K_{PW} in an equilibrium experiment and k_2 in a clearance experiment, then deduce k_1, although k_1 can be determined from the initial rate of uptake. Regrettably, there is often no easy way to deduce k_1 or k_2 from first principles.

The Mass Transfer Approach

In this case the equation is written as a version of Fick's first law, in which diffusion is regarded as the key process.

$$V_P dC_P/dt = k_M A(C_W - C_P/K_{PW}) \tag{4}$$

where k_M is a mass transfer coefficient or a diffusion velocity and A is the surface area of the particle. The group $(C_W - C_P/K_{PW})$ is a "departure from equilibrium" or the "driving force" for diffusion. Integrating as before for uptake gives:

$$C_P = K_{PW}C_W(1 - \exp(-k_M At/V_P K_{PW})) \tag{5}$$

It follows that k_1 is $k_M A/V_P$, k_2 is $k_M A/V_P K_{PW}$, and τ is $V_P K_{PW}/k_M A$.

Since k_M is a velocity and A the area, the group $k_M A$ is an effective flow rate, e.g., in cubic meters per second of water, to the particle surface. It can be viewed as a "water encounter" rate. This is significant from a bioavailability viewpoint because the maximum rate at which chemical can reach the particle is $k_M A C_W$, which may have units such as grams per second. The actual rate will generally be lower, but this is a limiting value which is useful to know. For fish exposed to chemical in water, it is the quantity of water or associated chemical which the fish can contact per unit time.

Whereas it is not possible to estimate k_1 from first principles, it is possible to estimate the likely magnitude of k_M. For diffusion in a stagnant medium, i.e., a motionless particle, it can be shown that the so-called Sherwood No.[2] is

$$k_M d/D = 2$$

where d is the particle diameter and D is the diffusivity of the chemical in water. In flow systems the quantity 2 becomes larger, e.g., 10, but for small particles in stagnant systems 2 is a reasonable value. Now D is typically 3×10^{-6} cm²/s or 0.01 cm²/h.

If the particle is a sphere, A is πd^2 and V is $(\pi/6) d^3$, then $\tau = V_P K_{PW}/A k_M = \pi/6 d^3 \cdot K_{PW}/(\pi d^2 \cdot 2D/d)$ and $= K_{PW} d^2/12D = K_{PW} d^2/0.12 = 8 K_{PW} d^2$ hours, with d in centimeters. Clearly, τ is very sensitive to d. Small particles reach equilibrium rapidly because d is small. Part of this is the high area/volume ratio, but k_M is also larger.

It is also useful to quantify the "encounter volume" $k_M A$ in terms of the number of particle volumes encountered per unit time, i.e., $k_M A/V_P$. This is $1/8d^2$. Alternatively, the time taken for a particle to encounter its own volume of water is $8d^2$ hours, with d expressed in centimeters. Thus, it takes 8 h for a 1-cm particle to contact its own volume of water. A 1-mm particle takes only 8/100 h, or 5 min.

The uptake rate constant, k_1, is simply $k_M A/V_P$ or $1/8d^2$. Mass transfer or diffusion theory thus helps by giving an estimate of k_1, and it provides an insight into its physical meaning.

The reason that the uptake or clearance time τ is $K_{PW} \cdot 8d^2$ now becomes obvious. It takes $8d^2$ hours to contact 1 volume of water, thus if the approach to equilibrium requires extraction of K_{PW} volumes of water, the time must be $K_{PW} \cdot 8d^2$ h.

Table 1. Table of $8K_{PW}d^a$ Showing Three Bands of

(I) Kinetic control (lower left)
(II) Kinetic and equilibrium control (middle)
(III) Equilibrium control (upper right)

DIAMETER d

	1 cm	1 mm	100 mm	10 mm	1 mm
1	8 h	5 min	3 s	30 ms	0-3 ms
10^2	3 days	8 h	5 min	3 s	30 ms
10^4	1 year	3 days	8 h	5 min	3 s
10^6	100 years	1 year	3 days	8 h	5 min

K_{PW} (row label)

ªIn this expression d is in centimeters.

It is interesting to construct a table for time (τ) as a function of d and K_{PW} as shown in Table 1. There are apparently three regions:

1. To the upper right, kinetics are so fast that they are difficult to measure and they may be irrelevant because equilibrium is assured. We must know K_{PW}, but we may not need an accurate k_1.
2. To the lower left, kinetics control and equilibrium is unlikely to be reached in practice. Equilibrium may be irrelevant. We must know k_1, but we need not know K_{PW}.
3. In the middle band both kinetics and equilibrium are important. We must know K_{PW} and k_1.

In environmental situations, it is useful to know which region applies because the location determines the information needs. This, in turn, dictates the nature of the experimental approach. As is discussed later, when analyzing experimental data there is often a poor fit when only one rate constant is used. Systems often display "two-compartment" behavior, with a fast or labile rate constant and a slow or resistant rate constant. This is usually conceived as rapid surface equilibration and slow diffusive penetration. The result is violation of the basic assumption of homogeneity within the particle, i.e., it is no longer well mixed. A particle may thus exhibit behavior as a region in Table 1 extending over two or more bands.

The Equilibrium Criterion (Fugacity) Approach

If the problem was to determine the kinetics of heat uptake from water by a particle which was initially cold, we would write the equation in the form:

$$V_P H \, dT_P/dt = hA(T_W - T_P) \tag{6}$$

where H is a heat capacity of the particle (e.g., $J/cm^3 \cdot K$), T_P and T_W are the temperatures of the particle and the water, and h is a heat transfer coefficient or a thermal conductivity term ($J/cm^2 \cdot K \cdot s$). Integrating from an initial temperature of zero would give

$$T_P = T_W (1-\exp(-hAt/V_PH)) \tag{7}$$

The rate constant is hA/V_PH or the time τ is V_PH/hA, which is clearly analogous to the mass transfer expression. The significant point is that the "driving force" for heat transfer is now expressed as $(T_W - T_P)$, the difference in the relevant applicable equilibrium criterion, in this case temperature. When examining systems of this type it is instructive to inspect the relative values of T_W and T_P to detect the extent of approach to equilibrium.

In the chemical uptake expression this is equivalent to examining C_W and C_P/K_{PW}. This suggests using the equilibrium criterion which is analogous to temperature, but which applies to mass transfer. It is fugacity (f), or in some cases, activity.

In the fugacity formalism f, (expressed in units of pressure, Pa), replaces concentration using the relationship C equals Zf where C has units of moles per cubic meter and Z, with units of moles per Pascal cubic meter, is a capacity term. Details have been given by Mackay.[3]

The fugacity equation is

$$V_P Z_P df_P/dt = D(f_W - f_P) \tag{8}$$

in which D is a "D value" or transport parameter (not to be confused with D the diffusivity) which yields the uptake equation

$$f_P = f_W (1 - \exp(-Dt/V_PZ_P)) \tag{9}$$

The rate constant, k_2, is clearly D/V_PZ_P. It can be shown that D is k_MAZ_W and K_{PW} is Z_P/Z_W.

The key conclusion is that the fugacity approach directly addresses the equilibrium status. When the particle fugacity is observed to be less than that of the water we presume that kinetics are important and that the system is migrating towards equilibrium, but it has not yet reached it. If the fugacities are equal, equilibrium is attained.

This approach is particularly useful for biotic particles. A simple first hypothesis is that the fugacity of a chemical in an organism will equal that of its environment, i.e., simple bioconcentration applies. Often, the organism's fugacity is less than that of the environment. This may be due to metabolism, growth, or insufficient time of exposure. Kinetic considerations then control the extent of departure from equilibrium.

In some cases the organism's fugacity exceeds that of the water, which implies biomagnification. This is equivalent in the temperature analogy to the organism being 30°C in water of 20°C. Thermodynamically, this is only possible if there is some other energy dissipation or entropy generation process at work. It is likely that this process occurs in the digestive tract of the consuming organism as lipids are hydrolyzed and absorbed.

This fugacity approach can be applied in the water column and in sediments in which equilibrium implies that lipid normalized chemical concentrations and organic carbon normalized concentration will be similar. The ratio of these concentrations is becoming known as the sediment bioconcentration factor.

SUMMARY

In this discussion the three approaches to quantifying the kinetics of uptake have been described and compared. All three approaches are correct and, as has been shown, are algebraically equivalent, but by viewing the processes from a different perspective we can identify different features. The rate constant approach is simplest. Mass transfer theory gives us the concept of the "encounter volume" and provides an estimate of k_1. The fugacity or equilibrium criterion approach gives a direct expression of equilibrium status.

When interpreting concentration data from the environment it is particularly illuminating to compare fugacities, since this reveals the equilibrium status directly and shows when kinetic considerations are playing an active role in determining concentrations.

Some Complications or Deviations

When we view the environment we find it to be much more complex than is expressed by this simple model. These complications usually affect the time τ.

1. Organisms usually demand more oxygen than simple, natural diffusive rates can supply. They thus increase their encounter volumes, expressed as flow rates by various physiological mechanisms. It is important to know these encounter volumes. For a small fish, it is typically increased to 100 body volumes per hour.

2. Organisms demand energy and nutrients, thus they also establish another "encounter volume" for food. The food may also be contaminated, so this acts as a parallel source of contaminant for bioconcentration. Typically, this encounter volume is about 0.001 body volumes per hour or 0.024 body volumes per day, i.e., 2.4% of body weight. It follows that the ratio of water to food encounter volume is of the order of 10^5. When the ratio of contaminant concentrations in water is 10^5, the organism will encounter equal quantities of contaminant from both sources. Thus, highly hydrophobic chemicals with ratios of 10^6 or greater thus mainly reach organisms in food. Another method of expressing this is to calculate the time necessary for an organism to contact one body volume of water or food. This is about 0.01 h for respiration and 1000 h (or 40 days) for food. Of course, there are differences in uptake efficiency from food and water, and some benthic organisms ingest sediment, thus the accurate calculation of relative *uptake* rates is more complex. This does not affect the general conclusion that to understand bioaccumulation we must know these encounter rates as one step in quantifying the uptake process. It is in this area that the work of the fish physiologist and ecologist becomes critically important. If we are to understand contaminant uptake, it is first essential to know the organism's behavior—its "encounter rate" of water, sediment, and various foods.

3. Not all the chemical in these encounter volumes may be "available". This applies especially to hydophobic chemicals in water and sediment. Conventionally, a "dissolved fraction" is calculated as discussed elsewhere in these proceedings.

4. There is often a diffusive resistance into or onto the particle. Examples are gill and gut membrane resistances and pharmacokinetic processes in organisms, especially into poorly perfused lipids. Especially important in any discussion of bioavailability is the issue of sorption of organic contaminants to natural organic matter. This is important in the water column, in bottom sediments, and in terrestrial soils. Soil scientists have devoted considerable effort to elucidating the kinetics and equilibrium of sorption as the sorbate diffuses within and between the accumulations of organic matter which play such a vital role in soil fertility. Notable are the studies by Chiou et al.[4,5] and those of Rao and Jessup,[6] Brusseau and Rao,[7a,7b] Brusseau et al.,[8] and Karickhoff.[9] These concepts have been applied to intraparticle diffusion in aquatic systems by Wu and Gschwend.[10] It is possible that partitioning to external surfaces of small biotic and abiotic particles is particularly important as a mechanism of reducing bioavailability. Partitioning is even observed at air-water interfaces in small bubbles and fog droplets. Obviously, quite complex multicompartment models can be assembled describing the kinetics of such processes but for many purposes a simpler approach may be adequate. It is possible to assign an approximate time to these processes. For hydrophobic contaminant transport from water to lipid storage in goldfish, a time of about 200 h has been deduced by Mackay and Hughes.[11] For sediment particles, a time of about 500 hours or 20 days seems typical. It may be helpful to assign different times to labile and resistant sorption processes to gain an appreciation of when equilibrium is expected to apply and when kinetics are important.

5. The particle may be growing and is thus "diluting" the chemical. This is important when the growth doubling time is similar to, or smaller than, the uptake time. This is observed for fast-growing algae and for very hydrophobic contaminants which take a long time to reach equilibrium in fish because of the high partition coefficients or bioconcentration factors.

6. There may be transformation of chemical in the particle which reduces the concentration. This is observed for PAHs in fish. Again, a characteristic time can be assigned.

No doubt there are other complications which require consideration in real environmental situations.

CONCLUSIONS

When interpreting observed concentrations in biotic and abiotic compartments from the environment we must have thermodynamic equilibrium information, e.g., K_{PW} or Z values. This information is a prerequisite for making any deductions about the role of kinetics.

When interpreting observed concentrations from the environment, it is instructive to examine the equilibrium status. This is best done using fugacity or activity, but can be done by comparing concentrations and partition coefficients, e.g., comparing C_W with C_P/K_{PW}. Near-equilibrium implies that the contact time exceeds the uptake time, and knowledge of rates may not be necessary.

When considering kinetics, it is useful to estimate encounter rates of organisms in terms of body volumes per hour or as its reciprocal, τ, the time to contact one body volume. The uptake time for a contaminant will be of the order of $K_{PW}\tau$. For small fish, the respiration τ is of the order of 0.01 h. For food, τ is about 40 days. The K_{PW} which is multiplied by the τ is the partition coefficient between the receiving "particle" and the transferring medium. For water it is K_{PW}, and is of the order of the octanol-water partition coefficient for organic chemicals. For food it is K_{PF}, the particle fish/food ratio. Since the fish and food often have similar lipid contents, K_{PF} is of the order of unity.

Ultimately, both for scientific and regulatory purposes, it is essential that the bioconcentration phenomenon be expressed quantitatively. This is a challenging task which will require that models be developed, tested, improved, and often discarded. This process is essential to the unraveling of the complex assembly of kinetic and equilibrium processes which occur as chemical contaminants migrate through our environment, and especially as they migrate into organisms such as fish.

REFERENCES

1. Landrum, P.F., Lee, H., and Lydy, M.J., Toxicokinetics in aquatic systems. Model comparisons and use in hazard assessment, *Environ. Toxicol. Chem.*, 11, 1709–1725, 1992.
2. Bird, R.B., Stewart, W.E., and Lightfoot, E.N., *Transport Phenomena*, John Wiley & Sons, New York, 1960.
3. Mackay, D., *Multimedia Environmental Models: The Fugacity Approach*, Lewis Publishers, Boca Raton, FL, 1991.
4. Chiou, C.T., Peters, L.J., and Freed, V.H., A physical concept of soil-water equilibria for nonionic organic compounds, *Science*, 206, 831–832, 1979.
5. Chiou, C.T., Porter, P.E., and Schmedding, D.W., Partition equilibria of nonionic organic compounds between soil organic matter and water, *Environ. Sci. Technol.*, 17, 227–231, 1983.
6. Rao, P.S.C. and Jessup, R.E., Sorption and movement of pesticides and other toxic organic substances in soils, in *Chemical Mobility and Reactivity in Soil Systems*, Soil Science Society of America, Madison, WI, 1983, chap. 13.
7a. Brusseau, M.L. and Rao, P.S.C., Influence of sorbate structure on nonequilibrium sorption of organic compounds, *Environ. Sci. Technol.*, 25, 1501–1506, 1991.
7b. Brusseau, M.L. and Rao P.S.C., Sorption kinetics of organic chemicals. Methods, models and mechanisms, in *Rates of Soil Chemical Processes*, SSSA Spec. Publ. No. 27, Soil Science Society of America, Madison, WI, 281–302, 1991.
8. Brusseau, M.L., Jessop, R.E., and Rao, P.S.C., Nonequilibrium sorption of organic chemicals: elucidation of rate limiting processes, *Environ. Sci. Technol.*, 25, 134–142, 1991.
9. Karickhoff, S.W., Semi-empirical estimation of sorption of hydrophobic pollutants on natural sediments and soils, *Chemosphere*, 10, 833–846, 1981.
10. Wu, S.C. and Gschwend, P.M., Sorption kinetics of hydrophobic organic compounds to natural sediments, *Environ. Sci. Technol.*, 20, 717–725, 1986.
11. Mackay, D. and Hughes, A.I., A three parameter equation describing the uptake of organic compounds by fish, *Environ. Sci. Technol.*, 18, 439–444, 1984.

Chapter 2

Physiological and Biochemical Mechanisms That Regulate the Accumulation and Toxicity of Environmental Chemicals in Fish

James M. McKim

INTRODUCTION

The field of environmental toxicology is complex and multidisciplinary in its approach, and must involve physical, chemical, and biological understanding in order to resolve the present and future environmental perturbations caused by toxic chemicals. The latest Pellston Workshop on bioavailability and chemical interactions was concerned primarily with understanding the relationships between toxic chemicals and the physical-chemical and biological processes inherent to the aquatic environment. In order to link these important environmental interactions to animal responses, a transition had to be made from environmental availability to accumulation and toxicity. This involved knowledge of the important exchange surfaces and the mechanisms that control the flux of chemicals across such surfaces. Figure 1 presents this transition in the context of ecological risk assessment inputs such as *hazard, exposure,* and biological *receptors.* This description further summarizes the processes that control the movement of chemicals through the environment; the *source* of contamination, *transport* through the system, *transformation* from one form to another by either biotic (i.e., biodegradation by microorganisms and/or biotransformation by higher organisms) or abiotic (i.e., photooxidation, hydrolysis) processes, and the *fate* as determined by bioaccumulation, sorption to sediment, or evaporation. These natural processes ultimately determine a chemical's bioavailability to aquatic organisms. The animals exposed to concentrations of specific chemicals represent the receptor. The impact of chemical exposure on the receptor organism depends on a number of major ecological variables such as *species, lifestage, habitat,* and *trophic level.* The environmental exposure of the animal to the chemical provides both a total dose received (in milligrams per kilogram per day) and an accumulated dose [body burden (in milligrams per kilogram)]. *Modifying factors* (physiological and biochemical mechanisms) in the animal control the magnitude of chemical uptake and accumulation and, thus, the toxic response of the animal. Toxic response is tied to a specific internal dose (in milligrams per kilogram) or exposure concentration (waterborne, food) and describes the *hazard.* Toxic *responses (acute and chronic)* are measured in single-species toxicity tests to determine potential hazard and to establish dose-response relationships. The proper collection and synthesis of these data inputs (Figure 1) will provide aquatic toxicologists with a better mechanistic approach to understanding the impacts of chemical toxicants on the environment.

The objectives of this review paper are to:

1. Describe major physiological and biochemical mechanisms that regulate the transfer of bioavailable chemicals from the aqueous environment to the animal across the gills and skin,
2. Review simple physiologically based mechanistic surface exchange models that can predict the toxicity and bioaccumulation of chemicals to specific species,
3. Demonstrate the ability of these prototype models to deal with environmental variables that impact the flux of chemicals at important exchange surfaces, and
4. Discuss the importance of an animal's capability to control and/or internally modify the toxicity and bioaccumulation of an absorbed chemical through biotransformation.

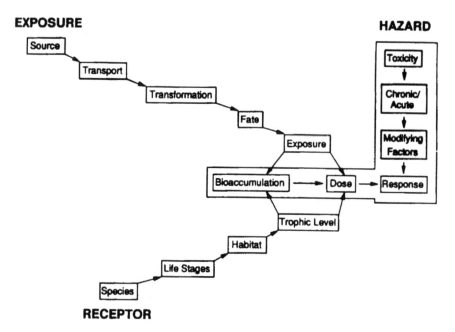

Figure 1. The major environmental and toxicological factors impacting the toxic response of animals to environmental toxicants.

BIOAVAILABILITY AND INTERNAL DOSE

Environmental bioavailability refers to that fraction of total chemical in the animals' exposure environment (water, food, sediment) that is available for absorption. In general, the properties of a chemical, the physiology of the species in question, and the environmental conditions of the exposure determine the bioavailability of a chemical, as outlined in Figure 1. These same factors also have a strong influence on the route of chemical absorption (inhalation/branchial, dermal, oral). In general, it is believed that very hydrophobic [log octanol/water partition coefficient (log K_{ow}) > 6.0] compounds are absorbed by fish primarily from food, while compounds exhibiting low to moderate hydrophobicity (log K_{ow} < 4.0) are absorbed directly from water, primarily at the gills.[1-3] Consistent with this suggestion is the observation that hydrophobic compounds exhibit a high affinity for organic substrates (dissolved organic material and suspended sediment) and, as a result, appear to be relatively unavailable for uptake across fish gills.[4] Moreover, fish prey, including invertebrates and smaller fishes, can accumulate hydrophobic compounds to very high levels relative to the water concentrations, providing a rich source of chemical for intestinal absorption. For compounds in the intermediate (>4.0 but <6.0) log K_{ow} range, both food and water probably contribute significantly to chemical uptake. Finally, the skin may represent an important route of chemical uptake and elimination for very small fish or for juveniles of larger species, because of the much larger skin surface area to body volume ratio in small fish.

Toxicological or pharmacological bioavailability refers to that portion of the applied dose or exposure concentration in the environment (water, food, sediment) that actually reaches the systemic (arterial) circulation for distribution throughout the animal's body.[5] In toxicology, the magnitude of toxic response depends on the concentration of chemical (or metabolite) at the site of toxic action. The applied dose in a toxicity test controls only the maximal concentration that can be attained at the site of action. The total amount of the applied dose that actually reaches the target organ is the net result of many internal processes including: (1) absorption, (2) transport, (3) biotransformation to active and/or inactive metabolites, and (4) excretion. Figure 2 summarizes the impact that these processes have on the amount of parent chemical reaching the site of action

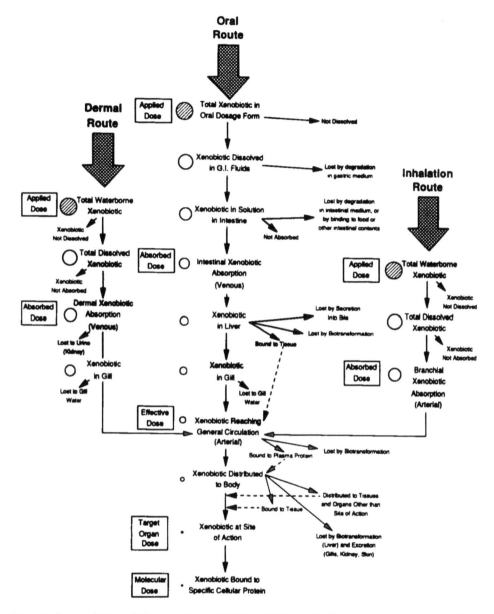

Figure 2. Factors that modify the amount of xenobiotic reaching the site of action following inhalation, oral and dermal exposure (● applied dose, ○ progressive reduction in applied dose) (Modified from McKim, J.M. and Nichols, J.W., *Aquatic Toxicology: Molecular, Biochemical, and Cellular Perspectives*, Ostrander, G.K., and Malins, P.C., Eds., CRC Press, Boca Raton, FL, 1994.)

for the three major exposure routes (inhalation/branchial, oral, and dermal). Toxicokinetics deal with the movement of chemical mass within an organism and therefore encompasses all of the internal processes listed above.

If we are to link mechanistic toxicology to effects seen in the whole animal, the difference between applied dose, absorbed dose, effective dose, and the biologically effective dose actually reaching the target organ, must be understood (Figure 2).[6] In addition, the route of exposure can strongly impact the effective dose and the disposition of a chemical within an animal, as demonstrated in Figure 2. Clearly, large differences can exist between an environmental exposure concentration and the concentration of chemical reaching the systemic circulation. An even greater discrepancy may occur between the environmentally available chemical and the portion that is actually involved in target organ and molecular toxicity.

Aquatic toxicologists generally have not dealt with any sort of an internal chemical dose, but rather with the concentration of chemical in the exposure water (LC50, MATC, NOEC).[7] This allows a relative comparison of effects between chemicals and species, but does not provide an understanding of the actual absorbed dose received by the animal. Recent estimates by McCarty[8-10] of lethal body burdens in fish have focused attention on the importance of internal dose, and provided toxic residue estimates for use in risk assessments with narcotic-type chemicals. However, this approach does not provide a complete understanding of the mechanisms controlling internal biological dose-response relationships, nor does it deal with estimates for more specific-acting chemicals. More importantly, these predictive studies give no information on dose received at the site of action (the **target organ dose** is the fraction of the total absorbed dose distributed to a specific organ or tissue that elicits a toxic response; the **molecular dose** is the fraction of total absorbed dose bound to cellular constituents that causes a specific set of responses leading to cellular changes or cell death).

The need to obtain an environmentally relevant measure of chemical absorption is of critical importance to the development of predictive dosimetry models.[6] Quantitation of chemical absorption across the gills of fish has provided the tools necessary to monitor the extraction efficiency of waterborne xenobiotics as they flow across the gills.[11-13] These studies provided a direct measure of the total absorbed dose (e.g., milligrams per kilogram per hour) and made available techniques to further explore the mechanisms controlling both gill and dermal flux of xenobiotic chemicals.

To fully develop a predictive capability for waterborne xenobiotic uptake and distribution by fish, it will be necessary to incorporate an improved understanding of gill and dermal flux into models that represent the entire animal. However, for these models to achieve maximum potential they must also incorporate accurate mechanistic descriptions of GI tract exposure from contaminated food, so that all relevant uptake and elimination routes can be included in the final model description of fish tissue dosimetry.[6]

CHEMICAL FLUX ACROSS FISH GILLS

Gill Morphology and Physiology

The structure of fish gills reflects their primary function as a gas exchange, osmoregulatory, and excretory organ. Gill surface area is maximized by an intricate arrangement of thin plate-like structures termed secondary lamellae, each of which is oriented longitudinally to the direction of water flow (Figure 3).[14,15] Deoxygenated blood flows through these lamellae in a posterior to anterior direction, providing for counter current exchange with water. Water flow over the gills (ventilation) and blood flow through the gills (perfusion) fluctuate with activity, temperature, and the level of environmental oxygen available.[16-19] The general anatomy of the fish gill and its unique adaptations for diffusive exchange of respiratory oxygen under various environmental conditions was recently reviewed.[20] Further, normal osmoregulatory and excretory functions of the gill in response to fluctuating environmental conditions has also been summarized.[21,22]

Gill Uptake of Organic Chemicals

Knowing the general physiology of the fish gill, the route that a waterborne organic chemical takes as it moves from the water into a fish can be described as follows: (1) movement of water with the dissolved chemical through the gill lamellar "sieve" by the branchial pump; (2) diffusion of the chemical across the water flow channel, the gill epithelium, and into the blood; and (3) removal of the chemical from the gill by the blood.[23,24]

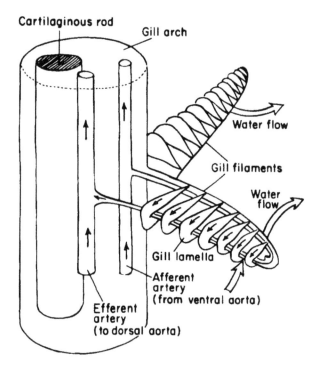

Figure 3. Diagram of fish gills showing countercurrent blood and water flows. (From Wedemeyer, G.A., Meyer, F.P., and Smith, L., *Environmental Stress and Fish Disease*, TFH Publications, Neptune, NJ, 1976. With permission.)

Xenobiotics are assumed to move across biological membranes by simple passive diffusion.[25] The physical-chemical and biological mechanisms which influence the qualitative and quantitative movement of chemicals across biological membranes include: charge on the molecule, molecular weight, lipid solubility, molecular volume, concentration in the water, surface area, and overall metabolic rate.[26] In addition, another very important consideration in the uptake of nonpolar chemicals is the plasma binding of these xenobiotics as they cross the gill epithelium and enter the blood. This rapid binding allows for maintenance of the diffusion gradient necessary for rapid passive flux.[23,24]

Using an *in vivo* system, McKim et al.[27] measured the gill flux of a heterogeneous group of 20 organic chemicals by adult rainbow trout at concentrations that caused no demonstrable physiological effect. These measurements (Figure 4a) revealed a striking relationship between gill flux and the chemical's octanol:water partition coefficient (K_{ow}). Figure 4a also shows rate constants based on other experiments by McKim and co-workers,[28-31] which further substantiated this relationship. Uptake rates are low for log K_{ow} less than 1, increase about 4-fold for log K_{ow} in the range of 1 to 3, show a plateau for log K_{ow} in the range of 3 to 6, and decrease for log K_{ow} greater than 6.

Saarikoski et al.[32] showed a similar relationship between the absorption rates of a series of phenols, anisoles, and carboxylic acids and log K_{ow} in the guppy (*Poecilia reticulata*) (Figure 4b). Both authors[27,32] attributed increases in absorption seen between log K_{ow} of 1 and 3 to greater membrane permeability for the more lipophilic chemicals. The maximum absorption reached at the higher log K_{ow} was believed to be related to the limitations imposed by diffusion across aqueous boundary layers. Finally, the drop in absorption shown above log K_{ow} of 6[27] was thought to be related to molecular weight, molecular volume, or other physical-chemical problems disturbing membrane transport. Similarly, other aquatic researchers have noted a downward deflection of BCFs at high log K_{ow} values of 6 to 10.[1,33-37] Again, explanations for this occurrence were

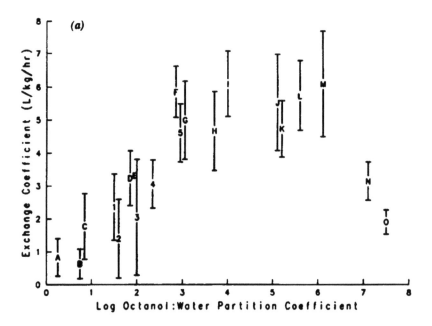

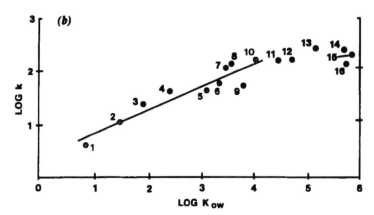

Figure 4. Relationship between rate of gill absorption in rainbow trout (the rate constant as h^{-1} from pH 7.8 water) and guppies, (the rate constant as h^{-1} from acidic water, pH < pK_a), and the octanol:water partition coefficient. (a): Rainbow trout sublethal exposures. A, ethyl formate; B, ethyl acetate; C, 1-butanol; D, nitrobenzene; E, p-cresol; F, chlorobenzene; G, 2,4-dichlorophenol; H, 2,4,5-trichlorophenol; I, 1-decanol; J, 1-dodecanol; K, pentachlorophenol; L, hexachlorobenzene; M, 2,5,2',5'-tetrachlorobiphenyl; N, fenvalerate; O, mirex. Rainbow trout lethal exposures: 1, benzaldehyde; 2, 2,4-dinitrophenol; 3, MS-222; 4, malathion; 5, 1-octanol. Error bars denote ± SD among replicate fish. (From McKim, J., Schmieder, P., and Veith, G., *Toxicol. Appl. Pharmacol.*, 77, 1–10, 1985. With permission.) (b): Guppy sublethal exposures: 1, butyric acid; 2, phenol; 3, benzoic acid; 4, 4-phenylbutyric acid; 5, 2,4-dichlorophenol; 6, 2-sec butyl-4,6-dinitrophenol; 7, 3,4-dichlorobenzoic acid; 8, 2,6-dibromo-4-nitrophenol; 9, 2,4,5-trichlorophenol; 10, 2,4,6-trichlorophenol; 11, 2,3,4,6-tetrachlorophenol; 12, tetrachloroverathrol; 13, pentachlorophenol; 14, pentachloroanisol; 15, 2,4,6-trichloro-5-phenylphenol; 16, DDT; and 17, 2,3,6-trichloro-4-nitrophenol. (Reported 95% error range about mean varied from 0.1 to 0.2 log units.) (From Saarikoski, J., et al., *Ecotoxicol. Environ. Saf.*, 11, 158–173, 1986. With permission.)

related to the structural aspects of the chemical in question and the physical-chemical limitations on gill absorption.

Various physiological and chemical factors account for the trends in gill flux described above. The importance of diffusion through lipid membranes and aqueous phases in the epithelium, and of chemical partitioning relationships (plasma binding) which define diffusion gradients, has already been noted. Gobas et al.[38] evaluated how gill exchange may be regulated by partitioning into lipid phases and diffusion through lipid membranes and adjacent aqueous layers. Lipid membranes may represent a distinct barrier for some chemicals. Membranes may inhibit transport of chemicals above a certain molecular volume,[39] although many such chemicals also rapidly bind to high molecular weight dissolved organic matter and are thus effectively excluded from crossing membranes.[4] Organic ions also exhibit a limited ability to cross lipid membranes, as demonstrated by Saarikoski et al.[32]

However, as noted earlier, the process of gill exchange also involves water and blood flows to and from the gill. These flows are essential for maintaining diffusion gradients. Norstrom et al.,[40] Neely,[41] and Bruggeman et al.[42] suggested that gill uptake could be directly related to the rate of water flow into the gill, and thus to the respiration rate. Gobas and MacKay[43] considered multiple diffusion and flow steps as parts of the exchange process. Barber et al.[44] presented a hydrodynamic-based exchange model that incorporates effects of advection and diffusion in the water flow through gills. Hayton and Barron[23] noted that blood flow, water flow, and diffusional barriers can all have a major influence on exchange depending on the organism, chemical, and environmental conditions of interest. It was also noted by Erickson and McKim[24] that the data for log K_{ow} below 6 in Figure 4b closely followed limits imposed by the equilibrium between water and blood flows. All of these factors must be considered in trying to understand the mechanisms that control gill flux.

Gill Uptake of Inorganic Chemicals

The mechanisms controlling the flux of toxic inorganic chemicals, such as metals, across the fish gill epithelium are less well understood.[45] Acquiring the mechanistic details to accurately describe these gill metal fluxes are a future prerequisite to any detailed gill uptake modeling attempts. Just as previously described for organic chemicals, early investigators[46,47] assumed that the mechanism of uptake for metals across the gills was by passive diffusion driven by the gradient from water to blood. More recently, however, investigators have observed that saturable processes seem to be controlling the flux of divalent metals across the respiratory epithelium. Perry and Wood[48] showed active transport of calcium across the trout gill in response to low environmental calcium and cortisol injection. Further investigations on zinc, cadmium, and copper showed very low gill flux rates as compared with previous work on calcium, sodium, and chloride.[45,49-51] The uptake of these metals did not support earlier assumptions on passive diffusion. The uptake pattern indicated a saturable gill uptake mechanism or selective pore, which is not necessarily active or even carrier mediated. An example of this saturable gill flux for metals is presented here in Figure 5 for zinc.[45] These authors also confirmed that the presence or absence of plasma binding proteins had no effect on the gill flux of zinc, which lends further support to a saturable process that has no dependence on a strong blood:water gradient. Furthermore, they found the maximum affinity (K_m) of the gill uptake mechanism for zinc was well into the acutely toxic range, which suggested that zinc transport is coincidental and could be an "accidental" uptake by a specific uptake mechanism (i.e., a calcium uptake mechanism). In contrast, Spry and co-workers[52] have recently shown that the competitive action of calcium, and a zinc-deficient diet can both alter the affinity of the zinc uptake mechanism, which may not be entirely "accidental".

Recently, Verbost et al.[53] demonstrated that cadmium gill flux in rainbow trout was transcellular, with evidence that strongly indicated the chloride cell calcium channels were responsible for the transcellular movement. Since they obtained no evidence for ATP-driven translocation

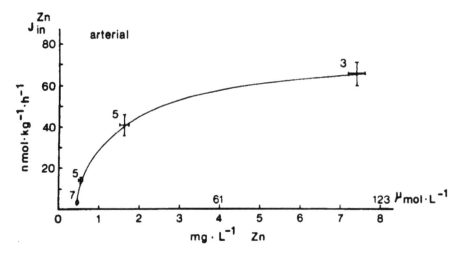

Figure 5. Zinc flux across the gills of an isolated perfused head preparation of a rainbow trout vs. waterborne zinc concentration. (From Spry, D.J. and Wood, C.M., *Can. J. Fish. Aquat. Sci.*, 45, 2206–2215, 1988. With permission.)

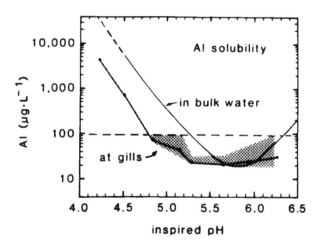

Figure 6. Aluminum solubility in bulkwater and Al solubility predicted at rainbow trout gills. Shaded area is area of Al precipitation on gills. (From Playle, R.C. and Wood, C.M., *J. Comp. Physiol. Biol.*, 159, 539–550, 1989. With permission.)

through the basolateral membrane to the blood, and because cadmium influx was too high to be explained by simple diffusion, some form of facilitated diffusion was suggested. Further substantiation of chloride cell-calcium channel cadmium flux in the gills of a second aquatic species, the oyster (*Crassostrea virginica*), was completed by Roesijadi and Unger.[54] In contrast to the earlier trout findings, these investigators found cadmium flux was partially inhibited by 2,4-dinitrophenol, an uncoupler of oxidative phosphorylation. This indicated that while cadmium flux may not depend on active transport, ATP-dependent mechanisms may be responsible for facilitating its movement through the calcium channels.

Aluminum has also received considerable attention because of its linkage with acid rain. This metal was shown to be toxic to fish, not through uptake into the systemic blood, but by precipitation on the gill surface, where it caused epithelial damage followed by respiratory and ionoregulatory upset.[55-58] The solubility of aluminum is closely controlled by environmental pH, and the precipitation on the gills is dictated by the pH (4.8 to 6.2) created in the microenvironment of the fish's gill (Figure 6) through the excretion of carbon dioxide, ammonia, and base.[58] The

GILL MODEL SCHEMATIC

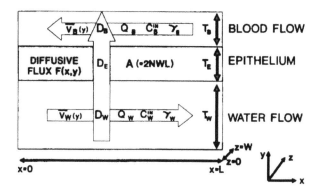

Figure 7. Gill model schematic showing major components and variables. (From Erickson, R.J. and McKim, J.M., *Aquat. Toxicol.*, 18, 175–198, 1990. With permission.)

chemistry of the microenvironment of the fish gill is extremely important in gill flux of both inorganics, and organics, and will continue to be an important area of research in the next decade.

MECHANISTIC GILL UPTAKE MODEL

Prediction of the exchange of organic and inorganic chemicals at fish gills requires a good quantitative assessment of the specific factors regulating the exchange. Models should be based on realistic mechanistic descriptions of only those factors identified as being important in regulating uptake, and should be parameterized independently of actual exchange measurements.

Inorganics

In the case of inorganics, a better understanding of the mechanisms involved in the uptake of toxic metals is required before it will be possible to accurately define the major rate-limiting barriers to gill uptake. Therefore, attempts to accurately model the uptake of metals across the gill surface must await future enlightenment on the specific mechanisms controlling metal flux.

Organics

Conversely, a better understanding of the rate-limiting barriers to organic chemical flux makes it possible to construct simple mechanistic gill uptake models for organic chemicals. Erickson and McKim[24] presented a simple model that demonstrated that exchange of organic chemicals at the gills of large rainbow trout appears to be limited largely by the flows of water and blood to the respiratory surface. With estimation only of water flow, blood flow, and the equilibrium partitioning of the chemical between water and blood, this 'flow-limited' model closely predicted the observed uptake rates and trends. However, discrepancies between observations and predictions remained and the applicability of this model to other organisms and conditions was uncertain. Consequently, they expanded their first model to account for the importance of diffusion through water, blood, and epithelial cells, and to better accommodate the effects of chemical speciation on exchange.[59]

Model Development

The conceptual model of Erickson and McKim[59] is shown in Figure 7, and represents a cross section of half of a secondary lamella—half of the water and blood channels abutting the epithelium on one side of the lamella. In general, this model accounts for the transport of chemical into the gill in water and blood flows, the diffusive flux across the water flow, epithelium, and blood flow, and the transport out of the gill in the water and blood flows.

They assumed that the gill consists of an array of N lamellae, which are parallel, uniform, rectangular plates of length L and width W. Water flows through channels of thickness $2T_W$ between the plates and blood flows counter current to the water within the lamellae in channels that are $2T_B$ thick. The water and blood flows are separated by an epithelial layer of thickness T_E.

Total water flow of Q_W was assumed to be distributed evenly among the lamellar channels and to have a quadratic, laminar velocity profile (V_W) perpendicular to the lamellar surface. Total blood flow of Q_B was also assumed to be uniformly distributed among the lamellae. Exchange occurs due to chemical gradients defined by the concentration and speciation of the compound throughout the gill, and the consequent diffusion perpendicular to the lamellar surface, with effective diffusivities D_W, D_E, and D_B in the respective layers of the model. Diffusion parallel to the lamellae, either longitudinal or transverse to the flow, was considered negligible.

It was also assumed that changes in afferent water and blood concentrations are slow relative to approach to steady-state within the gills, so that exchange rates can be based on steady-state concentration fields within the gill. Finite element analysis was used to compute this steady-state concentration field[59] and the exchange rate was calculated by multiplying efferent concentrations by flow velocities. More complete information on model formulation is reported elsewhere.[59]

Model Parameterization

Model parameters that must be estimated include water and blood flows, chemical diffusivities, chemical speciation coefficients, and various gill morphometric measures. Details on parameterization techniques can be found in the literature,[59] but an overview as presented earlier[60] is also given here.

Required information on gill morphometry includes average lamellar length (L), total gill area (A = 2NWL), and thicknesses of the water (T_W), epithelial (T_E), and blood (T_B) layers. Much of this information has been summarized[16] and scaling relationships were used to estimate total gill area, average lamellar bilateral area, and average lamellar density. Combined with information on epithelial thickness, reasonable estimates can be made of the required model parameters for a species by using data for that species or, alternatively, averages for all reported species.

The water flow (Q_W) in the model is only that portion of water which flows between lamellae. It thus excludes water that bypasses lamellae and flows around the ends of gill filaments ('anatomical dead space'). It should include water from which a chemical is not absorbed due to diffusional constraints, but still passes through the lamellar channels ('physiological dead space').[61] Estimates for Q_W are made by dividing oxygen consumption for the fish of interest by the afferent oxygen concentration and the estimated efficiency for oxygen removal in the lamellar channels. Blood flow through the lamellae (Q_B) is assumed to be the total cardiac output, and may be estimated from relationships reported for cardiac output vs. temperature[18] and weight.[62] Extrapolation of this information to smaller fish and to other species represents a major uncertainty in model parameterization. However, as more physiological and biochemical information on different species becomes available these uncertainties will be reduced.

Because blood and water flows were assumed to be laminar, the diffusivity in the water channels was set to the molecular diffusivity, which is based on molecular volume relationships. The diffusivity in blood was assumed to be similar to that in water. The epithelial diffusivity was assigned to be half of the molecular diffusivity in water to approximately account for the greater viscosity and tortuosity of transport in the epithelium.[59]

The speciation of the compound is a critical component for defining the diffusion gradient within the gill, especially across the epithelium. For lipophilic chemicals, binding by components

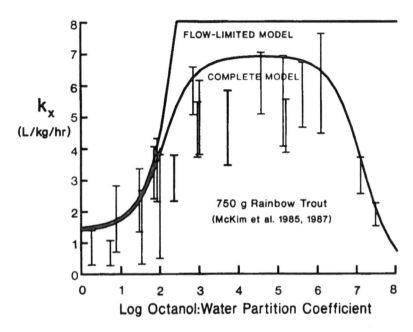

Figure 8. Gill model predictions for rainbow trout uptake vs. log K_{ow} with data as described in Figure 4a. Two model simulations are shown—one for the flow-limited model of Erickson and McKim[24] and the second for the complete model as described in the text. (From Erickson, R.J. and McKim, J.M., *Aquat. Toxicol.*, 18, 175–195, 1990. With permission.)

of the blood reduce the concentration of free chemical (which defines the diffusion gradient) and allow total concentrations to accumulate in excess of that in the water. Based on measurements of blood binding,[63,64] a relationship of blood binding to K_{ow} was derived[59] and is used in the model. As discussed earlier, high molecular weight dissolved organic matter (DOM) in water can also complex a significant fraction of highly lipophilic organic xenobiotics. Where available, measured partition coefficients are used in model calculations; otherwise, an empirical relationship of the DOM partition coefficient vs. the octanol:water partition coefficient was used.[65] Finally, the speciation of ionizable compounds (and thus their permeability through membranes) is affected by pH, and to a lesser extent by temperature and other chemical parameters. The concentration of the neutral form of the compound can be calculated based on the pK_a. However, whereas chemical bound to DOM and suspended solids in water and blood reasonably can be considered not to cross the epithelium, some transport of ions must be considered.[60] Application of the present model to simultaneous transport of both neutral and ionic forms will be further discussed in the section on pH effects below.

Model Validation

To illustrate the utility of this model, Figure 8 shows its application to the data set presented earlier regarding the dependence of uptake in large rainbow trout on the octanol-water partition coefficient.[59] Two simulations are shown. In the first, diffusional and chemical kinetic constraints are considered to be negligible ('flow-limited model'), while in the second, reasonable diffusion steps are included following the above parameterization guidelines ('complete model'). These simulations follow the pattern and the magnitude of the data quite well. The similarity between the two simulations for log K_{ow} less than 6 indicate that limitations imposed by water and blood flows are of primary importance for this system, with diffusional constraints having less influence. Trends of uptake with K_{ow} are largely attributable to relative flows of water and blood and the partitioning of chemical between these fluids. For log $K_{ow} < 1$, the blood:water partition coefficient is near 1, so the efferent blood becomes nearly saturated relative to the afferent water after removing only a small fraction of the chemical in the water. The uptake rate constant is therefore roughly equal to the blood flow rate. As log K_{ow} increases, increased binding of chemical to blood constituents gives the blood more capacity to carry chemical away from the gills, and the uptake coefficient rises correspondingly until the water passing between perfused lamellae has been essentially depleted of all chemical. Then, efferent water nears equilibrium with afferent blood and the uptake rate constant is roughly equal to the water flow rate. The rate stays approximately constant with further increases in log K_{ow} until organic complexation becomes important at log K_{ow} greater than 6, and reduces the availability of chemical in the water.

While these modeling efforts[59] were based on simple, realistic principles, their validation relied solely on empirical uptake rates.[27] The underlying physiological mechanisms were not directly tested, but were assumed to be correct because of model success. Recently, however, Schmieder and Weber[66] directly tested the assumption that blood and water flow limitations and partitioning are major factors in the gill exchange of organic chemicals in fish. *In vivo* techniques were applied that permitted direct measurement of cardiac output and ventilation volume, and focused on the influence of physiological flows of blood and water through the gill on chemical uptake rates.

Schmieder and Weber[66] developed an experimental protocol that used hypoxia and post-hypoxia conditions to manipulate gill water and blood flows at specific times. As the trout varied gill blood and water flows, gill flux rates were measured for butanol (logP = 0.88) and decanol (logP = 4.51). Changes in chemical uptake rate were then interpreted in relation to the observed variations in gill water and blood flows. Butanol (hydrophilic) removal from gill water increased 70% as cardiac output was elevated, while the uptake rate of decanol (hydrophobic) increased to the greatest extent (100%) with maximally elevated gill water flow. These researchers demonstrated for the first time that gill uptake of hydrophilic chemicals (butanol) by the trout is blood flow-limited, while the uptake of hydrophobic chemicals (decanol) is water flow-limited.[66] These direct measurements of uptake agree well with the flow-limited gill model[24] predictions of blood and water flow limitations on chemical uptake.

Further verification of the counter current gill model[24] structure and the specific inputs to the model was made possible through the successful use of the gill model in a physiologically based toxicokinetic (PB-TK) inhalation model for tetra, penta, and hexachloroethane in rainbow trout.[67,68]

CHEMICAL FLUX ACROSS FISH SKIN

The major function of the skin of fish is to provide a tough, resilient barrier that protects the fish from abrasive injury and disease, and is impermeable enough to maintain osmotic balance. The skin of fish also contributes to gas exchange and has important sensory functions.[69] Chemical flux across the dermal surface has also been shown to occur in several species in varying amounts, depending on the chemicals, species, and life stages being studied.[70-72]

Mechanistic Dermal Models

Two fish dermal models are presently being developed. The earlier work of McKim and Nichols[73] involved a large rainbow trout PB-TK dermal model that predicted the uptake and disposition of three chloroethanes. Chemical uptake across the skin of these large rainbow trout was modeled as a Fick diffusion process, and dermal permeability coefficients (K_d) were obtained by fitting model simulations to observed arterial blood data.[73] All other model inputs were unchanged from previous inhalation studies with trout.[68] Parameterized in this manner, the model

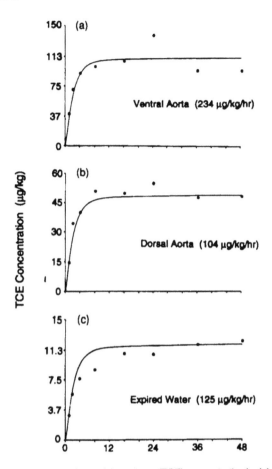

Figure 9. The 48-h time course for 1,1,2,2-tetrachloroethane (TCE) concentration in: (a) venous blood, (b) arterial blood, and (c) expired water. Model simulations are represented as solid lines, the observed values as individual points. Trout were dermally exposed to a waterborne TCE concentration of (mean) 1367 µg TCE/l. Each point represents the mean of five fish. (From McKim, J.M. and Nichols, J.W., *Aquatic Toxicology: Molecular, Biochemical, and Cellular Perspectives*, Ostrander, G.K. and Mallins, D.C., Eds., CRC Press, Boca Raton, FL, 1993.)

accurately described the kinetics of uptake and disposition of all three chloroethanes. Further, the kinetics of accumulation of tetrachloroethane in both venous and arterial blood, and its elimination in expired water (Figure 9), provided additional verification of the inputs to the gill model of Erickson and McKim,[59] discussed previously. The more recent work by Lien and McKim[74] deals with the dermal uptake of 2,5,2',5'-tetrachlorobiphenyl (TCB) in fathead minnows and Japanese medaka.

The chemical exchange across the skin of large rainbow trout is relatively low (2 to 5%) in comparison to the much higher gill exchanges observed. However, dermal exchange of chemicals relative to gill exchanges in small bioassay fish (<4 g) like the fathead minnow and Japanese medaka are quite substantial (40 to 50%). Earlier work on the guppy, another small bioassay fish, also demonstrated a considerable uptake across the dermal surface (25 to 35%).[32] Likewise, Rombough and Moroz[75] observed as much as 80% of the oxygen uptake of larval salmon to be absorbed across the dermal surface.

The surface area of the skin in small fish is often equal to or greater than that of the gills, while in large (1 Kg) trout the gill surface area can be as much as twofold greater than the skin area. A large skin area to volume ratio in the small fish, coupled with a much thinner skin, can result in high chemical exchange rates across the dermal surface. The recent study by Lien and McKim[74] suggests that 30-day-old fathead minnows weighing approximately 74 mg have a skin

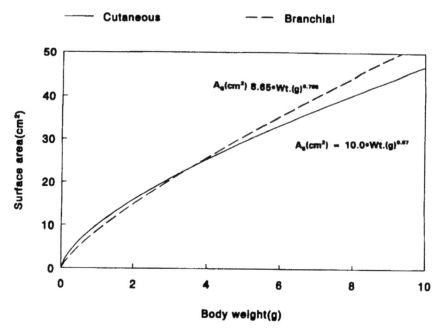

Figure 10. Calculated area for total skin and gill surfaces in small bioassay fish. (Modified from Lien, G.J. and McKim, J.M., *Aquat. Toxicol.*, 27, 15–32, 1993. With permission.)

surface area that is slightly greater than the total lamellar surface area (Figure 10). The influence of skin exchange on total exchange is even more pronounced when one considers the functional gill area (i.e., that portion of the lamellar surface in direct contact with respiratory water), which can vary from 36% of anatomical at rest to 70% during maximum activity.[76] In addition, the skin thickness (epidermis and dermis) in 30-day-old fatheads (68 µm) was shown to be much less than the thickness of rainbow trout skin (1.2 mm, epidermis and dermis), which greatly favors dermal exchange in the smaller fish.[74]

Model Development and Parameterization for Small Fish

The dermal model for small fish was structured with blood flow through the dermis (perfusion), and diffusion across the epidermis and dermis (Figure 11).[74] The exchange of chemical at cutaneous surfaces was estimated using a diffusion: perfusion conductance ratio derived from the approximate analytical solution presented by Erickson and McKim.[59]
 Important model parameters include:
1. Blood flow to the skin
2. Blood: water partition coefficient
3. Diffusivity of chemical across dermal tissue[59]
4. Diffusion distance across the dermal surface
5. Surface area of the skin, calculated using the allometric equation[77] $As(cm^2) = 10.0 \cdot Wt\ (g)^{0.67}$.

Dermal Model Validation for Small Fish

The model developed by Lien and McKim[74] describes the simultaneous uptake of TCB across the gills and skin of 30-day-old fathead minnows as being linear for the time period studied (Figure 12). There was excellent agreement between observed and predicted uptake of TCB from the combined gill and skin exposure in these fish (Figure 12, solid line). The model, parameterized independently of residue measurements, predicts that gill exchange alone would account for roughly one half the observed uptake of TCB in these fish (Figure 12, dashed line).

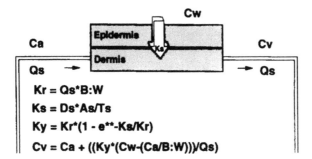

$$Kr = Qs^*B:W$$

$$Ks = Ds^*As/Ts$$

$$Ky = Kr^*(1 - e^{**}-Ks/Kr)$$

$$Cv = Ca + ((Ky^*(Cw-(Ca/B:W)))/Qs)$$

Where:

D = **Diffusivity**   a = **arterial**

A = **Area**   v = **venous**

T = **Thickness**   w = **water**

Q = **Blood Flow**   s = **skin**

C = **Concentration of Chemical**

B:W = **Blood:Water Partition Coefficient**

Figure 11. Conceptual dermal model for 2,5,2',5'-tetrachlorobiphenyl. (Modified from Lien, G.J. and McKim, J.M., *Aquat. Toxicol.*, 27, 15–32, 1993. With permission.)

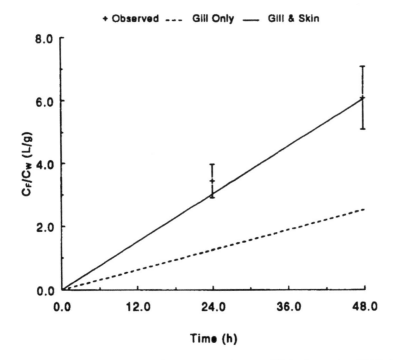

Figure 12. Predicted (solid line) and observed rate constants (mean ± SD) for the uptake of TCB via branchial and cutaneous exchange surfaces in the fathead minnow. The dashed line is the estimated rate constant for gill uptake only. (Modified from Lien, G.J. and McKim, J.M., *Aquat. Toxicol.*, 27, 15–32, 1993. With permission.)

The fact that most of the acute and chronic aquatic toxicity databases are built on data generated from small bioassay fish and studies on the early life stages of larger fish make the dermal contribution to chemical uptake important in predictive and mechanistic toxicology.

ENVIRONMENTAL AND BIOLOGICAL VARIABLES THAT IMPACT CHEMICAL ACCUMULATION AND TOXICITY

Physiological Mechanisms

The following environmental variables were selected to show their impact on the physiological mechanisms that control the uptake of waterborne chemicals. Simulations made possible by the use of the counter current gill model[59] were used to measure the impacts of some of these variables on the uptake of selected environmental chemicals where the necessary data sets existed.[60] The chemicals involved in the simulations and the conditions of the experiments are described and referenced in Figure 13.

Dissolved Oxygen

Observed chemical uptake increases with decreasing oxygen at high oxygen concentrations but flattens out at low oxygen concentrations (Figure 13a). The adherence of model predictions to the data reflects the relative importance of flows and diffusion barriers in controlling uptake.[60] At the highest oxygen concentration, the uptake is apparently largely water flow limited. As oxygen concentration declines, the water flow rate increases to maintain oxygen consumption and provides more chemical for absorption. However, as the flow increases even more, diffusion becomes relatively more important and makes it more difficult to maintain a high efficiency of chemical removal from water.

Temperature

The effects of temperature changes within the physiological range for the animal, impacting the uptake of chemicals primarily by altering the metabolic rate and, correspondingly, the oxygen requirements.[60] This changes the flow of water across the gills and for water flow-limited chemicals alters the chemical uptake (Figure 13b). The adherence of the model predictions to the data is primarily due to changes in oxygen consumption. Oxygen consumption decreases a little more than twofold from 20 to 5° C, but the higher oxygen concentration at the lower temperature makes the decrease in water flow more than threefold. Thus, changes in chemical uptake with temperature can exceed changes in oxygen uptake simply because of the relationship of water flow to oxygen demand and availability.

pH

Increased ionization of weak acids with increasing pH results in lower accumulation, apparently because of the inability or limited ability of the ionized compound to move across lipid membranes in the gill epithelium (Figure 13c). The observed decrease in accumulation is higher than would be predicted based on the assumption that only the neutral form is absorbed (lowest curve on Figure 9c). Three possible explanations for the greater than expected uptake at high pH exist:[60]

1. As the neutral compound is absorbed at the gill surface rapid acid-base equilibration results in the ionized compound forming more of the neutral compound, which can be further absorbed,

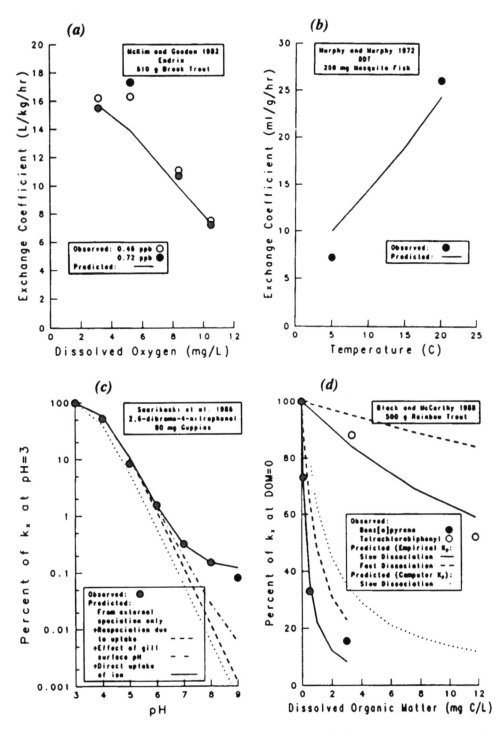

Figure 13. Gill model predictions vs. observations for the effects of: (a) dissolved oxygen concentrations on the exchange of endrin across the gills of brook trout at 12°C;[11] (b) the effect of temperature on the exchange of DDT across the gills of the mosquito fish at 5 and 20°C;[78] (c) the effect of pH on the exchange of 2,6-dibromo-4-nitrophenol across the gills of the guppy at 20°C;[32] (d) the effect of dissolved organic material (DOM) on the exchange of tetrachlorobiphenyl and benz[a]pyrene across the gills of rainbow trout at 18°C.[4] (From McKim, J.M. and Erickson, R.J., *Physiol. Zool.*, 64, 39–67, 1991. With permission.)

2. Respiration causes a reduction of the pH in the gill water near the gill surface and more of the ionic form is converted to the neutral form for rapid uptake, and

3. The ionic form may be taken up directly from the water at a much slower rate than the neutral form.

Each of these three scenarios was shown in Figure 13c to improve the model simulations, using the countercurrent gill model.[59]

Dissolved Organic Matter

DOM partition coefficients were used in model simulations, which assumed no significant dissociation of the complexes during passage through the gills and assumed instantaneous dissociation to maintain the equilibrium at all points.[60] The model simulations based on the assumption of slow dissociation clearly are best supported by the data in Figure 13d. The second model simulation is based on the relationship of the DOM partition coefficients to K_{ow} computed from a different source of DOM.[65] Simulations were poor, and suggested that caution be used in extrapolating DOM partition coefficients to other chemicals and sources of DOM.

Biochemical Mechanisms

Environmental variables such as temperature, stress, diet type and quantity, age, sexual maturity, species, etc., are also known to significantly vary the biochemical mechanisms of these lower vertebrates that biotransform organic chemicals.[79] In order to truly understand the toxicity of chemicals to aquatic animals, and to make tissue dosimetry models feasible in predictive toxicology, the internal distribution of both parent compounds and their major metabolites must be qualitatively and quantitatively understood. Biochemical reactions in fish that are responsible for the biotransformation of toxic organic chemicals have only recently been recognized for the impact they have on accumulation and toxicity. Through the use of chemical promoters and inhibitors of these reactions, researchers are now able to show the contribution that biotransformation has on final residue accumulation and toxicity.

An early example of the importance of biotransformation on toxicity was provided by Lech,[80] who experimented with 3-trifluoromethyl-4-nitrophenol (TFM), a lampricide developed for sea lamprey control in the Great Lakes. He inhibited TFM glucuronide formation with salicylamide (a glucuronyl transferase inhibitor) and increased toxicity twofold. As expected, the disposition of TFM was also altered, with a decrease in biliary elimination of TFM-glucuronide and increased levels of TFM in specific tissues. A similar study was carried out by Erickson et al.[81] on the effects of both a P-450 inhibitor and promoter on the toxicity of rotenone to the rainbow trout. They found a striking 60% reduction from controls in the 120-h LC50 for rotenone caused by the metabolic inhibitor (PBO), while a 23% increase in the LC50 was observed when the promoter (BNF) was applied. A recent example of the importance of biotransformation in the bioaccumulation and toxicity of 2,8-dichlorodibenzo-*p*-dioxin (DCDD) was reported by Sijm and Opperhuizen.[82] DCDD is normally not accumulated nor is it toxic to fish; however, when a P-450 inhibitor was introduced during exposure, DCDD accumulated rapidly and became almost as toxic as 2,3,7,8-tetrachloro-*p*-dioxin (TCDD). Other examples of the importance of promotion and inhibition of biotransformation reactions are available in a recent review by Kleinow et al.[79]

While bioconcentration of organic chemicals in fish is controlled in large part by animal fat content and chemical lipophilicity, it is becoming well recognized that biotransformation can play a key role in regulating the accumulation of metabolizable chemicals. This is especially true for bioconcentration factors (BCFs) estimated from water solubility data, where the estimated and observed values can vary considerably—as in the case of an alkylbenzene compound reported to have an observed BCF of only 35, yet the estimated BCF was 6300.[83] Similar studies on PCP[84] showed the calculated BCF to be 44% higher than the actual observed BCF. In both of these examples the chemicals underwent considerable biotransformation.

It should be recognized at this point that both the physiological mechanisms discussed previously, and the biochemical mechanisms responsible for biotransformation are major controlling factors in the accumulation and toxicity of organic and inorganic chemicals in aquatic animals. Furthermore, these controlling mechanisms are continually impacted by environmental variables (physical, chemical, biological).

CONCLUSIONS

The major physiological and chemical factors that regulate the uptake of waterborne, bioavailable, organic chemicals have been identified, are relatively simple, and can be easily incorporated into mathematical models that are truly predictive of chemical accumulation and toxicity. For gill models, these factors include respiratory water volume, cardiac output, blood:water partition coefficients, and gill morphometry. In addition to the above, dermal models require skin thickness, diffusivity, surface area, and percentage of cardiac output to the skin. With a knowledge of these factors and the simple mechanistic models discussed here, it is feasible to account for the general magnitude of xenobiotic exchange at gills and skin in various species under specific environmental conditions. Further evaluation of these models with inorganic chemicals, other species, and unique environmental situations is needed to better understand the mechanisms controlling xenobiotic gill and dermal fluxes.

The surface exchange models reviewed here, along with the future development of a mechanistic gut exchange model, will provide the link between the environmentally available chemical and animal accumulation. When these exchange models are coupled with species-unique physiological, biochemical, and behavioral attributes, and with pertinent ecological and physical-chemical information, they will be capable of mechanistically describing and predicting bioaccumulation and toxicity in aquatic animals.

RESEARCH NEEDS

1. Document physiological information on important species, especially those on which effects data exist.
2. Develop a mechanistic gut uptake model that can account for carrier, chemical structure, and species (determine rate-limiting barriers).
3. Determine blood:water partition values for nonvolatile organics.
4. Develop innovative ways to collect accurate biotransformation rate and capacity data for specific chemical classes.
5. Document physiological and morphological information on the exchange surfaces (gills, dermal, and gut) of specific life stages.
6. Develop and parameterize an invertebrate uptake model (determine rate-limiting barriers).
7. Establish toxic end points that can be effectively modeled.
8. Use available information outlined above to build physiologically-based models that allow the extrapolation of toxicity and disposition of chemicals across species, regardless of chemical, dosing regime, or route of exposure.

REFERENCES

1. Bruggeman, W.A., Opperhuizen, A., Hutzinger, A., and Wijbenga, O., Bioaccumulation of super-lipophilic chemicals in fish, *Toxicol. Environ. Chem.*, 7, 173–189, 1984.
2. Thomann, R.V. and Connolly, J.P., Model of PCB in the Lake Michigan lake trout food chain, *Environ. Sci. Technol.*, 18, 65–78, 1984.
3. Oliver, B.G. and Niimi, A.J., Bioconcentration factors of some halogenated organics for rainbow trout: limitations in their use for prediction of environmental residues, *Environ. Sci. Technol.*, 19, 842–847, 1985.

4. Black, M.C. and McCarthy, J.F.. Dissolved organic macromolecules reduce the uptake of hydrophobic organic contaminants by the gills of the rainbow trout (*Salmo gairdneri*), *Environ. Toxicol. Chem.*, 7, 593–600, 1988.
5. Winter, M.E., Katcher, B.S., and Koda-Kimble, M.A., *Basic Clinical Pharmacokinetics*, Applied Therapeutics, Spokane, WA, 1980.
6. McKim, J.M. and Nichols, J.W., Use of physiologically based toxicokinetic models in a mechanistic approach to aquatic toxicology, in Ostrander, G.K. and Malins, D.C., Eds., *Aquatic Toxicology: Molecular, Biochemical, and Cellular Perspectives*, CRC Press, Boca Raton, FL, 1994.
7. Rand, G.M. and Petrocelli, S.R., Introduction, in Rand, G.M. and Petrocelli, S.R., Eds., *Fundamentals of Aquatic Toxicology: Methods and Applications*, Hemisphere, Washington, D.C., 1985, 1–28.
8. McCarty, L.S., The relationship between aquatic toxicity QSARs and bioconcentration for some organic chemicals, *Environ. Toxicol. Chem.*, 5, 1071–1080, 1986.
9. McCarty, L.S., Relationship between toxicity and bioconcentration for some organic chemicals. I. Examination of the relationship, in Kaiser, K.L.E., Ed., *QSAR in Environmental Toxicology*, Vol. II. D. Reidel, Dordrecht, 1987, 207–220.
10. McCarty, L.S., Relationship between toxicity and bioconcentration for some organic chemicals. II. Application of the relationship, in Kaiser, K.L.E., Ed., *QSAR in Environmental Toxicology*, Vol. II, D. Reidel, Dordrecht, 1987, 221–230.
11. McKim, J.M. and Goeden, H.M., A direct measure of the uptake efficiency of a xenobiotic chemical across the gills of brook trout (*Salvelinus fontinalis*) under normoxic and hypoxic conditions, *Comp. Biochem. Physiol.*, 72C, 65–74, 1982.
12. McKim, J.M. and Heath, E.M., Dose determinations for waterborne 2,5,2'5'-14 C tetrachlorobiphenyl and related pharmacokinetics in two species of trout (*Salmo gairdneri* and *Salvelinus fontinalus*) A mass-balance approach, *Toxicol. Appl. Pharmacol.*, 68, 177–187, 1983.
13. Black, M.C., Millsop, D.S., and McCarthy, J.F., Effects of acute temperature change on respiration and toxicant uptake by rainbow trout (*Salmo gardneri*), *Physiol. Zool.*, 64, 145–168, 1991.
14. Satchell, G.H., Respiratory toxicology of fishes, in Weber, L.J., Ed., *Aquatic Toxicology*, Vol. 2, Raven Press, New York, 1984, 1–50.
15. Wedemeyer, G.A., Meyer, F.P., and Smith, L., *Environmental Stress and Fish Diseases*, TFH Publications Neptune, NJ, 1976.
16. Hughes, G.H., General anatomy of the gills, in Hoar, W.S. and Randall, D.J., Eds., *Fish Physiology*, Academic Press, Orlando, FL, 1984.
17. Randall, D.J. and Daxboeck, C., Oxygen and carbon dioxide transfer across fish gills, in Hoar, W.S. and Randall, D.J., Eds., *Fish Physiology*, Academic Press, Orlando, FL, 1984, 263–307.
18. Barron, M.G., Tarr, B.D., and Hayton, W.L., Temperature-dependence of cardiac output and regional blood flow in rainbow trout (*Salmo gairdneri*) Richardson, *J. Fish Biol.*, 31, 735–744, 1987.
19. Nilsson, S., Control of gill blood flow, in Nilsson, S. and Holmgren, S., Eds., *Fish Physiology, Recent Advances*, Croom Helm, Bristol, U.K., 1987, 86–101.
20. Laurent, P. and Perry, S.F., Environmental effects on fish gill morphology, *Physiol. Zool.*, 64, 4–25, 1990.
21. Randall, D.J., Lin, H., and Wright, P.A., Gill water flow and the chemistry of the boundary layer, *Physiol. Zool.*, 64, 26–38, 1991.
22. Wood, C.M., Branchial ion and acid-base transfer: environmental hypoxia as a probe, *Physiol. Zool.*, 64, 68–102, 1991.
23. Hayton, W.L. and Barron, M.G., Rate-limiting barriers to xenobiotic uptake by the fish gill, *Environ. Toxicol. Chem.*, 9, 151–157, 1990.
24. Erickson, R.J. and McKim, J.M., A simple flow-limited model for exchange of organic chemicals at fish gills, *Environ. Toxicol. Chem.*, 9, 159–165, 1990.
25. Hunn, J.B. and Allen, J.L., Movement of drugs across the gills of fishes, *Annu. Rev. Pharmacol.*, 14, 1–27, 1974.
26. Spacie, A. and Hamelink, J.L., Alternative models for describing the bioconcentration of organics in fish, *Environ. Toxicol. Chem.*, 1, 309–320, 1982.
27. McKim, J., Schmieder, P., and Veith, G., Absorption dynamics of organic chemical transport across trout gills as related to octanol-water partition coefficient, *Toxicol. Appl. Pharmacol.*, 77, 1–10, 1985.

28. McKim, J.M., Schmieder, P.K., and Erickson, R.J., Toxicokinetic modeling of (14C) Pentachlorophenol in rainbow trout *(Salmo gairdneri)*, *Aquat. Toxicol.*, 9, 59–80, 1986.

29. Bradbury, S.P., Coats, J.R., and McKim, J.M., Toxicokinetics of fenvalerate in rainbow trout *(Salmo gairdneri)*, *Environ. Toxicol. Chem.*, 5, 567–576, 1986.

30. McKim, J.M., Schmieder, P.K., Carlson, R.W., Hunt, E.P., and Niemi, G.J., Use of respiratory-cardiovascular responses of rainbow trout *(Salmo gairdneri)* in identifying acute toxicity syndromes in fish. I. Pentachlorophenol, 2,4-dinitrophenol, tricane, methanesulfonate, and 1-octanol, *Environ. Toxicol. Chem.*, 6, 295–312, 1987.

31. McKim, J.M., Schmieder, P.K., Niemi, G.J., Carlson, R.W., and Henry, T.R., Use of respiratory-cardiovascular responses of rainbow trout *(Salmo gairdneri)* in identifying acute toxicity syndromes in fish. II. Malathion, carbaryl, acrolein, and benzaldehyde, *Environ. Toxicol. Chem.*, 6, 313–328, 1987.

32. Saarikoski, J., Lindstrom, M., Tyynila, M., and Viluksela, M., Factors affecting the absorption of phenolics and carboxylic acids in the Guppy *(Poecilia reticulata)*, *Ecotoxicol. Environ. Saf.*, 11, 158–173, 1986.

33. Zitko, V., Uptake of chlorinated paraffins and PCB from suspended solids and food by juvenile Atlantic salmon, *Bull. Environ. Contam. Toxicol.*, 12, 406–412, 1974.

34. Zitko, V. and Hutzinger, O., Uptake of chloro- and bromobiphenyls, hexachloro-and hexabromobenzene by fish, *Bull. Environ. Contam. Toxicol.*, 16, 665–673, 1976.

35. Tulp, M. and Hutzinger, O., Some thoughts on aqueous solubilities and partition coefficients of PCB, and the mathematical correlation between bioaccumulation and physico-chemical properties, *Chemosphere*, 10, 849–860, 1978.

36. Sugiura, K., Ito, N., Matsumoto, N., Mihara, Y., Murata, K., Tsukakoshi, Y., and Goto, M., Accumulation of polychlorinated biphenyls and polybrominated biphenyls in fish: limitation of "correlation between partition coefficients and accumulation factors", *Chemosphere*, 9, 731–736, 1978.

37. Konemann, H. and Van Leeuwen, K., Toxicokinetics in fish: accumulation and elimination of six chlorobenzenes by guppies, *Chemosphere*, 9, 3–19, 1980.

38. Gobas, F.A.P.C., Opperhuizen, A., and Hutzinger, O., bioconcentration of hydrophobic permeation, *Environ. Toxicol. Chem.*, 5, 637–646, 1986.

39. Opperhuizen, A., van der Velde, E.W., Gobas, F.A.P.C., Liem, D.A.K., and van der Steen, J.M.D., Relationship between bioconcentration in fish and steric factors of hydrophobic chemicals, *Chemosphere*, 14, 1871–1896, 1985.

40. Norstrom, R.J., McKinnon, A.E., and DeFreitas, S.F., A bioenergetics based model for pollutants accumulation by fish. Simulation of PCB and methyl mercury residue levels in Ottawa River yellow perch *(Perca flaescens)*, *J. Fish. Res. Board. Can.*, 33, 248–267, 1976.

41. Neely, W.B., Estimating rate constants for the uptake and clearance of chemicals by fish, *Environ. Sci. Technol.*, 13, 1506–1510, 1979.

42. Bruggeman, W.A., Martron, L.B.J.M., Kooiman, D., and Hutzinger, O., Accumulation and elimination kinetics of di-, tri-, and tetra chlorobiphenyls by goldfish after dietary and aqueous exposure, *Chemosphere*, 10, 811–832, 1981.

43. Gobas, F.A.P.C. and Mackay, D., Dynamics of hydrophobic organic chemical bioconcentration in fish, *Environ. Toxicol. Chem.*, 6, 495–504, 1987.

44. Barber, M.C., Suarez, L.A., and Lassiter, R.R., Bioconcentration of nonpolar organic pollutants by fish, *Environ. Toxicol. Chem.*, 7, 545–558, 1988.

45. Spry, D.J. and Wood, C.M., Zinc influx across the isolated perfused head preparation of the rainbow trout *(Salmo gairdneri)* in hard and soft water, *Can. J. Fish. Aquat. Sci.*, 45, 2206–2215, 1988.

46. Pentreath, R.J., The accumulation and retention of ^{65}Zn and ^{54}Mn by the plaice, *Pleuronectes platessa* L., *J. Exp. Mar. Biol. Ecol.*, 12, 1–18, 1973.

47. Rankin, J.C., Stagg, R.M., and Bolis, L., Effects of pollutants on gills, in Houlihan, D.F., Rankin, J.C., and Shuttleworth, T.J., Eds. *Gills*, Soc. Exp. Biol. Semin. Ser. 16, Cambridge University Press, London, 1982, 207–220.

48. Perry, S.F. and Wood, C.M., Kinetics of branchial calcium uptake in the rainbow trout: effects of acclimation to various external calcium levels, *J. Exp. Biol.*, 116, 411–433, 1985.

49. Part, P. and Svanberg. O., Uptake of cadmium in perfused rainbow trout *(Salmo gairdneri)* gills, *Can. J. Fish. Aquat. Sci.*, 38, 917–924, 1981.

50. Laure, D.J. and McDonald, D.G., Influence of water hardness, pH, and alkalinity on the mechanisms of copper toxicity in juvenile rainbow trout (*Salmo gairdneri*), *Can. J. Fish. Aquat. Sci.*, 43, 1488–1496, 1986.

51. Spry, D.J. and Wood, C.M., A kinetic method for the measurement of zinc influx *in vivo* in the Rainbow Trout and the effects of waterborne calcium on flux rates, *J. Exp. Biol.*, 142, 425–446, 1989.

52. Spry, D.J., Hodson, P.V., and Wood, C.M., Relative contributions of dietary and waterborne zinc in the rainbow trout, *Salmo gairdneri*, *Can. J. Fish. Aquat. Sci.*, 45, 32–41, 1989.

53. Verbost, P.M., van Rooij, J., Flik, G., Lock, R.A.C., and Wendelaar Bonga, S.E., The movement of cadmium through freshwater trout branchial epithelium and its interference with calcium transport, *J. Exp. Biol.*, 145, 185–197, 1989.

54. Roesijadi, G. and Unger, M.E., Cadmium uptake in the gills of the mollusc *Crassostrea virginica* and inhibition by calcium channel blockers, *Aquat. Toxicol.*, 24, 195–206, 1993.

55. Wood, C.M., Playle, R.C., Simons, B.P., Goss, G.G., and McDonald, D.G., Blood gases, acid-base status, ions, and hematology in adult brook trout (*Salvelinus fontinalis*) under acid/aluminum exposure, *Can. J. Fish. Aquat. Sci.*, 45, 1575–1586, 1988.

56. Wood, C.M., McDonald, D.G., Booth, C.E., Simons, B.P., Ingersoll C.G., and Bergman, H.L., Physiological evidence of acclimation to acid/aluminum stress in adult brook trout (*Salvelinus fontinalis*). I. Blood composition and net sodium fluxes, *Can. J. Fish. Aquat. Sci.*, 45, 1587–1596, 1988.

57. Playle, R.C. and Wood, C.M., Water pH and aluminum chemistry in the gill micro-environment of rainbow trout during acid and aluminum exposures, *J. Comp. Physiol. Biol.*, 159, 539–550, 1989.

58. Playle, R.C. and Wood, C.M., Water chemistry changes in the gill micro-environment of rainbow trout: experimental observations and theory, *J. Comp. Physiol. Biol.*, 159, 527–537, 1989.

59. Erickson, R.J. and McKim, J.M., A model for exchange of organic chemicals at fish gills: flow and diffusion limitations, *Aquat. Toxicol.*, 18, 175–198, 1990.

60. McKim, J.M. and Erickson, R.J., Environmental impacts on the physiological mechanisms controlling xenobiotic transfer across fish gills, *Physiol. Zool.*, 64, 39–67, 1991.

61. Randall, D.J., Gas exchange in fish, in Hoar, W.S. and Randall, D.J., Eds., *Fish Physiology*, Vol. IV, Academic Press, New York, 1970, 253–292.

62. Wood, C.M. and Shelton, G., Cardiovascular dynamics and adrenergic responses of rainbow trout *in vivo*, *J. Exp. Biol.*, 87, 247–270, 1980.

63. Schmieder, P.K. and Henry, T.R., Plasma binding of 1-butanol, phenol, nitrobenzene and pentachlorophenol in the rainbow trout and rat: a comparative study, *Comp. Biochem. Physiol.*, 91C, 413–418, 1988.

64. Hoffman, A., Bertelson, S., and Gargas, M., An *in vitro* gas equilibration method for determination of chemical partition coefficients in fish, *Comp. Biochem. Physiol.*, 101A, 47–51, 1992.

65. McCarthy, J.F., Jimenez, B.D., and Barbee, T., Effect of dissolved humic material on accumulation of polycyclic aromatic hydrocarbons: structure-activity relationships, *Aquat. Toxicol.*, 7, 15–24, 1985.

66. Schmieder, P.K. and Weber, L.J., Blood and water flow limitations on gill uptake of organic chemicals in the rainbow trout (*Oncorhynchus mykiss*), *Aquat. Toxicol.*, 24, 103–122, 1992.

67. Nichols, J.W., McKim, J.M., Andersen, M.E., Gargas, M.L., Clewell, H.J., III, and Erickson, R.J., A physiologically based toxicokinetic model for the uptake and disposition of waterborne organic chemicals in fish, *Toxicol. Appl. Pharmacol.*, 106, 433–455, 1990.

68. Nichols, J.W., McKim, J.M., Lien, G.J., Hoffman, A.D., and Bertelsen, S.L. Physiologically based toxicokinetic modeling of three waterborne chloroethanes in rainbow trout (*Oncorhynchus mykiss*), *Toxicol. Appl. Pharmacol.*, 110, 374–388, 1991.

69. Van Oosten, J., The skin and scales, in Brown, M.E., Ed., *The Physiology of Fishes*, Vol. I, Academic Press, New York, 1957, 207–244.

70. Tovell, P.W.A., Howes, D., and Newsome, C.S., Absorption, metabolism and excretion by goldfish of the anionic detergent sodium lauryl sulphate, *Toxicology*, 4, 17–29, 1975.

71. Ferreira, J.T., Schoonbee, H.J., and Smit, G.L., The uptake of the anaesthetic benzocaine hydrochloride by the gills and skin of three freshwater fish species, *J. Fish. Biol.*, 25, 35–41, 1984.

72. Varanasi, U., Uhler, J., and Stranahan, S.I., Uptake and release of naphthalene and its metabolites in skin and epidermal mucus of salmonids, *Toxicol. Appl. Pharmacol.*, 44, 277–289, 1978.

73. McKim, J.M. and Nichols, J.W., Physiologically based toxicokinetic modeling of the dermal uptake of three waterborne chloroethanes in rainbow trout (Oncorhynchus mykiss), *Toxicologist,* 11(Abstr.), 35, 1991.

74. Lien, G.J. and McKim, J.M., Predicting branchial and cutaneous uptake of 2,5,2′,5′-tetrachlorobiphenyl in fathead minnow (*Pimephales promelas*) and Japanese medaka (*Oryzias latipes*): rate limiting factors, *Aquat. Toxicol.,* 27, 15–32, 1993.

75. Rombough, P.J. and Moroz, B.M., The scaling and potential importance of cutaneous and branchial surfaces in respiratory gas exchange in young chinook salmon (*Oncorhynchus tshawytscha*), *J. Exp. Biol.,* 154, 1–12, 1990.

76. Gehrke, P.C., Cardio-respiratory morphometrics of spangled perch, *Leiopotherapon unicolor* (Gunther, 1859), (Percoidei, Terponidae), *J. Fish. Biol.,* 31, 617–623, 1987.

77. Schmidt-Nielson, K., Scaling: why is animal size so important?, Cambridge University Press, Cambridge, U.K., 1984.

78. Murphy, P.G. and Murphy, J.V., Correlations between respiration and direct uptake of DDT in the mosquito fish (*Gambusia affinis*), *Bull. Environ. Contam. Toxicol.,* 6, 581–588, 1971.

79. Kleinow, K.M., Melancon, M.J., and Lech, J.J., Biotransformation and induction. Implications for toxicity, bioaccumulation and monitoring of environmental xenobiotics in fish, *Environ. Health Perspect.,* 71, 105–119, 1987.

80. Lech, J.J., Glucuronide formation in rainbow trout—effect of salicylamide on the acute toxicity, conjugation and excretion of 3-trifluoromethyl-4-nitrophenol, *Biochem. Pharmacol.,* 23, 2403–2410, 1974.

81. Erickson, D.A., Goodrich, J.S., and Lech, J.J., The effect of B-naphthoflavone and piperonyl butoxide on hepatic monooxygenase activity and the toxicity of rotenone to rainbow trout, *Salmo gairdneri, Toxicologist,* 6(Abstr.), 160, 1986.

82. Sijm, D.T.H.M. and Opperhuizen, A., Biotransformation, bioaccumulation and lethality of 2,8-dichlorodibenzo-*p*-dioxin: a proposal to explain the biotic fate and toxicity of PCDD's and PCDF's, *Chemosphere,* 17, 83–99, 1988.

83. Werner, A.F. and Kimerle, R.A., Uptake and distribution of C12 alkylbenzene in bluegill (*Lepomis macrochirus*), *Environ. Toxicol. Chem.,* 1, 143–146, 1982.

84. Stehly, G.R. and Hayton, W.L., Disposition of pentachlorophenol in rainbow trout (*Salmo gairdneri*): effect of inhibition of metabolism, *Aquat. Toxicol.,* 14, 131–148, 1989.

Chapter 3

Synopsis of Discussion Session on the Kinetics Behind Environmental Bioavailability

Peter F. Landrum (Chair), William L. Hayton, Henry Lee II, Lynn S. McCarty,
Donald Mackay, and James M. McKim

INTRODUCTION

Kinetics are involved in virtually all environmental processes. The time frame (frequency) can very from extremely brief, say, billionths of seconds, to very long, on the order of billions of years. However, only those rate processes which are within a few orders of magnitude of the examined biological rates are of practical importance. One of the most important determinants is the influence of body size on the different metabolic rates at which organisms live.[1,2] That is to say, internal processes and interactions with the external environment occur at a faster rate for small organisms than for larger organisms. Since this body-size effect can encompass several orders of magnitude, factors that influence environmental bioavailability may vary considerably between organisms of substantially different body sizes. Thus, the processes that are kinetically important for microbes will likely be different from those that are important for fish.

Rather than continuing to examine fundamental influencing factors, we propose to discuss kinetics within a toxicological framework, and on a process-orientated basis within that framework. Kinetics come into play in all three of the basic areas defined in classic toxicology/pharmacology:

1. Exposure—those factors external to the organism that control contact with the chemical;
2. Toxicokinetics—those physiological and behavioral factors controlling uptake, distribution, metabolic processing, elimination and, ultimately, delivery to the site(s) of toxic action; and
3. Toxicodynamics—those processes which constitute the biochemical and physiological sequelae resulting from the effects of the chemical at the site(s) of action.

For this report, discussion will be limited to the first two categories.

FUNDAMENTALS

Why Kinetics?

Many interesting and useful kinetic relations exist in the environment. Our objective is to describe those kinetic processes that alter exposure regimes and the toxicokinetic factors that result in toxicologically significant accumulation of chemicals on or in the body of exposed organisms. Once such factors are understood, it may then be possible to modify exposure regimes or avoid circumstances that create exposures so that adverse effects do not occur. This represents the essence of the current approach for the application of sound scientific principles to the development of exposure-based regulatory criteria for environmental protection. Within this scope, several areas of specific interest exist.

Description and Prediction of the Accumulation Process

Kinetic prediction of accumulation allows estimates not only of the steady-state condition, but also the time required to reach steady state and the shape of the accumulation-time curve.

1-56670-086-8/94/$0.00+$.50
© 1994 by CRC Press, Inc.

Current models use thermodynamic relations to empirically predict steady state, i.e., the relation between the bioconcentration factor (BCF) for nonpolar organic compounds and the octanol-water partition coefficient (K_{ow}).[3] Recent models for bioaccumulation include simple empirical mass-transfer models[4] and gill exchange models,[5,6] where accumulation of organics is controlled largely by hydrophobicity (usually approximated by log K_{ow}), as well as sophisticated physiologically based pharmacokinetic (PB-PK) models where not only body burden, but individual organ accumulation is determined.[7,8] These models are largely successful for nonpolar organic compounds where hydrophobicity dominates partitioning, somewhat successful for polar organic chemicals, and not currently viable for readily metabolized and/or reactive organics[9] and metals.

When organisms accumulate chemicals via multiple routes of exposure, such as occurs from a combination of water respiration and food ingestion, concentrations are often greater than for accumulation from water exposure alone (see Reference 10). This phenomenon is termed bio-magnification, and forces the use of kinetic models in place of simple thermodynamic potential models. Similarly, chemical accumulation from mixtures requires an explicit consideration of the individual kinetics of the mixture components.[11]

Examination of the Mechanisms Affecting the Accumulation and Loss of Contaminants in Organisms

Mechanistic investigations focus on the rate-limiting barriers to accumulation and loss (see Session 6, Chapter 2). Several physiological measures for fish have been compared with the accumulation or loss processes (e.g., ventilation rate and accumulation from water).[12,13] Similar studies with invertebrates have been few[14,15] and were unable to establish predictive relationships. Furthermore, kinetic models facilitate study and interpretation of changes in physiology and environmental conditions that influence accumulation (e.g., growth, reproductive state, temperature, species). For example, the influence of seasonal changes on toxicant accumulation may be incorporated when the time to steady state is long enough to observe such environmental and physiological cycles.[16]

Prediction of Adverse Biological Responses using the Tissue Residue Approach

Toxicity assays, as actually reported, do not always ensure that the end points under consideration represent response at steady state, and rarely provide the kinetics so that the time-response curve can be interpreted, despite advice to the contrary.[17] Effects are generally measured for environmental contaminants; although, toxicity has been predicted for some classes of compounds using log K_{ow}.[18–23] Current efforts allow prediction of effects for nonpolar organics acting by a narcotic mechanism based on body-burden residue and should be predictable from accumulation models.[25,26] Such relations are not well established for metals or for organics with a specific mechanism of toxic action.

What Is Bioavailability?

The term bioavailability is used by a variety of investigators in various fields of research. The following series of definitions is proposed to help clarify the situation, especially since the well-established toxicological/pharmacological definition of bioavailability is not compatible with some other uses of this term.

Environmental availability can be defined as the portion of the total material in a compartment (or compartments) of the environment that is involved in a process or group of processes and is subject to physical, chemical, and biological modifying influences. This represents the total pool of material that is potentially available to organisms and represents the broadest sense of the concept.

Bioavailability can be viewed as a special case of environmental availability in which organisms are involved as a target rather than as part of a process. The exclusive use of the prefix "bio" indicates this. Thus bioavailability, by necessity, incorporates not only the characteristics of the chemical and its environmental speciation, but also the behavior and physiology of the organism. Incorporation of the biological component increases the complexity of the term. However, depending on the environment and specificity of the question of interest, groups of organisms with similar characteristics may be combined through the use of clear, up-front assumptions. Two definitions below reflect two related uses and are concerned with some portion of the material defined above as being environmentally available.

Toxicological bioavailability is the fraction of the total available dose absorbed by organisms which is distributed by the systemic circulation, and ultimately presented to the receptors or sites of toxic action. This description works well for systemic toxicants, but is of less value for topically active materials and is the classical toxicological definition of bioavailability. It is often estimated as the ratio of the absorbed dose, area under the blood concentration time curve of a drug under test conditions, compared with the absorbed dose under standard conditions were the administered doses are the same. Generally, the standard condition for reference is an intravenous dose.

The difficulty in applying this concept to the general bioavailability of contaminants in the environment is the definition of dose and the appropriate reference condition. Thus, to avoid confusion with the term bioavailability as employed in pharmacology and toxicology, the term environmental bioavailability should be employed. Environmental bioavailability (EBA) constitutes the fraction of the environmentally available compound which an organism accumulates when processing a given medium.

One approach to resolving the practical difficulties in estimation of environmental bioavailability is to employ the efficiency of accumulation from each matrix (e.g., water, sediment) as a representative measure of EBA. This would be formulated as the ratio of the uptake clearance divided by the rate at which as organism encounters a given contaminant for each matrix being processed.

$$\text{EBA} = \frac{\text{Uptake Clearance } (Lkg^{-1}\,h^{-1})}{\text{Encounter Rate (Volume/Mass)}(Lkg^{-1}\,h^{-1})}$$

EBA should be used for comparing the relative availability of a contaminant between systems and/or experimental conditions. However, as with any tool, it should not be used out of context. Thus, the exposure concentrations and the uptake clearance should always be provided, because under some special conditions the flux into an organism may increase even though the efficiency decreases. This will be found under conditions where the encounter volume increases faster than the uptake coefficient.

When sediment or food exposures are considered, it is likely that the encounter mass or volume may not always be well defined, particularly in the sediment case. Thus, the best comparison between exposure conditions that provides insight into the EBA may be comparison of the uptake coefficients of two exposures.[27] Hence, while the foregoing discussion attempts to describe the efficiency of accumulation, it will not describe the effective dose at the receptor. For that, toxicological bioavailability is suggested, and further discussion of this term is beyond the scope of this workshop.

When considering a framework for discussing the processes that may kinetically limit the EBA, it is useful to consider the following simple empirical kinetic model of accumulation by an organism.

$$\frac{dC_a}{dt} = k_1 C_w + k_f\, C_f - (k_2 + k_e + k_m + k_g)\, C_a$$

where k_1 is the uptake clearance for accumulation from water (in milliliters per gram per hour); C_w is the contaminant concentration freely dissolved in the water (in micrograms per milliliter); k_f is the uptake clearance from diet (in grams per gram per hour); C_f is the contaminant concentration in the ingested food (in micrograms per gram); C_a is the contaminant concentration in the organism (in micrograms per gram); k_2 is the elimination rate constant (in hours) across diffusional membranes, e.g., the gill; k_e is the egestion (fecal, urinary, and biliary elimination) rate constant (in hours); k_m is the rate constant for metabolism (in hours); and k_g is the rate constant for growth dilution (in hours).

The uptake clearance from water incorporates the ventilation rate, the blood flow, and the membrane transfer efficiency for a contaminant flowing across the gill. Similarly, k_f incorporates the ingestion rate and assimilation efficiency for the selected food. It is not currently possible to identify all the factors that influence the assimilation efficiency for organisms ingesting sediment. Thus, while C_w represents the freely dissolved compound, which is presumably the biologically available portion of the compound in water, C_f represents the concentration in/on the ingested particles or food which is currently the best representation of the biologically available fraction that can be described. For organic contaminants in sediments, it is not sufficient to use whole-sediment concentration for C_f, as animals selectively, to varying degrees, ingest specific organic fractions of sediment,[28] nor will normalization to organic carbon completely solve the problem of EBA for organic contaminants.[29] Polar organics and metals have not been examined to date.

PHYSICAL FACTORS AFFECTING EXPOSURE

No clear distinctions can be made between the processes that control the mass transport of pollutants in an environment and those that directly affect bioavailability. If the pollutant is not transported to the environment where an organism lives, there can be no bioaccumulation or toxicity; in that sense, the transport processes are the rate-limiting steps. However, in this section, we will concentrate on the processes that affect the immediate environment of the organism and/or deviations from the exposure predicted from thermodynamic considerations. Based on physical factors, the potential or actual uptake of toxicants are affected by changes of either the concentration of the pollutant in the organism's immediate environment or the environmental bioavailability of the pollutant.

Temperature

All kinetic processes—physical, chemical, and biological—are influenced by the temperature of the system. Even homeotherms, with relatively constant body temperatures that reduce the effect of varying environmental temperatures on the kinetics of internal processes, are subject to the effects of temperature on the kinetics of external processes. Clearly, changes in environmental temperature will alter EBA under kinetically limited conditions.

Advection

Advection is the mass movement of a fluid past a point. The rate of advection can limit the rate of accumulation if it exceeds the rate of pollutant sorption/desorption so that equilibrium is not reached between, say, the water- and solid-phase compartments. In cases where the sediment is the source, the water concentration would be below that expected at equilibrium. Such disequilibria may be common in environments where relatively rapid currents of clean water flow over contaminated sediment, as in the case of the contaminated sediment surrounding the Los Angeles sewage discharge. A nonequilibrium distribution may also occur at a relatively low current speed if the sediment contamination is limited to a localized "hot" spot and substantial mixing occurs within the water body.

Another possible case where advection may drive a nonequilibrium condition is the advection of clean overlying water into the surface layer of contaminated sediments. Based on the advection of oxygen needed to maintain a 1-cm oxic layer, Lee[27] calculated that the interstitial water in the surface sediment layer would have to turn over about 80 times a day. Although this speculation needs to be tested, it raises the question of whether the exchange of interstitial water in some sediments exceeds the desorption rate of the more strongly sorbed compounds (i.e., highly hydrophobic compounds). In higher energy environments, turbulence with the bottom may be the advective mechanism that mixes the bottom waters and increases the volume of interstitial water to be equilibrated, creating an advective-limited system.

Sedimentation and Resuspension

Sedimentation is the major pathway for transport of many, if not most, hydrophobic organic pollutants from the pelagic to the benthic environment. Considerable progress has been made in predicting sedimentation in natural systems, and the near-field sedimentation around sewage discharges.[30,31] While whole-sediment pollutant concentrations appear relatively stable, at least over nonseasonal time frames, benthic organisms could experience substantially altered uptake because of short-term fluctuations in sedimentation if the deposited material was preferentially ingested. For example, *Diporeia* spp. shift their feeding from sediment detritus seasonally, and ingest the spring and fall diatom bloom as a major source of nutrition.[32] As discussed here, the kinetics of sedimentation are unlikely to be the rate-determining factor for the bioavailability of contaminants per se, although it will be one of the major transport processes controlling the ultimate sediment concentration of pollutants and organic matter in depositional environments.

Resuspension can have an immediate rate-enhancing effect on exposure concentrations and bioavailability. Resuspension and subsequent ingestion of contaminated sediment will greatly enhance the exposure of filter-feeding organisms, such as mussels or oysters, living above or surrounding a contaminated sediment. The flux of sediment into the water column may also increase the total concentration of aqueous-phase neutral organics, which would increase exposure to nonfilter-feeding organisms as well. Generally, this effect would be counterbalanced by simultaneous fluxes of dissolved organic matter (DOM) that could sorb the organic pollutants. For the metals, both the total concentrations and the bioavailability may increase in the water column with resuspension because reduced metals, such as those precipitated with acid volatile sulfide (AVS), could be oxidized and become bioavailable. Resuspension processes are generally short term and the additional exposure would need to be described through a kinetic model.

Transition Environments

Certain transition environments can undergo profound changes in key environmental factors in time frames faster than many of the corresponding physical/chemical processes. Bioaccumulative processes for organisms living in such environments may be subject to rapidly changing exposure conditions. Examples of such transitional environments include the zone of dilution surrounding sewage or industrial discharges and the turbidity maximum zone in estuaries. In a sewage discharge into marine waters, the freshwater effluent will be diluted by seawater on the order of 100 to 1 times or more within minutes. During this time, a host of changes take place including dilution of particles and dissolved materials, desorption, changes in pH and ionic strength, and flocculation of particles. Each of these processes occurs independently at its own rate. With so many factors occurring simultaneously, these environments are not at equilibrium and may be better thought of as zones of complex kinetics, at least in terms of bioavailability.

PHYSICOCHEMICAL FACTORS AFFECTING EXPOSURE

Dissolution

In general, chemicals must be dissolved in the exposure water before they can diffuse across absorbing membranes. When a chemical is present in the solid state, dissolution must occur prior to absorption, then dissolution is a potential rate-controlling process. The dissolution rate of chemicals is described by the Noyes-Whitney equation:[33]

$$DR = \left(\frac{DA}{h}\right)(C_s - C)$$

where D is the diffusion coefficient of the chemical in the aqueous stagnant boundary layer adjacent to the solid, A is the surface area of the solid, h is the thickness of the boundary layer, C_s is the solubility of the chemical in water, and C is the concentration of the chemical already dissolved in the water. The diffusion rate (DR) is increased by an increase in agitation intensity (via a decrease in h), and an increase in surface area (via a reduction in particle size). Also, DR is directly related to water solubility; among a series of chemicals, the slowest to dissolve will be those that are least soluble. Thus, dissolution rate-limited uptake will be more likely encountered with poorly water-soluble chemicals, some of which are relatively lipophilic. This process is particularly important for ingested materials.

Desorption (Complex Dissociation)

The role of sorbing phases is currently modeled as an equilibrium process in which the freely dissolved material is considered to be bioavailable. For nonpolar organics, organic carbon is considered to be the sorbing phase and the sorption is represented by the log-linear relation between K_{ow} and K_{oc}.[34] This simplification of sorption does not account for observed variability in the sorptive capacity of various molecular compositions of the organic matter. Similar relations remain elusive for metals and organic molecules with functional groups that interact with organic matter by other than a simple partitioning process.

Kinetics generally focus on the desorption rate of compounds from the sorbed phase. The desorption process for nonpolar organics is apparently sufficiently slow that it does not occur from dissolved organic matter during transition across the gill.[35] Therefore, for chemicals that accumulate from water the equilibrium representation appears to adequately describe the reduction in bioavailability. The kinetic data are not available for sorbed polar organics or for metals, so whether the desorption kinetics are fast enough to alter the EBA for uptake from water remains in question.

The kinetics for the sorption of nonpolar organic compounds to organic matter is of greatest importance in the sediment. The rate of desorption from particles will dictate the rate of accumulation from interstitial water because organisms rapidly accumulate freely dissolved compounds from interstitial water and diffusion is limited within sediments. The importance of the interstitial water accumulation route for infaunal benthos depends on the relative flux from all routes of accumulation. The influence of desorption on the encountered interstitial water concentration can be altered by the organism's movement to undisturbed interstitial water. However, this desorption process apparently limits the accumulation of poorly water-soluble polycyclic aromatic hydrocarbons by benthic organisms.[36,37] Thus, for these compounds, the uptake process from water is sufficiently fast that it is no longer the kinetically determinative step in the sediment environment. Rather, the uptake rate from interstitial water is determined by the desorption process from particles. The relative magnitudes of the particle desorption rates and the uptake rates from water are not known for polar organic molecules and metals. Therefore, whether desorption processes for these compounds are kinetically rate-limited is unknown.

A second phenomenon that significantly influences this desorption process is the contact time of the toxicant with the particle. For organic compounds that have low aqueous solubility, the sorption process can be described as a two-phase process. As contact time increases between the compounds and particles, the chemical extractability and EBA are both reduced.[38] Furthermore, the EBA declines faster than the chemical extractability.[36,39] The rates for these sorption processes are not well studied. Neither do we know whether this process occurs with polar organics or metals, or the extent of influence on EBA in the field.

A third issue involves sorption across a range of particles sizes. For example, hexachloro-biphenyl distribution does not always correlate with the organic carbon distribution among sediment particles.[70] Thus, sorption processes appear to be influenced by both particle size and composition and the influence on EBA remains in question.

The above discussion focused on sorption to an organic matrix. Sorption to the inorganic matrix of particles and sediment is well recognized for metals, but the importance of such phases for organic compounds needs study. We do not know how the kinetics of these sorption process will influence EBA. To summarize, we need to know more about both the time needed to reach equilibrium and desorption kinetics for sorption processes, so that significant kinetic limitations can be identified and properly modeled. Further, the sorption relations for polar organic compounds[40] and the complex sorption of metals are not fully defined, particularly in terms of the influence on EBA. Both the thermodynamics and kinetics of these processes relative to EBA should be investigated. Finally, the influence of the composition of the sorbing or complexing phase on the extent of sorption also needs study.

Diffusion

Some potential rate-controlling steps in bioaccumulation involve diffusion processes, i.e., chemical movement within sediments, movement across unstirred layers, and movement across membranes and epithelia. The process of diffusion involves random movements of the diffusant down an activity gradient which is usually described in terms of a concentration gradient. The rate of diffusion is described by Fick's Law:[41]

$$\text{Flux} = -D \left(\frac{dC}{dX} \right)$$

where D is the diffusion coefficient, and dC/dX is the concentration gradient across the diffusion barrier. The magnitude of the diffusion coefficient is influenced by the environmental variables in the Stokes-Einstein equation:[41]

$$D = \frac{kT}{6\pi\eta r}$$

where k is Boltzman's constant, T is absolute temperature, η is viscosity of the matrix in which the chemical is diffusing, and r is the radius of the diffusant. The diffusion coefficient will increase with an increase in temperature (due more to a decrease in η than an increase in T), a decrease in viscosity, and a decrease in molecular weight. For small molecules and metal species (< 1000 Da) molecular weight and temperature effects on D are small. In some matrices, there may be binding sites that can interact with the diffusant and reduce D, thereby creating an effective resistance to diffusion. For example, the presence of functional groups (e.g., hydroxyl and carboxyl) that can hydrogen bond to sites in a membrane will reduce the value of D. Similar interactions will be expected to occur with π bonding, including induced dipole interactions and charge-charge interactions.

Diffusion of a chemical in sediments occurs when an organism in a sediment has taken up chemical in its immediate vicinity, setting up a concentration gradient between the absorbing surface of the organism and the sediment away from the absorbing surface. If the uptake of chemical by the absorbing surface is fast relative to diffusion of the chemical to the absorbing surface, the diffusion process can control the rate of uptake of the chemical. Such diffusion in sediments is also limited by the twisted path among sediment particles (tortuosity), as well as strict diffusion limits.

Similarly, at an absorbing membrane such as the gill epithelium, depletion of chemical from water immediately adjacent to the epithelium can set up a gradient in the water layer bounding the epithelium.[42] This layer is a "stagnant boundary layer"; it is stationary relative to the epithelium even though water farther from the surface is flowing. Diffusion across this boundary layer can be a rate-controlling step in the absorption of a chemical by the membrane, particularly when the permeability of the membrane is high relative to that of the boundary layer. Whether this boundary layer operates in the fish gill is not known with certainty. It may be that the close spacing of secondary lamellae precludes the development of an aqueous diffusion layer of sufficient resistance to control the rate of chemical uptake by the gill.

Where an aqueous diffusion layer offers significant resistance to chemical transfer (e.g., across the integument or to nonmobile particles), the resistance offered by the boundary layer can be influenced by its thickness which, in turn, is affected by the agitation intensity of the water or the rate at which water flows past the absorbing surface. When an aqueous diffusion layer controls the absorption rate, an increase in the ventilation rate should reduce the thickness of this layer and increase the rate of absorption. Thus, an increase in organism activity can reduce such diffusional resistance.

When an aqueous boundary layer controls the rate of exchange (uptake or elimination) at an adsorbing or absorbing surface (or a desorbing surface), binding or complexation of the chemical can reduce the resistance offered by the boundary layer in the absence of the binding or complexing agent.[43] The mechanism involves release of bound chemical from the complex to the membrane/water interface. This becomes another source of chemical entry into the stagnant layer, in addition to diffusion from the bulk-water phase. For this mechanism to operate, dissociation/desorption must be rapid relative to diffusion and uptake at the membrane/water interface. This mechanism was not observed in the fish gill for high the K_{ow} compounds, benzo(a)pyrene and tetrachlorobiphenyl, in the presence of the macromolecular binding agent, Aldrich humic acid.[35]

The diffusive flux of chemicals across a biological membrane is also described by Fick's Law, but in this case a term (K_m) is added to account for the difference in capacity between the membrane and the water:[44]

$$Flux = -DK_m \frac{dC}{dX}$$

The membrane is generally composed primarily of lipid, and the membrane/water distribution coefficient tends to correlate with distribution coefficients measured between water and other lipoidal phases, including common organic solvents. Because partition coefficients for solvents are much easier to measure than membrane/water coefficients, it is common to use K_{ow}, for example, to indicate the permeability of a membrane toward a chemical. The permeability of absorbing membranes generally increases for a series of chemicals as their K_{ow} values increase. As K_{ow} increases, membrane permeability increases and absorption is then limited mostly by blood flow through the absorbing surface and ventilation.[12,13] For relatively polar and charged chemicals (low K_{ow}; e.g., <1 for fish gill epithelium), diffusion across the absorbing membranes is the dominant rate-determining step.

Speciation

The rate at which the speciation of metals and reactive organics (weak acids and bases) can change forms may be fast enough to alter EBA or toxicity for externally acting materials. This process is generally not an issue in the bulk environment, but it may play a significant role at the gill surface or other membrane surfaces, depending on transit time and the gill or gut chemistry. Under the conditions of transit, a metal (i.e., Al) may experience a microenvironment that results in a change in speciation. Such changes may be fast enough for precipitation to occur on the membrane surface, and it has been suggested as an important process for Al toxicity at the gill of fish.[45,56] Similarly, neutralization of an organic acid or base may occur in a micro-environment at a rate faster than the transit time. Such a change in speciation could make the compound more available for biological accumulation than would be predicted from the bulk environment. The importance of these processes has not been well studied.

Degradation

Processes that can compete kinetically for the biologically available compound can effectively reduce the EBA to the organism. Among the chemical processes, hydrolysis and photolysis can effectively reduce the available concentration before an organism can reach its thermodynamic limit. The mechanistic understanding of hydrolysis is good, and predictive models are under development.[3] For both direct and sensitized photolysis, predictive relations remain elusive. If sufficient sunlight is available, these processes are more than fast enough to alter the biologically available compound and reduce accumulation.

It is well recognized that redox processes control the speciation of metals. Often the rate-controlling factor is the diffusion of oxygen or production of sulfide. For organic compounds such as the polychlorinated biphenyls, reductive dechlorination alters the chemical composition and, thus, the particle sorption characteristics of compounds in sediments which, in turn, alter the bioavailability. However, this process is generally slower than other processes such as ingestion, which likely controls the EBA of highly sorbed sediment contaminants.

BIOLOGICAL FACTORS AFFECTING EXPOSURE

Diagenesis

As organic matter decays, the particle size decreases and total surface area per unit weight increases, at least until most of the carbon is decomposed. As DOM is released into the water, the chemical composition of the organic matter changes both in response to changes in the organic matrix and from colonization by microbes. The mass declines as carbon is respired by the microbes and multiple metabolic products are released by the colonizing bacteria and fungi. The rate of release of DOM from the carbon matrix and from the colonizing microbes will affect the percentage of "free" vs. bound organic contaminants in the overlying or interstitial water. This chain of physical and chemical alterations has major effects on the bioavailability of organics and metals, especially in sediments where most diagenesis occurs.

Sorption kinetics may change in response to the changes in the size distribution of the particles, whereas the equilibrium distribution between water and solids will change in response to changes both in the mass of particulate carbon and organic composition as well. As the particle size decreases and the surface area:volume ratio increases, the rate of sorption should increase. Presumably, the smaller the particles, the faster both the rapidly and slowly reversible contaminant pools should reach equilibrium.[38] This sorption appears to be multicompartmental, with the initial sorption rapidly reversible, and after extended contact between the contaminant and the particle the reversibility slows.[38]

Production of reducing end products by microbes during decomposition of organic matter[47] affects metal speciation and, hence, metal bioavailability. When the rate of production of these products exceeds the rate of oxygen transport into the sediment, the sediment becomes anaerobic and many of the metals form insoluble metal precipitates. One end product of recent interest is acid volatile sulfide (AVS), which is hypothesized to be the primary regulator of the bioavailability of metals in sediments.[48]

The diagenesis of organic matter is largely controlled by bacteria and fungi, though marine and freshwater macroinvertebrates can play an important role in the initial physical breakdown of the organic matter. Because of the dominant biological control of diagenesis, the rates of diagenesis are sensitive to factors controlling microbial and macroinvertebrate activity. Temperature, especially, has a dominant influence. With the relatively high particulate and dissolved carbon contents of most sediments, short-term changes in the diagenesis rate are unlikely to have a major effect on bioavailability. However, changes over the course of a season can affect bioavailability from a sediment. For example, it is thought that the reduced production of AVS during the winter in certain sediments can result in substantial changes in metal bioavailability and toxicity.[48] With its lower concentrations of dissolved and particulate carbon, pelagic systems should be more sensitive to short-term changes in the rates of diagenesis.

Organic Matter Loading

The rate of organic matter input directly affects the particulate and dissolved organic content of both the water column and sediment, specifically, aqueous and solid-phase geochemistry and, hence, bioavailability. One end of the organic loading spectrum is oligotrophic lakes and low-productivity marine systems. Areas undergoing phytoplankton blooms and/or areas surrounding sewage discharges represent examples at the organic-rich end of the spectrum. In general, the lower the organic content the larger will be the fraction of aqueous-phase pollutants that are freely dissolved and bioavailable due to the reduced binding to DOM. Conversely, the high particle loads in organic-rich systems will tend to scavenge pollutants from the water column and transport a greater percentage to the near-field sediments.[49] During a phytoplankton bloom, the organic matter associated with phytoplankton and excreted DOM increases, thereby substantially altering the amount of freely dissolved chemical available for accumulation by biota that do not ingest the algae.

Bioturbation

Biological mixing of the sediment (bioturbation) results in significant alteration of the three-dimensional sediment structure. The extent and rate of bioturbation of the sediment matrix can profoundly affect the exposure environment, both of the species modifying the sediment and other species within and outside the benthic community.[50]

Bioturbation often results in the transport of deeper sediment to the surface. Such transport results in a mixing of the surface and subsurface materials.[38,51,52] This mixing prolongs the residence time of deposited materials and the exposure of benthic organisms in the bioactive zone to sediment-associated contaminants.[53] As sediment-associated contaminants are buried by the deposition of less contaminated sediments, a pollution maximum will move downward through the sediment column, as is observed for DDT off the Palos Verdes Peninsula.[54] As the contaminated sediment is buried, the exposure regime for organisms feeding at any particular depth varies over time. The rate of such burial and, hence, the rate of change in the exposure concentration depends on the rate of bioturbation, as well as the rate of net sediment deposition. In biologically active sediments, substantial changes in pollutant concentrations may occur within the life span of longer-lived benthic species.

Intense bioturbation can also increase the water content of muddy sediments and reduce their cohesion, such that they are easily resuspended. Once resuspended, as previously discussed, the contaminated sediments are potentially available to both infaunal filter-feeding organisms and epifaunal filter-feeding organisms above the sediment. Resuspension can also result in substantial off-site transport of the sediment-associated pollutants. Such transport can result in increasing concentrations of particle-associated contaminants through sediment focusing into depositional basins.[49,53]

TOXICOKINETIC FACTORS AFFECTING EXPOSURE

Physiology and behavior control the volume or mass of the source compartment encountered by the organism over time. Further, the physiology also controls the internal compound distribution required to maintain concentration gradient for passive accumulation processes.

Encounter Volume

The maximum possible uptake rate of a chemical by an organism is controlled by the rate at which the chemical is presented to the absorption sites. For chemicals with high membrane permeability that are rapidly removed from the inside surface of the absorbing membrane (i.e., highly bound to blood), the rate of supply of a chemical can control its rate of uptake. Since both membrane permeability and blood binding increase with the lipophilicity, the probability of this happening increases with the lipophilicity of a compound. For fish, chemicals having a log $K_{ow}>3$ generally have encounter-volume rate-limited uptake.[55] In fish and other aquatic species with gills, the encounter volume is dominated by the volume of water that passes across the respiratory surfaces of the gill epithelium. This flow amounts to about 60 to 80% of the total flow of water across the gills. This concept also applies to organisms living in sediments, where the encounter volume is the volume of sediment-interstitial water adjacent to the organism and the volume transected while the organism moves through the sediment. Some sediment organisms (e.g., the clam, macoma) have very little contact with the interstitial water because they respire overlying water.

The encounter volume is strongly influenced by the activity of the organism, with an active fish able to encounter more than ten times the volume water of a fish at rest. For aquatic species with gills, the encounter volume is largely controlled by the ventilation rate and is sensitive to the concentrations of oxygen and carbon dioxide; low O_2 or high CO_2 in the water increases the encounter volume. The encounter volume also tends to increase with temperature, because dissolved oxygen decreases and metabolism increases.

The distribution of the chemicals to tissues in the organism by the blood or by diffusion will also be involved in the kinetics of accumulation for encounter-volume rate-limited uptake at the absorbing surface. This happens because encounter-volume rate-limited uptake is generally found for relatively lipophilic chemicals. Such chemicals also tend to have high affinity for poorly perfused tissues, like the adipose tissues. In such cases, the accumulation of chemicals by highly perfused tissues like muscle tissue may be controlled by the encounter volume, but the filling of the poorly perfused tissues may be controlled by blood flow to them. Further, the poorly perfused tissues tend to hold more lipophilic chemicals than the highly perfused tissues, often by a large amount. When this happens, accumulation of most of the chemical may not be controlled by the encounter volume, but by the smaller blood flow to the poorly perfused storage tissues, as observed with trifluralin accumulation by fish.[56]

Encounter Mass For Ingested Materials

Ingestion Rate

Gut uptake becomes increasingly important with higher K_{ow} compounds, and probably dominates for compounds with a log K_{ow} of about 6 and above.[10] This route will also likely dominate other compounds strongly sorbed to particles. The ingested dose from food/sediment is a function of both the total mass of pollutant ingested, which is a function of the ingestion rate and the pollutant concentration in the food/sediment, and gut uptake efficiency for the pollutant. Any of these factors—ingestion rate, pollutant concentration, or assimilation efficiency—can become the rate-determining factor for bioaccumulation.

In fish, the ingestion rate of prey is related to the organism's energy or oxygen requirements. If the pollutant concentrations in the prey are known or can be predicted, the dose and resulting tissue residues can be predicted with a reasonable degree of accuracy from bioenergetic model equation (see Reference 57). The prediction of ingestion rates and ingested doses for sediment-ingesting invertebrates is complicated by the type of feeding mode (i.e., filter feeding or deposit feeding), the current speed, selective ingestion of particles based on size or composition, and the wide range of food "quality". As pointed out by Lopez and Levinton,[58] the range in food quality varies more widely among different particle types within a sediment than among prey items for herbivores or carnivores. Models predicting sediment ingestion have been developed based on a constant ingestion rate of carbon[59] and on optimal foraging strategy,[60] though none is as general or accurate as the fish ingestion model.

Assimilation Efficiency

A key factor in determining the kinetics of pollutant uptake from food or ingested sediment is the efficiency with which the pollutants are absorbed from the gut. Assimilation efficiencies in fish for neutral organics with relatively large K_{ow} values range from a few percent to about 90%.[61-63]

Assimilation efficiencies for compounds with large K_{ow} values have been reported in fish both as independent of, and varying with, the concentration of the pollutant in the food.[64,65] Some of this variation may be due to different study techniques, such as bolus doses vs. chronic exposures, and use of a lipid vehicle vs. dosing the food source. Nonetheless, it seems likely that assimilation efficiencies will vary, at least at extreme concentrations, though it is unclear whether this is an important source of uncertainty for the concentrations found in prey.

Determination of pollutant assimilation efficiencies is problematic for sediment-ingesting invertebrates. Most deposit feeders selectively ingest specific sizes or types of particles. Usually the finer, high TOC particles are ingested, so that the whole-sediment concentrations underestimate the actual pollutant concentration of the ingested particles. Furthermore, because of this selection, assimilation efficiencies cannot be calculated from the difference in the pollutant concentrations in the feces and the whole sediment. Various tracers are used to correct for the selection process,[28,66,67] although there are limits to all of these techniques. Even with these limitations, gut assimilation efficiencies for PCB congeners appear to decline in a marine clam, with the increase in K_{ow} from over 80% for a trichlorobiphenyl to only a few percent for decachlorobiphenyl.[71] The very low assimilation efficiencies for the high K_{ow} PCB congeners could limit the accumulation of these compounds by deposit feeders.

Metabolism

Metabolic breakdown or degradation can influence toxicity in two ways: (1) the biotransformation results in detoxication, and (2) the biotransformation results in a material of similar or greater toxicity. In the first case, the internal concentration of the absorbed chemical is reduced in proportion to the rate of degradation and any toxic effects are similarly reduced (e.g. diethylhexylphthalate[68]). Because metabolism is treated as an elimination constant in the simple model discussed earlier, the time to achieve steady state for the parent compound and the steady-state concentration will decrease in proportion to the metabolic rate. For toxicity bioassays, the exposure concentration must be increased to compensate for the amount absorbed and metabolized before reaching the target sites. This makes the chemical appear less toxic than a compound of similar toxicity that is not metabolized. Typically, the rate must approach a significant portion of the uptake clearance to substantially influence the system. Conversely, in the second case, the rate of metabolism of the parent compound may determine the rate of appearance of the toxic agent in the organism and may well be the rate-determining step in production of adverse effects. Although considerable work has been carried out to estimate the metabolism of compounds as a function of chemical or physical properties, the rate of metabolic breakdown or transformation of chemicals cannot be reliably predicted for either terrestrial or aquatic organisms.

Growth

Organism growth can affect apparent accumulation of a chemical through growth dilution. As an organism grows, it increases in volume and, therefore, in capacity. If, for example, the rate of growth, which appears as a component of elimination in the simple model discussed earlier, is of the same magnitude as the uptake clearance rate, the body concentration will not change with time and will not achieve the thermodynamically predicted steady-state accumulation. For organic chemicals this effect has been noted to be seasonally significant in algae[69] and fish,[61] and can be estimated as a function of the size of the organism and the log K_{ow} of the chemical in question (see Section 6, Chapter 1). In addition to simple growth, as organisms grow they may also increase the proportion of lipid relative to other tissue types, which also increases the capacity of the animal to store nonpolar organics.

SUMMARY

The kinetics working group and workshop participants recognized that kinetics are involved as a part of all the topics of the workshop. Thus, this discussion paper reflects a more comprehensive discussion of kinetics and their potential effect than were covered under the two discussion initiation papers. The question was not whether the bioaccumulation processes had kinetic aspects, but whether the temporal scale was of a magnitude that they would alter bioaccumulation. The answer to this question depends on the size of the organism of interest and remains one of the most prominent for directing future research. In general, more insight and information on kinetic processes were available for nonpolar organic contaminants than for the polar-reactive organics and metals. Yet, despite the relatively large information base, kinetically limiting processes are not that well defined, and minimal information is available for the polar organic compounds and the metals.

Although there were many definitions of bioavailability presented at the workshop, the kinetics working group thought that the potential use of the environmental bioavailability concept, i.e., the efficiency of accumulation from a matrix of consideration, would provide a useful quantitative tool for comparisons between various experimental conditions and exposure scenarios. Other quantitative formulations will likely be developed with further research. Further, there may be situations where this concept would prove unworkable. However, research on the methodology and conceptual approach should be enhanced by having an initial definition against which to develop hypotheses. Finally, there are several recommendations within the manuscript for further research. Overall, improved recognition of the conditions that result in kinetic limitations and better predictive modeling could greatly improve our understanding of organism exposure and accumulation of contaminants.

REFERENCES

1. Boxenbaum, H., Interspecies scaling, allometry, physiological time, and the ground plan of pharmacokinetics, *J. Pharmacokinet. Biopharm.*, 10, 201-227, 1982.
2. Peters, R. H., *The Ecological Implications of Body Size*, Cambridge University Press, Cambridge, U.K., 1983.
3. Lyman, W. J., Reehl, W. F., and Rosenblatt, D. H., *Handbook of Chemical Property Estimation Methods*, American Chemical Society, Washington, D.C., 1990.
4. Spacie, A. and Hamelink, J., Alternative models for describing the bioconcentration of organics in fish, *Environ. Toxicol. Chem.*, 1, 309–320, 1982.
5. Barron, M. G., Stehly, G. R., and Hayton, W. L., Pharmacokinetic modeling in aquatic animals. I. Models and concepts, *Aquat. Toxicol.*, 17, 187–212, 1990.
6. Erickson, R. J. and McKim, J. M., A model for exchange of organic chemicals at fish gills: flow and diffusion limitations, *Aquat. Toxicol.*, 18, 175–198, 1990.
7. Nichols, J. W., McKim, J. M., Andersen, M. E., Gargas, M. L., Clewell, H. J., III, and Erickson, R. J., A physiologically based toxicokinetic model for the uptake and disposition of waterborne organic chemicals in fish, *Toxicol. Appl. Pharmacol.*, 106, 433–447, 1990.
8. Nichols, J. W., McKim, J. M., Lien, G. J., Hoffman, A. D., and Bertelsen, S. L., Physiologically based toxicokinetic modeling of three water borne chloroethanes in rainbow trout *(Oncorhynchus mykiss)*, *Toxicol. Appl. Pharmacol.*, 110, 374–389, 1991.
9. de Wolf, W., de Bruijn, J. H. M., Seinen, W., and Hermens, J. L. M., Influence of biotransformation on the relationship between bioconcentration factors and octanol-water partition coefficients, *Environ. Sci. Technol.*, 26, 1197–1201, 1992.
10. Connolly, J. P. and Pedersen, C. J., Thermodynamic-based evaluation of organic chemical accumulation in aquatic organisms, *Environ. Sci. Technol.*, 22, 99–103. 1988.
11. McCarty, L. S., Ozburn, G. W., Smith, A. D., and Dixon, D. G., Toxicokinetic modeling of mixtures of organic chemicals, *Environ. Toxicol. Chem.*, 11, 1037–1047, 1992.
12. Hayton, W. L. and Barron, M. G., Rate-limiting barriers to xenobiotic uptake by the gill, *Environ. Toxicol. Chem.*, 9, 151–157, 1990.
13. Erickson, R. J. and McKim, J. M., A simple flow-limited model for exchange of organic chemicals at fish gills, *Environ. Toxicol. Chem.*, 9, 159–165, 1990.
14. Landrum, P. F. and Stubblefield, C. R., Role of respiration in the accumulation of organic xenobiotics by the amphipod *Diporeia* sp., *Environ. Toxicol. Chem.*, 10, 1019–1028, 1991.
15. Landrum, P. F., Frez, W. A., and Simmons, M. S., Relationship of toxicokinetic parameters to respiration rates in *Mysis relicta*, *J. Great Lakes Res.*, 18, 331–339, 1992.
16. Landrum, P. F., Fontaine, T. D., Faust, W. R., Eadie, B. J., and Lang, G. A., Modeling the accumulation of polycyclic aromatic hydrocarbons by the amohipod *Diporeia* spp., in Gobas, F. A. P. C. and McCorquodale, J. A., Eds., *Chemical Dynamics in Fresh Water Ecosystems*, Lewis Publishers, Boca Raton, FL, 111–128, 1992.
17. Sprague, J., Measurement of pollutant toxicity to fish. 1. Bioassay methods for acute toxicity, *Water Res.*, 3, 793–821, 1969.
18. McCarty, L. S., Hodson, P. V., Craig, G. R., and Kaiser, K. L. E., The use of quantitative structure-activity relationships to predict the acute and chronic toxicities of organic chemicals to fish, *Environ. Toxicol. Chem.*, 4, 595–606, 1985.
19. Veith, G., Call, D., and Brooke, L., Structure-activity relationship for the fathead minnow *Pimephales promelas;* narcotic industrial chemicals, *Can. J. Fish. Aquat. Sci.*, 40, 743–748, 1983.
20. Veith, G D., DeFoe, D. L., and Knuth, M., Structure-activity relationships for screening organic chemicals for potential ecotoxicity effects, *Drug Metab. Rev.*, 15, 1295–1303, 1985.
21. Veith, G. D. and Broderius, S. J., Structure-activity relationships for industrial chemicals causing type (II) narcosis syndrome, in Kaiser, K. L. E., Ed.,*QSAR in Environmental Toxicology II*, D. Reidel, Dordrecht, Holland, 1987, 385–392.
22. Konemann, H., Quantitative structure-activity relationships in fish toxicity studies. I. Relationship for 50 industrial pollutants, *Toxicology*, 19, 209–221, 1981.
23. Konemann, H. and Musch, A., Quantitative structure-activity relationships in fish toxicity studies. II. The influence of pH on the QSAR for chlorophenols, *Toxicology*, 19, 223–228, 1981.

24. Hermens, J. L. M., Quantitative structure-activity relationships of environmental pollutants, in Hutzinger, O., Ed., *Handbook of Environmental Chemistry*, Vol. 2, Springer-Verlag, Berlin, 1989, 111–162.

25. McCarty, L. S., Mackay, D., Smith, A. D., and Dixon, D. G., Residue-based interpretation of toxicity and bioconcentration QSARs from aquatic bioassays: neutral narcotic organics, *Environ. Toxicol. Chem.*, 11, 917–930, 1992.

26. Mackay, D., Puig, H., and McCarty, L. S., An equation describing the time course and variability in uptake and toxicity of narcotic chemicals to fish, *Environ. Toxicol. Chem.*, 11, 941–951, 1992.

27. Lee, H., II, A clam's eye view of the bioavailability of sediment-associated pollutants, in Baker, R. A., Ed., *Organic Substances and Sediments in Water*, Vol. 3, Lewis Publishers, Boca Raton, FL, 1991, 73–93.

28. Lee, H., II, Boese, B. L., Pelletier, J., and Randall, R. C., A method to estimate gut uptake efficiencies for hydrophovic organic pollutants in a deposit-feeding clam, *Environ. Toxicol. Chem.*, 9, 215–219, 1990.

29. Lee, H., II, Models, muddles, and mud. Predicting bioaccumulation of sediment-associated pollutants, in Burton, G. A., Ed., *Sediment Toxicity Assessment*, Lewis Publishers, Boca Raton, FL, 1992, 267–293.

30. Bodeen, C. A., Hendricks, T. J., Frick, W. E., Baumgartner, D. J., Yerxa, J. E., and Steele, A., SEDDEP: A Program for Computing Seabed Deposition Rates of Outfall Particulates in Coastal Marine Environments, ERL-Narragansett Contrib. No. 109, U.S. Environmental Protection Agency, Narragansett, RI, 1989.

31. Tetra Tech, A Simplified Deposition Calculation (DECAL) for Organic Colloids Accumulated Near Sewage Outfalls, EPA Contract No. 68–01–6938, TC-3953–02, U.S. Environmental Protection Agency, Newport, OR, 1985.

32. Gardner, W. S., Quigley, M. A., Fahnensteil, G. L., Scavia, D. S., and Frez, W. A., *Ponoporeia hoyi*—a direct trophic link between spring diatoms and fish in Lake Michigan, in Tilzer, M. M. and Serruya, C., Eds., *Large Lakes: Ecological Structure and Function*, Springer-Verlag, New York, 1990, 632–644.

33. Carstensen, J. T., Theories of dissolution—simple particle systems, in Leeson, L. J. and Carstensen, J. T., Eds., *Dissolution Technology*, Academy of Pharmaceutical Science, Washington, D.C., 1974, 1–28.

34. DiToro, D. M., A particle interaction model of reversible chemical sorption, *Chemosphere*, 14, 1503–1538, 1985.

35. Black, M. C. and McCarthy, J. F., Dissolved organic macromolecules reduce the uptake of hydrophobic organic contaminants by the gills of rainbow trout (*Salmo gairdneri*), *Environ. Toxicol. Chem.*, 7, 593–600, 1988.

36. Landrum, P.F., Bioavailability and toxicokinetics of polycyclic aromatic hydrocarbons sorbed to sediments for the amphipod, *Pontoporeia hoyi*, *Environ. Sci. Technol.*, 23, 588–595, 1989.

37. Landrum, P. F. and Robbins, J. A., Bioavailability of sediment-associated contaminants to benthos invertebrates, in Baudo, R., Giesy, J. P., and Muntau, H., Eds., *Sediments: Chemistry and Toxicity of In-Place Pollutants*, Lewis Publishers, Boca Raton, FL, 1990, 237–263.

38. Karickhoff, S. W. and Morris, K. R., Impact of tubificid oligochaetes on pollutant transport in bottom sediments, *Environ. Sci. Technol.*, 19, 51–56, 1985.

39. Landrum, P. F., Eadie, B. J., and Faust, W. R., Variation in the bioavailability of polycyclic aromatic hydrocarbons to the amphipod *Diporeia* spp. with sediment aging, *Environ. Toxicol. Chem.*, 11, 1197–1208, 1992.

40. Jafvert, C. T., Westall, J. C., Greider, E., and Schwarzenbach, R. P., Distribution of hydrophobic ionogenic organic compounds between octanol and water. Organic acids, *Environ. Sci. Technol.*, 24, 1795–1803, 1990.

41. Flynn, G. L., Yalkowsky, S. H., and Roseman, T. J., Mass transport phenomena and models. Theoretical concepts, *J. Pharm. Sci.*, 63, 479–510, 1974.

42. Pedley, T. J., Calculation of unstirred layer thickness in membrane transport experiments. A survey, *Q. Rev. Biophys.*, 16, 115–150, 1983.

43. Weisinger, R. A., Pond, S. M., and Bass, L., Albumin enhances unidirectional fluxes of fatty acid across lipid-water interface. Theory and experiments, *Am. J. Physiol.*, 257, G904–G916, 1989.

44. Stein, W. D., *Transport and Diffusion Across Cell Membranes*, Academic Press, New York, 1986, 72–93.

45. Playle, R. C. and Wood, C. M., Water pH and aluminum chemistry in the micro-environment of rainbow trout during acid and aluminum exposures, *J. Comp. Physiol. B*, 159, 539–550, 1989.

46. Playle, R. C. and Wood, C. M., Is precipitation of aluminum fast enough to explain aluminum deposition of fish gills?, *Can. J. Fish. Aquat. Sci.*, 47, 1558–1561, 1989.

47. Berner, R. A., *Early Diagenesis: A Theoretical Approach*, Princeton University Press, Princeton, NJ, 1980.

48. DiToro, D. M., Mahony, J. D., Hansen, D. J., Scott, K. J., Hicks, M. B., Mayer, S. M., and Redmond, M. S., Toxicity of cadmium in sediments. The role of acid volatile sulfide, *Environ. Toxicol. Chem.*, 9, 1487–1502, 1990.

49. Vanderford, M. J. and Hamelink, J. L., Influence of environmental factors on pesticide levels in sport fish, *Pestic. Monit. J.*, 11, 138–145, 1977.

50. Lee, H., II and Swartz, R., Biological processes affecting the distribution of pollutants in marine sediments. I. Biodepositon and bioturbation, in Baker, R. A., Ed., *Contaminants and Sediments*, Vol. 2, Ann Arbor Science, Ann Arbor, MI, 1980, 555–606.

51. Kielty, T. J., White, D. S., and Landrum, P. F., Sublethal responses to endrin in sediment by *Stylodrilius heringianus* (Lumbriculidae) as measured by a [137]Cesium marker layer technique, *Aquat. Toxicol*, 13, 251–270, 1988.

52. Kielty, T. J., White, D. S., and Landrum, P. F., Sublethal responses to endrin in sediment by *Limnodrilus hoffmeisteri* (Tubificidae) and in mixed culture with *Stylodrilius heringianus* (Lumbriculidae), *Aquat. Toxicol.*, 13, 227–250, 1988.

53. Eadie, B. J. and Robbins, J. A., The role of particulate matter in the movement of contaminants in the Great Lakes, in Hites, R. A. and Eisenreich, S. J., Eds., *Sources and Fates of Aquatic Pollutants*, Adv. Chem. Ser., 216, American Chemical Society, Washington, D.C., 319–364, 1987.

54. Stull, J. K., Baird, R. B., and Heesen, T. C., Marine sediment core profiles of trace constituents offshore a deep wastewater outfall, *J. Water Pollut. Control Fed.*, 58, 985–991, 1986.

55. McKim, J., Schmieder, P., and Veith, G., Absorption dynamics of organic chemical transport across trout gills as related to octanol-water partition coefficient, *Toxicol. Appl. Pharmacol.*, 77, 1–10, 1985.

56. Hayton, W. L. and Archer, B. G., Kinetics of accumulation of trifluralin by fish, *Proc. West. Pharmacol. Soc.*, 24, 147–149, 1981.

57. Norstom, R. J., McKinnon, A. E., and deFreitas, A. S., A bioenergetic based model for pollutant accumulation by fish. Simulation of PCB and methylmercury residue levels in Ottawa River, *J. Fish. Res. Bd. Can.*, 33, 248–267, 1976.

58. Lopez, G. R. and Levinton, J. S., Ecology of deposit-feeding animals in marine sediments, *Q. Rev. Biol.*, 62, 235–260, 1987.

59. Cammen, L. M., Ingestion rate. An empirical model for aquatic deposit feeders and detritivores, *Oecologia* (Berlin), 44, 303–310, 1980.

60. Plante, C. J., Jumars, P. A., and Baross, J. A., Digestive associations between marine detritivores and bacteria, *Annu. Rev. Ecol. Syst.*, 21, 93–127, 1990.

61. Niimi, A. J. and Cho, C. Y., Elimination of hexachlorobenzene (HCB) by rainbow trout (*Salmo gairdneri*), and examination of its kinetics in Lake Ontario salmonids, *Can. J. Fish. Aquat. Sci.*, 38, 1350–1356, 1981.

62. Niimi, A. J. and Oliver, B. G., Biological half-lives of chlorinated dipenzo-p-dioxins and dibenzofurans in rainbow trout (*Salmo gairdneri*), *Environ. Toxicol. Chem.*, 5, 49–53, 1986.

63. Van Veld, P. A., Absorption and metabolism of dietary xenobiotics by the intestine of fish, *Aquat. Sci.*, 2, 185–203, 1990.

64. Opperhuizen, A. and Schrap, S. M., Uptake efficiencies of two polychlorinated biphenyls in fish after dietary exposure to five concentrations, *Chemosphere*, 17, 253–262, 1988.

65. Clark, K. and Mackay, D., Dietary uptake and biomagnification of four chlorinated hydrocarbons by guppies, *Environ. Toxicol. Chem.*, 10, 1205–1217, 1991.

66. Klump, J. V., Krezoski, J. R., Smith, M. E., and Kaster, J. L., Dual tracer study of the assimilation of an organic contaminant from sediment by deposit feeding oligochaetes, *Can. J. Fish. Aquat. Sci.*, 44, 1574–1583, 1987.

67. Lydy, M. J. and Landrum, P. F., Assimilation efficiency for sediment sorbed benzo(a)pyrene by *Diporeia* spp., *Aquat. Toxicol.*, 26, 209–224, 1993.
68. Barron, M. G., Schultz, I. R., and Hayton, W. L., Presystemic branchial metabolism limits to di-2-ethylhexylphthalate accumulation in fish, *Toxicol. Appl. Pharmacol.*, 98, 49–57, 1989.
69. Swackhamer, D. L. and Skoglund, R. S., The role of phytoplankton in the partitioning of hydrophobic organic contaminants in water, in Baker, R. A., Ed., *Organic Substances and Sediments in Water*, Vol 2, Lewis Publishers, Boca Raton, FL, 1991, 91–105.
70. Kukkonen, J., University of Joensuu, Finland, personal communication, 1993.
71. Lee, H., Unpublished data, 1993.

SESSION 7

CLOSING REMARKS

Summary and Conclusions

Kenneth L. Dickson (Chair), John P. Giesy, Rodney Parrish, and Lee Wolfe

Workshop Objectives

The principal workshop objective was to examine the current knowledge of the processes which govern the accumulation of chemicals by aquatic organisms. The workshop participants were challenged to address the following tasks:

- Provide a review of the present mechanistic understanding of the processes that affect bioavailability and bioaccumulation of toxic contaminants in aquatic environments.
- Identify the current limitations of methods currently used to assess the bioavailability of environmental contaminants.
- Develop research priorities for improving the understanding of factors that affect or control bioavailability.
- Explore incorporating considerations of environmental factors limiting bioavailability into regulations and environmental decision making.

To accomplish these objectives the workshop participants were divided into five subgroups, which addressed the following subject areas:

- Chemical and Physical Factors Affecting Bioavailability
- Bioavailability of Inorganic Chemicals
- Bioavailability of Organic Chemicals
- Effects of Variable Redox Potentials, pH and Light on Bioavailability in Dynamic Water-Sediment Environments
- Kinetics and Bioavailability

Each of the subgroups was asked to identify the important empirical relationships that are currently being used by environmental scientists to describe the behavior and/or predict the fate and effects of contaminants. Fundamental mechanisms that underlie these relationships were to be identified and evaluated. The subgroups were also charged with identifying problems, limitations, and exceptions to key empirical relationships, and to explain how these can be resolved. Finally, each subgroup was asked to identify key hypotheses and needs for research to advance understanding of the bioavailability of environmental contaminants. The following sections discuss the extent to which the above objectives were met by the workshop.

Bioavailability and Hazard Assessment of Chemicals

The basic strategy in chemical hazard evaluation is comparison of the expected exposure concentration (EEC) to a measured no-observed-effect concentration (NOEC). This principle originated and evolved during previous Pellston workshops[1-4] and is illustrated in Figure 1.

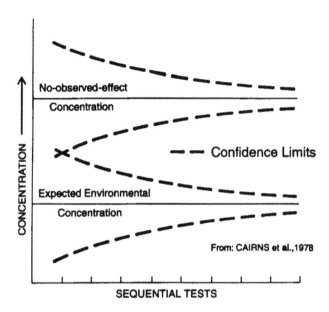

Figure 1. Conceptual model of hazard assessment approach.

Known as the quotient approach to hazard assessment, it is now used by environmental scientists and managers in industry and government to make decisions about the hazards of chemicals in freshwater and marine environments. Estimates of the concentration of a chemical in water and/or sediment are compared to results of toxicity tests which examine the concentration response relationship of the chemical to test organisms. If the expected environmental concentration exceeds the no-effect concentration, then an imminent hazard exists and decisions about the regulation of the chemical are straightforward. Likewise, if the expected environmental concentration is far below the no-effect concentration, decisions are relatively easy to make. However, when the EEC approaches the NOEC, decision-making is difficult. In that case, more testing is required to decrease the uncertainty that is associated with the estimates.

The quotient approach requires that the environmental concentration of a chemical be measured or estimated. The principal tools used to estimate the EEC are environmental fate models. The fourth Pellston Workshop, that was held in August, 1981 and titled "Modeling the Fate of Chemicals in the Aquatic Environment", examined the use of models to predict the EEC. Since that workshop, considerable advancements have been made in understanding the behavior of chemicals in water and sediments and in predicting and measuring the various forms of organic and inorganic chemicals in water and sediments. At the same time advances have been made in aquatic toxicology. Historically, aquatic toxicology was primarily concerned with the effects of chemicals on survival, growth, and reproduction of surrogate species suitable for laboratory testing. Little emphasis was placed on understanding the mechanisms of toxicity and the pharmacokinetics of chemicals in aquatic systems. However, as aquatic toxicology has developed as a science these areas have become more important, and has resulted in a better understanding of the interactions between organisms and chemicals in the environment.

All together, these advances have expanded our overall understanding of the bioavailability of chemicals. With this comes the possibility of better defining the expected environmental concentration aspect of the quotient method. This workshop examined current knowledge in the fields of environmental chemistry, fate modeling, and aquatic toxicology to further refine our understanding of the exposure concentration aspect of the quotient approach.

The ability of aquatic life to survive, grow, and reproduce depends on water and sediment quality, habitat suitability, productivity, biotic interactions, and hydrodynamics. This workshop focused primarily on water/sediment quality. Aquatic organisms have evolved tolerance ranges for environmental variables such as temperature, dissolved oxygen, pH, etc. Outside these tolerance ranges, organisms are stressed. Aquatic organisms have also evolved a variety of behavioral and physiological defense mechanisms in order to remain within their tolerance ranges of environmental variables, whether chemical, physical, or biological. However, there are limits to the effectiveness of these adaptive mechanisms, and human activities that cause excursions outside the optimal ranges often lead to environmental problems.

The chemical, physical, and biotic aspects of water/sediment quality are complex and dynamic. Fresh and marine waters are complex matrices of dissolved and particulate inorganic and organic chemicals. Many are naturally occurring, having been derived from the soils and vegetation of the lands being drained. Others are introduced from anthropogenic sources, and may even be chemicals or materials that are not naturally found in nature. These chemicals may interact with each other and with aquatic life. They may be changed in form, different compounds may be derived, and they be redistributed into various aqueous compartments such as bottom sediments, water column, surface microlayer, or biota. These interactions affect the spatial and temporal distribution of the chemicals and, thus, their available concentrations.

Whether or not a contaminant causes adverse effects to aquatic life is therefore a function of many factors. An assessment of the bioavailability of a chemical contaminant must consider the chemical and physical interactions in water and sediments that may, in turn, affect the form (soluble or sorbed to particles, ionized, or nonionized, etc.) and also the spatial and temporal distribution of the chemical in aquatic ecosystems.

In addition, consideration must be given to the interactions of the contaminant with the organism. Aquatic organisms also have evolved diverse and complex mechanisms to manage their interactions with chemicals, and are capable of further modifying the nature of a contaminant prior to actively or passively incorporating the contaminant into their bodies. Thus, an assessment of the bioavailability of a contaminant needs to consider not only the chemical and physical interactions that occur in water and/or sediments, but also the interactions that occur when an organism comes in contact with the contaminant.

Definitions of Bioavailability

Although many environmental chemists, toxicologists, and engineers purport to know what bioavailability means, the term eludes a consensus definition. The concept of biological availability was first proposed in 1975 at a National Science Foundation workshop on ecosystem processes and organic contaminants,[5] and was based mainly on physical chemistry. In the introduction to the workshop, bioavailability was defined as "the extent to which a toxic contaminant is available for biologically mediated transformations and/or biological actions in a aquatic environment". This definition came from a previous Pellston workshop dealing with biotransformation and fate of chemicals in aquatic environments.[3] Any discussion of bioavailability will be influenced by the working definition used by each of the discussants. Therefore, the definition used by different disciplines may vary depending on their perspective. Furthermore, bioavailability can be defined in either relative or absolute terms and in either chemical or biological terms. Each of these approaches strives to describe the mole-average activity of a compound that has the potential to interact with biological systems.

Each of these approaches has advantages and disadvantages, and there is no single "correct" definition or technique that will be the best choice under all conditions. Traditionally, chemists have defined bioavailability in terms of the chemical form in which the compound or element of interest occurs at a given time. Alternatively, definitions derived by biologists have assumed that the chemical form in the bulk phase is relevant only to the presence of a biological receptor; thus they have defined bioavailability based on the portion of the compound that could pass into an organism under a given set of conditions. The chemical form approach must be calibrated, based

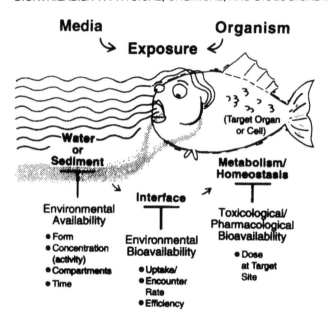

Figure 2. Conceptual model of bioavailability.

on the availability of the forms. The chemical approach is more theoretically appealing, while the biological approach is more directly relevant and more easily interpreted. The chemical approach offers the possibility of greater generality, while biologically based approaches can be used as rapid integrating systems, which avoid the need to make operationally defined separations or quantifications.

Since a number of questions can be posed relative to bioavailability and a number of uses of the measured values, a number of possible definitions could apply. A few that were presented during the workshop follow:

- The degree to which a contaminant in a potential source is free for uptake (movement into and/or *onto* an organism).
- The study of the interaction of compounds/phases. Studies are completed to define equilibrium between phases and kinetics of interactions.
- Although not rigidly defined, the idea of "encounters" by aquatic organisms was used to explain bioavailability. (Encounter volumes describe respiration, with concomitant toxicant uptake, and encounter rates describe food intake, with concomitant toxicant uptake.)
- Environmental bioavailability is the ratio of the uptake clearance divided by the encounter rate for a contaminant; toxicological bioavailability is that fraction of the absorbed toxicant which reaches the target site or the effective blood concentration.
- Bioavailability is equivalent to the activity of the freely dissolved organic compound in the water.
- Bioavailability is the degree to which a contaminant in a potential source is free for uptake (movement into or onto an organism).
- Sediment bioavailability is the weight-specific increase in tissue residue per unit time normalized to the sediment pollutant concentration.

Given the discussions during the workshop, it is clear that a diversity of definitions of bioavailability exist. While a consensus definition of bioavailability could not be reached by the workshop participants, a majority favored the concepts illustrated in Figure 2. Figure 2 portrays the concept of bioavailability as having three major components. One component, **Environmental Availability,** can be defined as that portion of the total material present in a compartment or compartments of the environment which actually participates in a particular process or group of processes and is subject to modifying influences. This represents the total amount of material potentially available, although the total is not necessarily involved in a given process. This definition represents the broadest sense of the concept. **Bioavailability** can be viewed as a special case of environmental availability in which organism(s) are involved in the process as a target rather than as a part of a process. The exclusive use of the prefix "bio" indicates this. The following definitions reflect two related uses. In these definitions, availability is "actual" rather than "potential" and represents a portion of the potential amount defined above.

- **Environmental Bioavailability**—This is the ratio of the uptake clearance rate divided by the rate at which an organism encounters a given contaminant in a medium being processed by the organism. This represents a measure of the general uptake efficiency.
- **Toxicological/Pharmacological Bioavailability**—Toxicological bioavailability is the fraction of the dose adsorbed/absorbed by the organism which reaches the target sites on/in the organism or the effective intercellular (blood, typically) concentration in the organism. This would be considered the classical definition of bioavailability.

Empirical Relationships

Environmental scientists have developed and used a number of empirical relationships to guide decision making regarding the fate and effects of organic and inorganic chemicals in the aquatic environment. These empirical relationships have been developed using scientific judgement and field and laboratory data. Examples of some of these relationships that have been used follow:

- The relationship between body residues, fat content, and the octanol:water partition coefficient,
- Water hardness adjustments for metals toxicity, and
- pH effects on metals bioavailability.

A better understanding of the underlying mechanisms could improve the utility of these relationships and assist in making better predictions about the fate and effects of chemicals in the aquatic environment. The workshop participants were asked to identify empirical relationships related to the bioavailability of chemicals to aquatic life that are used in decision making, and to evaluate the relationships in terms of the current understanding of the mechanism(s), how the empirical relationship is or can be used, important problems or exceptions with the relationships, and to identify research needs associated with each empirical relationship. Table 1 summarizes some of the more important empirical relationships identified during the workshop and reports the participant's evaluation of each relationship. More detailed discussion of these relationships are in the preceding chapters.

CRITICAL ISSUES

Chemical Activity as an Estimate of Exposure

To gain generality and, thus, predictive power, it was concluded by the workshop participants that a greater understanding of the availability of various operationally defined chemical forms is needed. This would provide the greatest flexibility and predictive power; but that unless there was some standardization of these techniques, they would not be very useful for making global predictions or for use in models. Furthermore, where possible the disciplines of Aquatic Toxicology and Chemistry should adopt the use of chemical activity rather than concentration. This is well defined for charged compounds, and relationships such as the Davies and Debye-Huckle equations have been derived to predict the chemical activity of these types of ions from the ionic strength of a solution. Analogous techniques can be visualized for charged organic compounds, but these relations are less developed.

Table 1.　Summary of Empirical Relationships

Empirical relationships	Understanding of mechanism	Use(s)	Problems or exceptions	Research needs
Hardness—organics no effect on toxicity or bioaccumulation once corrected for pH	Good	Extrapolation Regulations	Understanding the relationship between hardness and accumulation	Develop empirical relationship
Hardness— inorganics Hardness has no effect on toxicity	Limited	Limited	Few	Identify possible mechanisms
pH—organics calculate species using pKa and pH	Good	Prediction Extrapolation Regulation	Improve predictions pH effects on surfactants	Understanding surface events Better empirical toxicity data bases
pH—inorganics Toxicity decreases with pH for acids and increases with pH for bases	Fair	Extrapolation Prediction Regulation	Separation or speciation and better physiological effects data base	
Temperature— organics Acute toxicity increases with temp	Limited	Extrapolation	Varies for classes of cmpds	Establish mechanism increase
Temperature— inorganics Toxicity may increase or decrease with increasing temp.	Poor	Extrapolation	Many	Many
Salinity—organics Generally small effects	Poor	Extrapolation Regulations	Many	Many Better empirical data base
Salinity—inorganics	Limited	Extrapolation	Many	Relationship of hardness to effects
K_{oc} partitioning Describes interactions with sediments and water	Good	Estimate Exposure Concentrations	All organic carbon not equal	Define sorption/ desorption kinetics and add classes of comp.
Free metal ion Controls bioavailability	Poor	Predictive	Ion pair small organic ligands	Activity should be used instead of concentration
Photochemical, redox, hydrolysis processes generally detoxify organic chemicals	Limited	Regulation EPA requires product studies across all species	Some	Compilation of compounds that are photochemically active
Activities should be used instead of concentrations	Good	Used in research and modeling	Some	Availability of activity coefficients for organics
All metals cycle with Fe and Mn	Good	Prediction	Not always correct	Better empirical data needed
AVS control metal availability in sediments	Good	Prediction Potential regulations	No evidence from intact sediments	Better empirical data needed

The least developed area is chemical activity of uncharged organic molecules, but it should be possible to develop the theory and empirical relationships to allow the calculation of the chemical activity of these compounds. Normalization to total organic carbon in the bulk phases seems to work well as a first approximation for the chemical activity of these types of compounds; however, this relationship is limited by the heterogeneous nature of organic carbon in aquatic systems. To make better use of this relationship, the workshop participants felt definition of the range in variability of the properties of naturally occurring organic compounds that influence the distribution of synthetic organic compounds was necessary.

Naturally Occurring Organic Carbon

Recently, it has been observed that although organic carbon concentrations are similar among locations, not all organic carbon has the same binding/sorption capacity for toxic organic compounds. This raises a question about the practice of normalizing by dissolved organic carbon. The degree of uncertainty caused by this variation is unknown, but should be evaluated before a decision on the adequacy of the currently used K_{oc} methods for determining the activity of organic compounds is made.

Surface Phenomena

Prediction of bioavailability based on the form of chemicals in the bulk phase may be limited because of effects in the boundary layer. First, the lack of turbulent flow in this area results in different rates of reaction than would be predicted from rates in the bulk phase. Secondly, the ordering of water molecules at surfaces is known to affect the solubility of ions and molecules. Also, organisms may have the capacity to alter the microenvironment, such that measurements of chemical form in the bulk phase are not sufficient to allow prediction of the accumulation of ions or molecules.

Forms Used for Decision Making

Since the form in which a compound or element occurs can effect the exposure, and thus toxicity, accurate estimates of the effects of toxicants and nutrients require a knowledge of both the short-term and long-term exposure, which is a function, in part, of bioavailability. However, speciation or bioavailability is seldom used in decision making. There are several reasons for this, some of which can be addressed with additional information and understanding, while others cannot. The utility of bioavailability in decision making is limited by the fact that there is little uniformity in the operational techniques used to measure bioavailability. Thus, more uniform, validated techniques would be necessary before these corrections will be routinely used. Furthermore, the parameters required to make corrections for bioavailability are seldom quantified. Also, the degree to which correction for bioavailability would change the site-specific criteria is unknown for most chemicals.

In addition to these technical issues is the philosophical reason for not including bioavailability in predicting protective exposure concentrations. Specifically, any corrections of the exposure concentration are usually only applicable to fairly restricted geographic regions and time periods. For instance, changes in the bulk chemical/physical properties can change with daily and annual cycles and over geological time frames. Also, as masses of material are moved from one location to another, such as advective flow in rivers or estuaries or when sediments are dredged, the important bulk properties can greatly change the bioavailability of contaminants.

Environmental properties generally result in decreased bioavailability through binding, degradation, and sequestration. Therefore, if total concentrations of toxic chemicals and nutrients are used, there will be a degree of safety in all regulations and there will be no danger of exceeding the threshold for effect when environmental conditions change. The degree of overprotection is known for a number of chemicals. We suggest that a compilation of the ranges of possible exposures be estimated for a number of chemicals from several classes over a range of environmental conditions; in this way the relative error for the use of total concentrations can be assessed and if this margin of error appears to be excessive, actions could be undertaken to address those few exceptions in a most cost- and time-effective manner.

Equilibrium Partitioning

The participants concluded that there was no information to suggest that thermodynamic equilibrium partitioning relationships do not apply in determining the bioavailability of organic molecules to aquatic biota. The greatest limitation to the application of equilibrium partition concepts is disequilibrium. The errors introduced by disequilibrium were thought to be for compounds with log P values (log K_{ow}) >5.5 or <2. Metabolism or active accumulation or excretion of organic compounds could also result in concentrations in biota that are not predicted by simple partitioning coefficients. Site saturation and homeostatic regulation may limit the utility of speciation in the bulk media for predicting bioavailability of elements, especially required elements.

Several concepts which relate to the use of the equilibrium partitioning theory for prediction of accumulation of materials by and their effects on organisms have been the subject of intense debate and research. These include the use of K_{oc} to normalize the availability of neutral organic compounds from sediments, acid volatile sulfides (AVS) to normalize the availability of metals from sediments, and the solids concentration phenomenon.

One of the most well-developed and effective normalization procedures to account for speciation and bioavailability is the K_{oc}, where the activities of organic compounds are estimated from the total concentration and the concentration of naturally occurring organic carbon to which they can partition in water and sediments. While this method works well for the development of the technique, improved predictability will be dependent on a better description of the properties of the organic carbon compounds contained in the bulk phase, and on the influences of other cocontaminants like automotive fluids and synthetic polymers.

AVS has been demonstrated to be a useful covariate to predict the availability of metals in homogenized sediments in laboratory tests. If the molar ratio of AVS to metals in a sediment is >1, it is predicted that little or no free ions of transition metals will occur in the pore water of the sediments; thus, there will be little or no metal bioavailability. This seems to work well in anaerobic marine sediments for several metals. However, in systems where there is little AVS, and for certain metals such as copper, the method does not seem to work very well. In fact, total organic carbon may be a more important determinant of bioavailability in many situations. Furthermore, the normalization to AVS does not consider the availability of metals in the aerobic microenvironments of sediments that are created by biota. Finally, because the redox potential of the sediments can change with time, normalization to AVS will be of limited value in setting sediment quality criteria, which are normalized for bioavailability. Workshop participants emphasized that before the metals to AVS ratio is adopted for regulation of sediment quality, the relationship of AVS to metal bioavaiability must be demonstrated in intact (preferably in-place), structured sediments.

The solids concentration effect has been described as a situation where the equilibrium partitioning theory breaks down. This has been attributed to unique properties of sediment solutions, where concentrations of organic compounds in the pore water are greater than predicted. While this effect has been observed in experiments, and several possible mechanisms advanced to explain the effect, there has never been a definitive explanation of the phenomenon. The participants discussed this phenomenon at length and concluded that it was an artifact of the experiments. One popular opinion was that it was due to the presence of colloidal material in the freely dissolved phase which resulted in an artifact due to measurement techniques. Another opinion was that the effect was a kinetic phenomenon which arose because the laboratory experiments were not conducted for sufficiently long periods of time to allow true steady-state conditions to be reached.

Relationships between Thermodynamics and Pharmacokinetics

Bioavailability is a complex function of interactions of compounds with their chemical, physical, and biological environment, which can change as a function of both space and time. Often, the form in which a compound or element occurs will limit the instantaneous flux of materials into an organism. However, the ultimate concentration of the material accumulated by the organism may be independent of the form in the bulk environment. This can occur due to the complex interactions between and among factors such as the relative distribution constants for each pool, the rates of release from pools in the bulk phase, rates of accumulation by the organism, and both the relative and absolute aqueous solubilities of the compounds of interest. For instance, if a large but unavailable pool of material could supply the available pool with an inexhaustible supply of material, during short-term exposures there could be a great difference in the amount of material accumulated, but there may not be a large difference in the distribution between the available pool and the organism at steady state. Also, the relative availability and magnitude of distribution coefficients is critical to the relative importance of exposure and the error in estimating steady-state concentrations in the organism. For example, steady state will not be attained during any duration of exposure for compounds with relatively great partitioning coefficients ($K_{ow} > 6$); in this case the steady-state concentration could be underestimated. Comparison of the bioavailability of compounds from studies conducted for the same duration would result in an underestimate of the available portion of the total concentration of material in the bulk phase. For compounds with relatively small partitioning coefficients (log $K_{ow} < 2$), different trophic levels would potentially indicate different degrees of bioavailability.

FUTURE RESEARCH NEEDS

Knowledge of factors affecting the bioavailability of chemicals from water and sediments has greatly expanded in recent years. However, there are a number of areas where additional research is needed. The preceding chapters discuss the bioavailability of inorganic and organic chemicals, chemical and physical factors affecting bioavailability, and kinetics. Specific research needs are identified in each of the chapters. The major themes of bioavailability research can be summarized as follows:

- Research is needed to determine the biologically available forms of inorganic and organic chemicals and to develop methods for the analysis of these forms and for the prediction of their concentrations.
- The understanding of bioavailability must extend from the bulk environment, including both the water column and sediments, to the site(s) of uptake by or incorporation on organisms.
- A better understanding of chemical speciation in the water and sediments is essential to the understanding of bioavailability. Of particular need is information about the effects of dissolved and colloidal humic materials, composition of dissolved organic matter (DOM), temperature, and salinity on chemical speciation.
- Research to define the kinetics of chemical reactions important to the speciation of chemicals is also needed.
- Identification of the conditions that result in kinetic limitations is needed to understand organism exposure and accumulation of contaminants.
- While some information on kinetically limiting processes are available for nonpolar organic contaminants, minimal information is available for polar organic compounds and metals.

• Research is needed on the relation of physical/chemical factors in the water and sediment bulk phases and conditions (pH, permeabilities, and adsorption characteristics) at the organism exchange surfaces (cell walls, gills, skin, and gut lining). Research is also needed that elucidates the mechanism(s) which determine what forms of a chemical are bioavailable at exchange surfaces.

REFERENCES

1. Cairns, J., Jr., Dickson, K. L., and Maki, A. W., Eds., *Estimating the Hazard of Chemical Substances to Aquatic Life,* ASTM Tech. Publ. 657, American Society for Testing and Materials, Philadelphia, 1978.
2. Dickson, K. L., Maki, A. W., and Cairns, J., Jr., Eds., *Analyzing the Hazard Evaluation Process,* American Fisheries Society, Bethesda, MD, 1979.
3. Maki, A. W., Dickson, K. L., and Cairns, J., Jr., Eds., *Biotransformation and Fate of Chemicals in the Aquatic Environment,* American Society for Microbiology, Washington, D.C., 1980.
4. Dickson, K. L., Maki, A. W., and Cairns, J., Jr., Eds., *Modeling the Fate of Chemicals in the Aquatic Environment,* Ann Arbor Science Publishers, Ann Arbor, MI, 1982.
5. Pavlou, S. P., Dexer, R. N., Mayer, F. L., Fisher, C., and Hague, R. H., Chemical-ecosystem interface, in Neuhold, J. M. and Ruggerio, L. F., Eds., ecosystem processes and organic contaminants, in *Research Needs and An Interdisciplinary Perspective,* National Science Foundation, U.S. Government Printing Office, Washington, D.C., 1977, 21–26.

Index